AFRICA'S
THREATENED
RHINOS

AFRICA'S THREATENED RHINOS

A History of Exploitation and Conservation

KEITH SOMERVILLE

PELAGIC PUBLISHING

First published in 2025 by
Pelagic Publishing
20–22 Wenlock Road
London N1 7GU

www.pelagicpublishing.com

Africa's Threatened Rhinos: A History of Exploitation and Conservation

https://doi.org/10.53061/TCFB1570

A CIP record for this book is available from the British Library.

ISBN 978-1-78427-454-2 Hbk
ISBN 978-1-78427-455-9 ePub
ISBN 978-1-78427-456-6 PDF

Cover photograph © Mario Moreno Photography

Typeset in Minion Pro by S4Carlisle Publishing Services, Chennai, India

Dedicated to the memory of the great trade investigator Esmond Bradley Martin, to the courageous and incorruptible ranger Anton Mzimba, murdered because of his commitment to protecting rhinos, and to those who have given their lives or dedicated their careers to conserving rhinos and supporting the communities that live with them.

Black rhino. (Keith Somerville)

Contents

Introduction

The black and the white rhinoceros are two of the most charismatic megaherbivores left on our planet and have become flagship species for international conservation. They are significant not only for the continuation of a major evolutionary heritage but also as symbols for the protection of African savannahs. Africa's two rhinoceros species have a chequered history and the battle for their survival has been marked by some notable successes and sadly, many failures.[1]

I have always been fascinated by rhinos – their bulky, armoured appearance, their power and the tragedy of their slaughter at the hands of humans. In 1960, when I was three years old, there were an estimated 100,000 black rhino, and perhaps 1,200–2,400 northern white rhino.[2] Between 1948 and 1968, southern white rhino increased from 550 to 1,800.[3] As a child and then a young adult I saw black and white rhino in zoos, such as the Zoological Society of London (ZSL) zoos at Regent's Park and Whipsnade, and Marwell Zoo in Hampshire. I was always intrigued by how and why they had developed into the animal equivalent of a combat tank with horns, the front horn sometimes of amazing length and shape. They were charismatic and impressive, but it was always sad seeing them enclosed and reading about their decline, brought about by human desires and greed.

When I first worked in Africa in 1981–2, I smashed my BBC employer's saloon car to pieces in Kasungu National Park (NP) in Malawi but failed to see the few rhinos left there. I had more luck in April 1982, when I visited Hwange NP in Zimbabwe and saw white rhino at close quarters. Since then, I have seen them from vehicles and on foot in a variety of locations in East and southern Africa and even taken part, in a small way, in dehornings on John Hume's Buffalo Dreams Ranch in South Africa. Now, in my late sixties, it is sad for me to record that black and white rhino have declined catastrophically to under 23,000, and that in my lifetime the western black rhino has become extinct and just two female northern white rhino survive, so the species is functionally extinct in the wild. In October 2023, the African Rhino Specialist Group (AfRSG) of the International Union for the Conservation of Nature (IUCN) put the overall figure as 23,171 rhinos, an increase of 5.2% on 2021, not including nearly 1,300 in zoos and collections outside rhino range states. The total comprised 6,468 black and 16,801 white rhino, the latter registering the first increase in overall numbers since 2012.[4] But in January 2024, Pelham Jones of the Private Rhino Owners Association of South Africa (PROA) told me that around 600 rhinos had been poached in South Africa in 2023, 282 of these in KwaZulu-Natal (KZN), the location of Hluhluwe-iMfolozi reserve. This was nearly the final home of the last southern white rhino in the nineteenth century, but then bred the rhinos that now live in parks, reserves and on private land across southern Africa and have been translocated to Kenya, Rwanda, Uganda and the Democratic Republic of the Congo. When the official figures were released in late February 2024, they showed a slightly less extreme rise in poaching of rhinos overall but a bigger increase in rhinos poached in KZN. Barbara Creecy, the Minister of Forestry, Fisheries and the Environment, reported that there had been a rise of 51 to 499 killed for their horns – 406 in state parks or reserves and 93 on private reserves or game farms. The minister lamented that 'The pressure again has been felt in the KwaZulu-Natal (KZN) province with Hluhluwe-iMfolozi Park facing the brunt of poaching cases losing 307 of the total national poaching loss.' Kruger's rhinos had a slight respite, with 78 killed in 2023, down on 2022, when 124 were poached.[5] This national rise in poaching has shattered the optimism that numbers might slowly recover, and reminds us that the scourge of poaching is far from over – poaching that feeds demand in East Asia for medicinal uses of horn of dubious efficacy, the possession or giving of horn as a prestige product or to make Yemeni dagger handles.

My interest and fascination over the last decade have prompted me to find out everything I could about the animals themselves, their cruel treatment by humans and the attempts at conservation and recovery. The results are assembled here. This is a work of history, not biological science, but I hope that I have successfully woven scientific knowledge into my largely chronological account of the fate of Africa's rhinos and given a clear account of the trade in rhino horn. In compiling this account, I have climbed on to the shoulders of the leading rhino scientists and trade specialists, to get as clear and wide a view as possible, but any errors or misinterpretations are mine alone.

Chapter content

Chapter 1 deals with the evolution of the rhino and the basic details of size, distribution, diet, breeding, mortality, social behaviour and interactions with other large mammals, predators and humans. Chapter 2 looks at the development of human–rhino coexistence and exploitation, from early humans hunting rhinos for their meat and hides on a very small scale tens of thousands of years ago, to the end of the eighteenth century, detailing the development of demand outside Africa for rhino horn. Chapter 3 looks at East and Central Africa during the period of European penetration and then colonial rule, and the disastrous consequences of the rise in demand from outside Africa for horn. Chapter 4 looks at the same history in southern Africa. Chapter 5 examines the development of demand and the nature of the illegal trade in rhino horn from 1960 up to the present. Chapters 6 and 7 deal with the period from 1960 to 2005, which saw the recovery of the southern white rhino and the destructive waves of poaching across Africa. Chapters 8 and 9 chart conservation and poaching in the African rhino ranges, with emphasis on the tsunami of poaching that hit southern Africa from 2009 onwards. Finally, Chapter 10 examines approaches to management and the funding of rhino conservation, together with the key role of communities.

A note on sources

I have used peer-reviewed papers from a range of disciplines – conservation biology, zoology, ecology, palaeontology, archaeology, history, wildlife economics and the social sciences – to support my narrative. But when it comes to the history of human interactions with rhinos, hunting and early conservation attempts, there are major gaps prior to the twentieth century in the academic literature, and limited first-hand accounts by knowledgeable biologists or zoologists. To fill this gap I have used the extensive reports on wildlife contained in the *Proceedings of the Zoological Society of London* (thanks are due to the ZSL Library staff, especially the always helpful Emma Milnes), particularly the nineteenth- and early twentieth-century accounts by ZSL Fellows. The *Journal of the Society for the Preservation of the Fauna of the Empire* (now known as *Oryx*) has also been an invaluable source, especially for early twentieth-century surveys of wildlife in the colonies and the annual game reports from the game or wildlife departments established in European colonial territories (thanks are due to Flora and Fauna International for sending me the entire collection of the *Journal* on DVD).

Also important, though they are often sources that have to be weighed carefully and treated more as impressionistic than scientific or rigorously accurate accounts of African wildlife and the effects of African and European hunting, are the narratives of hunters and travellers such as Burchell, Oswell Cotton, Harris, Gordon-Cumming, Burton, Speke, Thomson, Stanley and Selous, and chroniclers of colonial life in Kenya such as Huxley and Blixen.

Throughout, except where directly quoting sources, I have opted to use rhino rather than rhinoceros, for brevity's sake. I have also used the most common and accepted usages for names and, again except where quoting or referring directly to a report or survey, have used the contemporary names for countries and regions rather than colonial ones. In South Africa, I have used iMfolozi for the KZN reserve linked to Hluhluwe, though in quotations and historical accounts Umfolozi and Imfolozi are retained. Throughout I have used the $ sign to signify US dollars – the dollars of any other country are identified as such.

In compiling my account I have been fortunate in having had guidance and advice from the leading experts on rhino conservation, ecology and the trade in rhino horn – and am hugely grateful to them. In particular, I'd like to thank the following: Lucy Vigne, who read, commented on and vastly improved Chapter 5 on trade in rhino horn; the late Esmond Bradley Martin; Sam Ferreira (Scientific Officer, AfRSG); trade expert Daniel Stiles; wildlife economist Michael 't Sas Rolfes; leading Kenyan conservationist Mordecai Ogada; Mike Knight and Richard Emslie of the AfRSG; Xolani Nicholus Funda, the head ranger of Kruger National Park (KNP); investigative journalist Julian Rademeyer; John Hanks; Ted and Mick Reilly of Big Game Parks in eSwatini; Pelham Jones of PROA; John and Tammy Hume of Platinum Rhino; Cedric Coetzee, head of rhino security at Hluhluwe-iMfolozi; John Jackson III of Conservation Force; David Cooper; Dr Marc Dupuis-Désormeaux; Dr Kees Rookmaaker, chief editor of the Rhino Resource Centre, for his advice, suggestions of papers to read, permission to use maps and images, and for his curation of the excellent Rhino Resource Centre; Cathy Dean, chief executive officer (CEO) of Save the Rhino for many years, who has always been a source of help, advice and encouragement, as has her successor Dr Jo Shaw. Thanks to Maxi Louis Pia for help with material relating to Namibia's community conservancies and to Gail Thomson Potgeiter of Conservation Namibia for advice and contacts.

Dr Helen Anderson, Africa Curator at the British Museum, was very helpful in getting permission to use images of rhinos from rock art in Namibia. Brianna Cregle of Princeton University Library helped in sourcing and getting permission for the reproduction of images in their collections.

My wife Liz and son Tom have offered advice and constant support, especially towards the end of writing this account, when 'rhino' became a dirty word in our house.

CHAPTER 1

Evolution, status and behaviour

Rhinoceros means nose horn in Greek. It is the rhino's horn that has been the defining characteristic of the animal and the cause of its progressive depletion across its African ranges, with the loss of the northern white rhino and western black rhino subspecies, and near extermination of two of the Asian species – the Sumatran rhino (*Dicerorhinus sumatrensis*) and Javan rhino (*Rhinoceros sondaicus*) – and the historical reduction in numbers and range of the Indian one-horned rhino (*Rhinoceros unicornis*), though the latter is now recovering in numbers in Nepal and India.

Rhino horn consists mainly of tiny keratin filaments or tubules, with a dense central area reinforced by calcium and melanin, which creates a strong core. Horn is chemically complex and contains large quantities of sulphur-bearing amino acids, including cysteine, tyrosine, histidine, lysine and arginine, and the salts calcium carbonate and calcium phosphate. The core has more calcium concentrated in it, making it resistant to wear. The softer exterior is worn down over time, leading to the development in older rhinos of long, tapering horns often ending in a sharp point. Growth occurs in layers from 'specialized skin cells which are keratinized and become hard and inert'.[1] The horns of African rhinos are part of their defensive capability, especially for female rhinos that protect young from predators such as lions and hyenas, and males that use them in territorial disputes, dominance competitions and fights over access to females for mating.

The African white rhino and black rhino have far larger horns, especially in terms of length, than the Asian species. Black rhinos have two horns – with the anterior (front) horn capable of growing to 130 cm while the posterior (rear) horn is 2–55 cm. The white rhino anterior horn is 94–102 cm and the posterior up to 55 cm. Horns grow constantly from the base, adding about 7 cm a year. The horns of young rhinos grow faster, with an average of 5.98 cm, than those of old rhinos.[2] White rhinos usually have longer horns than black rhinos. White males and females have similar overall lengths, but the males' horns are heavier and greater in girth.[3] Both species are very large: the black rhino can measure up to 1.8 m in height at the shoulder, and can weigh up to 1,350 kg; equivalent figures for the white rhino are 1.75 m and 3,600 kg. Both are larger than the three Asian species.[4]

Origins of the rhinoceros species

The five surviving rhino species are members of the family Rhincerotidae, which belongs to the Perissodactyla division of hoofed mammals, with three toes evenly spaced around the front of the foot. Contrary to popular belief, rhinos and elephants are not closely related. The rhinos (Rhinocerotidae) are one of the three surviving families of the Perissodactyla, the others being horses and asses (Equidae) and tapirs (Tapiridae). Fossils from North America, Europe, Africa and Asia show that species of rhinoceros have been on earth for 60 million years.[5] Numerous extinct species of rhino are known from the fossil record, having disappeared hundreds of thousands or even millions of years ago. Most Eurasian species became extinct in northern Eurasia when steppe habitat was largely replaced by dense coniferous forests, and with the spread of hominins and then *Homo sapiens* capable of hunting megafauna like the woolly rhinoceros and the mammoth, though there is surprisingly little evidence of early humans hunting rhinos, in contrast to the evidence of mammoth hunting.[6]

How did rhinos reach Africa?

Many African mammal species reached Africa in a series of migrations from Eurasia, and then evolved into the current African fauna. There are also many extinct species known only from fossils.[7] The timing of dispersals of species from Eurasia to Africa was dependent on the shifting of continental landmasses caused by tectonic forces in the Earth's crust. In the east, the African landmass and the Arabian peninsula were joined between 30 mya and 12 mya. Then Arabia broke from Africa and collided with Eurasia and the Iranian tectonic plate, with the Red Sea becoming part of the rift system from the Middle East down into the Horn of Africa, East and Central Africa. At that time, the narrowest gaps between Eurasia and Africa were at the Gulf of Suez and Sinai.[8] Many mammal species entered Africa across the narrow remaining strip of the land bridge with Asia,[9] including ancestors of the black and the white rhino.[10] In the middle of the Miocene (between 23 mya and 5.3 mya) there was an expansion of the northern coniferous forest in Eurasia and the retreat of some tropical flora owing to lower temperatures,[11] which prompted certain grassland and woodland species to move southwards, including to Africa.

Around 55–60 mya, rhinos evolved into a diversity of species distributed across Africa, Eurasia, and North and Central America. From the Middle Eocene, 47.8–38 mya, to the Late Pleistocene there is evidence of more than 41 genera and 142 species in Europe, Asia, North America and Africa.[12] The majority of these ancient rhinos disappeared during the late Miocene and early Pliocene.[13] During the Pliocene, rhinos became extinct in North America, and during the late Pleistocene they disappeared from Eurasia and Africa, with the exception of the extant five species.[14] The sole surviving family is the Rhinocerotidae, of which the five living species are members.

The first members of the Rhinocerotidae are thought to have appeared in Africa in the early Miocene, with evidence of both subfamilies, the Aceratheriinae (a group which lived from 33.9–3.4 mya) and Rhinocerotinae (including the surviving species and many that became extinct).[15] The presence of rhinos in the Miocene in Africa is known from fossils found across areas that had suitable habitat, from North Africa through the grassland/woodlands of north-west Africa and into the central and eastern regions as far south as Namibia, where the hornless, tapir-like Aceratheriinae were found.[16] In an attempt to trace the lineages of the extant rhino species, Shanlin Liu et al. sequenced the genomes of the five living species. This showed an 'an early divergence between extant African and Eurasian lineages … This early Miocene (16 mya) split post-dates the land bridge formation between the Afro-Arabian and Eurasian landmasses.'[17] In the late Miocene, three genera and four species were found in Africa, including the North African *Ceratotherium neumayri* (also described as *Diceros pachygnathus*).[18] Globally, only nine rhino species survived until the late Pleistocene in Eurasia and Africa; of these, four – the Siberian unicorn (*Elasmotherium sibiricum*), Merck's rhinoceros (*Stephanorhinus kirchbergensis*), its relative the narrow-nosed rhino (*Stephanorhinus hemitoechus*) and best known of all of them, the woolly rhinoceros (*Coelodonta antiquitatis*) – became extinct at the end of the Pleistocene.[19]

Some of the Eurasian species reached Africa before those that remained in Eurasia became extinct. Remains of Merck's rhinoceros dating from between 47 kya and 14 kya have been found in Cyrenaica in Libya. They are believed to have dispersed into North Africa between 250 kya and 130 kya but disappeared at the end of the last interglacial period or perhaps the beginning of the last glacial period about 115 kya, as a result of climatic changes.[20] Africa did not experience substantial megafaunal extinctions in the late Pleistocene, as other continents did. Some species died out as savanna came to cover about 40% of Africa's land surface. Grassland and open woodland, with abundant food for grazers (white rhino) and browsers (black rhino), ensured their survival. Owen-Smith estimates that during the megafaunal extinctions at the end of the Pleistocene between 15,000 and 10,000 years ago, the Americas lost 75% of their genera, Europe and Australia lost 45% but Africa lost only 13%.[21]

It is known, from fossils discovered in North Africa, that the rhino subfamily Elasmotheriinae dispersed into Africa in the Miocene.[22] Excavations on the southern slopes of the Central High Atlas in Morocco unearthed a nearly complete skull of an elasmotheriine from the late Miocene. Other remains of Rhinocerotidae there were not numerous, but included a mandible and teeth of a species of *Ceratotherium*.[23] The two African species that survived the periods of rhino extinction are part of the clade known as Diceroti, separate from the clade that included Merck's rhinoceros, the woolly rhinoceros and the extant Sumatran rhino, with a third clade containing the Javan and Indian one-horned rhino. The clades diverged around 16 mya as a result of the geographical division between the ancestors of modern rhinos that dispersed into Africa and those that remained in Eurasia.

The splitting of Africa and Eurasia isolated the rhinos in Africa, where they evolved into a number of species. One of these was *Paradiceros mukirii*, an ancestor of the black rhino, known from Kenya and Morocco.[24] Another descendant was *Ceratotherium neumayri*, the ancestor of both black and white rhinos, fossils of which have been found in the eastern Mediterranean as well as East Africa.[25] Their descendants split around the transition from Miocene to Pliocene (about 4.83 mya) – one branch that came from this ancestral species was *Diceros*, believed to be the direct ancestor of the black rhino, and the other branch was *Ceratotherium mauritanicum*, closely related to the white rhino.[26] The differences between those that evolved into *Diceros* and those that became *Ceratotherium* are related to diet and methods of foraging, with the former being browsers with prehensile lips like the black rhino, and the latter wide-mouthed grazers like the white rhino. Several species of *Diceros* occurred in Africa and Europe, with *Diceros bicornis* appearing about 4 to 5 mya. The lineage of the white rhino is much more recent; the current genus *Ceratotherium* first appeared during the Pliocene (5.33–2.58 mya).[27]

Guerin recorded that *Diceros bicornis* (the modern black rhino) first appears in the fossil record during the Pliocene about 4 to 5 mya, with remains found at more than 20 sites from the Pliocene up to the middle Pleistocene, including Hadar and Omo in Ethiopia, East Turkana in Kenya, and Laetoli and Olduvai in Tanzania.[28] Sánchez-Barreiro et al.'s examination of DNA of black rhino subspecies led them to propose 'central and/or eastern

Africa, east of the Congo basin, as the putative region of origin for the black rhinoceros. This is supported by the fossil record', with evidence of modern black rhinos at Koobi Fora in Kenya, 2.5 mya. The species' range appears to have expanded rapidly, with remains in South Africa (2 mya) and Ethiopia (1.6 mya).[29]

Guérin set out the evidence of the white rhino's early range: 'The genus *Ceratotherium* appears during the Pliocene with *C. praecox*, from Kanopol and Ekora in East Africa ... [and] abundant material of the same species from Langebaanweg in South Africa ... [it] is now known in 11 localities of East and South Africa'.[30] For the two recent forms, *simum* and *cottoni*, he wrote that he was able to find about 30 fossils: 'only 16 skulls and 8 postcranial skeletons were certainly from *cottoni*, and 8 skulls with 2 postcranial skeletons from *simum*'.[31] His conclusions suggest that white rhino were spread through East and southern Africa from the Pliocene until a point when they diverged geographically and became the two subspecies of *Ceratotherium simum*. The geographical split between the northern white rhino and southern white rhino, which became more complete than the divisions between the black rhino subspecies, occurred during the last 1 million years. Harley et al. believe that the mitochondrial DNA of northern and southern white rhino diverged somewhere between 0.46 and 0.97 mya, but that 'the dating of the actual lineages split between NWR [northern] and SWR [southern] is likely to be significantly more recent than the mitochondrial genome split ... it is possible that the two white rhinoceros lineages could have diverged even as recently as 200,000 years ago'.[32]

Status, distribution, ecology and behaviour

Black rhino
Species: *Diceros bicornis*

Subspecies:
Western (*Diceros bicornis longipes*) – declared extinct by the IUCN in 2011[33]
Eastern (*Diceros bicornis michaeli*)
South-western (*Diceros bicornis bicornis*) – sometimes labelled as *occidentalis*
South-central (*Diceros bicornis minor*)

The four recognised black rhino subspecies occur in different regions of Africa, though their respective ranges may overlap. Emslie and Brooks suggest it is likely that each has specific genetic or behavioural adaptations to its environment.[34] A combination of habitat loss and poaching has led to the decline and then extinction of the Western subspecies across its Central

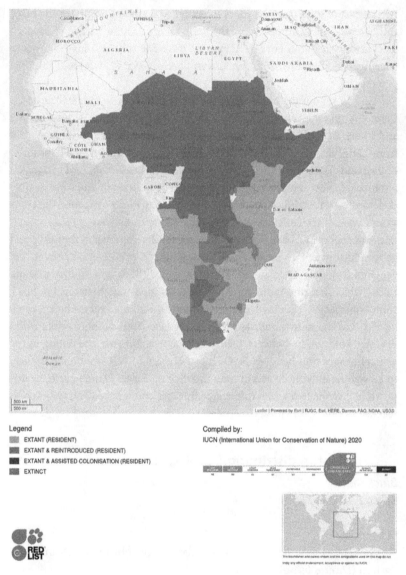

Legend

EXTANT (RESIDENT)

EXTANT & REINTRODUCED (RESIDENT)

EXTANT & ASSISTED COLONISATION (RESIDENT)

EXTINCT

Compiled by:

IUCN (International Union for Conservation of Nature) 2020

IUCN Red List map for black rhino. (Reproduced with kind permission of the IUCN)

and West African range.[35] There was also said to be another geographically defined subspecies, *Diceros bicornis chobiensis*, in northern Botswana, the Zambezi (formerly Caprivi Strip) region of Namibia and a small area off south-eastern Angola,[36] but DNA analysis of samples of rhinos from the region does not uphold the existence of a separate subspecies.[37]

Black rhino with trimmed horns, Etosha National Park, Namibia.
(Keith Somerville)

Size and appearance

The black rhino is a large, heavy, short-legged animal. Male black rhinos stand about 1.43–1.8 m tall at the shoulder and are 2.86–3.05 m in length; females are a bit smaller.[38] Adult males weigh up to 1,350 kg, and females up to 900 kg. They are not black in colour but grey, with variations according to local soil conditions, as rhinos regularly wallow in mud, leading to different shades in different locations. They have two horns, tae longer front or anterior horn and shorter posterior one. The horns grow continuously from the base throughout life.[39] The Namibian south-western black rhino (*Diceros bicornis bicornis*) is the largest of all the subspecies and is adapted to arid, semi-desert conditions, as in the Damaraland and north-west Kunene regions of Namibia; these rhinos are the only substantial population anywhere in the world living outside protected areas.[40] Black rhino are smaller than white rhino and lack the large hump behind the neck.

As browsers, eating from bushes or trees, they have a hooked upper lip, in contrast to the flat, wide lips of the white rhino. They have no incisor teeth, as cattle and most grazers have, and cannot clip grass. Instead they have tan extended upper lip, 'like a miniature elephant's trunk, about four inches in length, with which he feels out bush shoots and pulls them into his great mouth where the enormous back molars grind them into pulp'.[41]

Distribution and conservation status

Black rhinos were widely distributed across savanna, semi-desert thornbush and woodland/forest habitat in West, Central, East and southern Africa. While the subspecies have great similarity, the IUCN believes there are sufficient differences for them to be listed separately on the IUCN Red List and for their populations to be considered separately.[42] The eastern subspecies is listed as Critically Endangered, the south-central (also referred to as south-eastern) subspecies as Critically Endangered and the south-western subspecies as Near Threatened – while the western one is listed as extinct.

There are doubts about the westernmost extent of the original range of the western subspecies, with Rookmaaker noting unverified accounts suggesting it was once found in Senegal, Sierra Leone, Guinea, Ivory Coast and Ghana, and a black rhino was reportedly killed at Bouma in Ivory Coast in 1905.[43] What is verified is that the western black rhino was resident in Niger, Nigeria, northern Togo, Benin, Burkina Faso, Cameroon, Chad and the Central African Republic (CAR). The IUCN Red List recorded that by 2006 the western black rhino was believed to be extinct, mainly as a result of relentless hunting for its horn.[44] The last surviving remnant had been in northern Cameroon, but when it became extinct was not clear, as there was evidence of wildlife rangers and local rhino monitors faking tracks to justify their continued employment.[45]

The eastern subspecies (*Diceros bicornis michaeli*) was found in Rwanda, Uganda, Sudan, South Sudan, Ethiopia, Kenya and Tanzania, but is now limited to Kenya and northern Tanzania, and the Akagera NP in Rwanda, where it has been reintroduced, and to areas of South Africa to which it has been translocated. The south-central black rhino (*Diceros bicornis minor*) was once widely distributed across Angola, the southern Congo (Democratic Republic of Congo), Tanzania, Zambia, Malawi, Zimbabwe, Mozambique, South Africa, eSwatini and Botswana. It has been exterminated in much of its range but is still found in south-central Tanzania, Mozambique (in very small numbers), South Africa and Zimbabwe, and reintroductions have been carried out in Botswana, eSwatini, Malawi and Zambia.[46] The south-western black rhino is only found in South Africa, Namibia and, in small numbers, in southern Angola. The African wild black rhino declined in numbers and in the size of its range by over 80% in the last 50 years of the twentieth century, from over 100,000 in the 1960s down to about 2,345 in the mid-1990s,[47] and has only begun to pick up in the last few years; the population is now put at 6,487.[48]

Although habitat loss played some role in the dramatic fall in numbers, the overwhelming cause of the huge decline in the population was poaching for rhino horn to meet demand from Yemen, China and East Asia.

Breeding and natural mortality

Apart from chance encounters, bulls and cows only come together for mating. Cows come into oestrus at between four and a half and six years old. After mating, the gestation period is about 450 days. The interval between births for a female averages two to two and a half years, depending on the availability of browse and water, the density of rhinos in the area and the health of the cow. Calves stay with their mothers for two to four years, and cows may been seen with a following subadult of up to four years old and a young calf.[49] Cows generally end their reproductive life at 30–35 years of age, and may live for up to 45 years. The breeding age of bulls varies according to the density of bulls and availability of females in their ranges. Bulls compete with great violence for the chance to mate. There is no set breeding season, and mating occurs when a female comes into season and her scent attracts males. Males may spend a considerable time following a female in oestrus.[50]

Apart from predation, which is generally low, except for young calves or old, sick or injured rhinos predated by hyenas and lions, natural causes of death are mainly lack of food or water during droughts, and disease. A major cause of non-poaching mortality among rhinos is fighting between males or injuries received by females during mating. Sas Rolfes et al. found that data from wildlife authorities and rhino owners indicated fighting accounted for 40% of natural black rhino deaths, of which 70.7% were male, 13.8% were breeding females and 14.2% were female calves or subadults.[51] There was an increase in fighting in monitored populations when new males were introduced to improve the breeding stock in fenced reserves or game ranches. A study in Pilanesburg NP of deaths between 1985 and 2001 found that fighting was the major cause of death among adults, with some subadults or calves killed during fights in which they were innocent bystanders. Juvenile deaths were higher in years with low rainfall or at times when the density of rhinos increased in particular areas.[52] Rhinos may be affected by tick-borne diseases as well as tuberculosis (TB), anthrax and other diseases present among wild ungulates.

Diet and foraging

Black rhinos are browsers, not grazers of grass. Their hooked lip enables them to feed on the leaves and shoots of a wide variety of plants, even some such as the Damara milkbush (*Euphorbia damarana*) that are toxic to most animals. A study in the Galana-Yatta region of Kenya, north of Tsavo East NP, found that 22 different plants were eaten regularly and another 47 occasionally.[53] In Tanzania's Ngorongoro Crater and Olduvai Gorge, black rhinos were found to feed on nearly 200 different plants, with a particular preference for acacias.[54] Rhinos released in Malawi's Majete Wildlife Reserve were observed to feed on 59 different plant species, including poisonous euphorbia species, which are succulent and usually eaten during the dry season.[55]

Social behaviour and territoriality

Black rhinos have gained an unwarranted reputation for being aggressive, bad-tempered, dangerous animals – a result of the views of Europeans who hunted in southern, eastern and central Africa in the nineteenth and early twentieth centuries. The prolific nineteenth-century hunter William Cornwallis Harris called the black rhino 'a swinish, cross-grained, ill-favoured, wallowing brute', and many others referred to its short temper and generally belligerent attitude.[56] Often this was a result of hunters pursuing or taking unviable shots at rhinos, which then charged them, having been chased into thick bush from which they emerged in pain and angered by their injuries. However, as Hillman Smith notes, having had a wealth of experience tracking and observing these animals, they are

> surprisingly placid and peaceful. Their responses to humans
> and other animals are tempered by environment and
> circumstance. Rhinos have poor sight but very good hearing
> and sense of smell – they are also aided in being wary by the
> presence of oxpeckers foraging for insects on their hides –
> the oxpeckers will give alarm calls if they sense danger. Most
> of the time if it detects a sound or scent that disturbs it, it
> will move away. Sometimes it will move towards the sound
> or scent out of curiosity.[57]

Although rhinos are generally solitary as adults, females with young offspring may inhabit territories that overlap with adult bulls and slightly with other females. Subadult males are often tolerated by adult bulls but are chased off or attacked when they mature. It is not clear whether occupied areas should be viewed as well-delineated territories or as less-defined home ranges; for simplicity's sake I will refer to them as territories. Territorial fights between adult bulls may result in serious injuries or even death.[58] Territoriality may break down around waterholes in arid areas – as I have witnessed at the Okaukuejo Waterhole in Etosha NP, where I have seen several black rhino, male and female, using the waterhole together in relative harmony at night. Males and females are relatively sedentary and do not stray much outside their territories.

Hillman Smith believes that general range sizes have increased as rhino numbers have fallen, reducing population density and freeing up areas for rhino dispersal where poaching and other factors have reduced the population.[59] Ranges were 2.6–15.4 km^2 in Ngorongoro Crater, 5.6–22.7 km^2 in the Maasai Mara and 70–133 km^2 in the Serengeti in the early 1980s, but poaching has massively depleted numbers, and in the Maasai Mara and Serengeti increasing potential territory sizes. In areas such as Damaraland, where aridity and limited browse naturally limit numbers, there are very large ranges, which may be in excess of 200 km^2 and in some cases up to 500 km^2.[60] Sizes of territories may vary somewhat between wet and dry seasons and the associated changes in water and food sources, with dry-season ranges being smaller.[61]

The marking of territories is done with dung middens, the scraping and spreading of dung with the rhinos' feet, horn thrashing in prominent bushes and urine spraying. Urine may be sprayed and then stamped in to spread the scent. Middens are both markers and social points, with more than one rhino depositing dung at a site. The scent of the middens serves to indicate to a rhino the presence of other rhinos and when a female that has used the midden may be coming into season. Scent may also be released from scent glands to add to the marking. This olfactory communication may enable males to avoid violent territorial disputes and help to sort out territories when rhinos are released into new areas without existing territories, as part of translocation programmes. In addition to middens, rhinos use vocalisations to communicate – warning growls, bleats when alarmed or threatened (or when a subordinate male is signalling submission in a dispute), loud honks and squeaks.

White rhino
Species: *Ceratotherium simum*

Subspecies:
Southern (*Ceratotherium simum simum*)
Northern (*Ceratotherium simum cottoni*)

Size and appearance
The southern white rhino and northern white rhino are considerably larger than the black rhino. Males can be 3.7–4 m in length, whereas females are 3.4–3.65 m, and males stand 1.7–1.86 m tall at the shoulder, whereas females are 1.6–1.77 m tall. Adult weights are 1,350–2,200 kg for males and 1,350–1,600 kg for females. At birth, white rhinos weigh 40–60 kg. White rhinos have huge heads and massive muscled necks, needed to support the weight of the head. The skin is thick, hairless and grey – colour variations according to local soil conditions are common, resulting from regular wallowing in mud or rolling in dry soil or sand. In shape they differ from

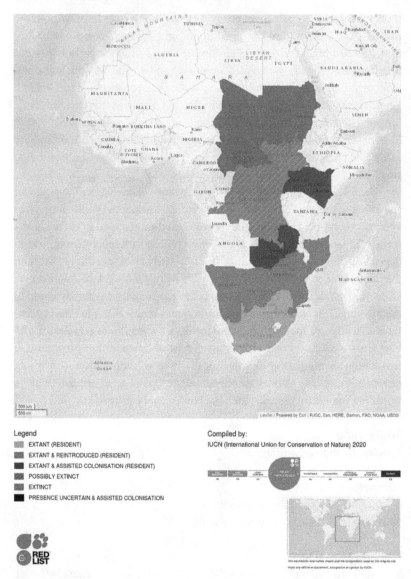

Legend

- EXTANT (RESIDENT)
- EXTANT & REINTRODUCED (RESIDENT)
- EXTANT & ASSISTED COLONISATION (RESIDENT)
- POSSIBLY EXTINCT
- EXTINCT
- PRESENCE UNCERTAIN & ASSISTED COLONISATION

Compiled by:
IUCN (International Union for Conservation of Nature) 2020

IUCN Red List map for white rhino. (Reproduced by kind permission of the IUCN)

black rhinos in having a very large hump on the back of the neck. The front horn is much larger than the rear one. White rhino horns may grow to greater lengths than those of black rhino. White rhinos have wide mouths, evolved to enable them to graze grass right down to the soil surface. This wide-mouthed appearance is the reason for their name – 'white' being a mistranslation of the word 'wyd', meaning wide-mouthed in Dutch/Afrikaans.[62]

White rhino, Hlane NP, eSwatini. (Keith Somerville)

Distribution and conservation status

Northern and southern white rhinos are essentially the same size, have the same bodily structures and breeding behaviour, and exhibit the same social and territorial behaviour. The northern white rhino was identified as a separate subspecies from the southern in 1903, when one that had been shot in the Lado Enclave, in what is now South Sudan, was exhibited in Europe.[63] As far back as 1876, Henry Stanley shot what he believed to be a white rhino near Chief Rumanika's territory in the Karagwe Kingdom of north-western Tanzania between Rwanda and the western shores of Lake Victoria, but he did not attempt to find out what species or subspecies the animal belonged to, beyond saying he thought it was a white rhino.

More thought was given to the taxonomy of rhinos shot or seen in French central African colonies by Professor Trouessart of the Museum of Natural History in Paris, who wrote to the ZSL in 1909 about the animals. He clearly thought them to be a northern form of the white rhino species and to be native to parts of Uganda, Sudan, Chad and the surrounding areas.[64] From then on it became clear that this was a white rhino that was geographically separated from the southern white rhino. The northern white, as far as could be gleaned from the accounts by European hunters, was found no further south than the Kagera River in Rwanda and Karagwe in north-western Tanzania.[65] Given the name *Ceratotherium simum cottoni*, they were present in southern Chad, northern and eastern CAR, south-western Sudan, north-eastern Congo and north-western Uganda, with no overlap for probably several tens of thousands of years with the southern subspecies.[66] Within recorded history, no white rhinos have been known to inhabit the area between the south-eastern limits of the northern subspecies (the River Nile, northern DRC, Rwanda, north-western Tanzania and Uganda) and the northern limits of the southern subspecies (the Zambezi River), a distance of some 2,000 km. Fossil records and cave paintings indicate that the two white rhino subspecies might have had a more continuous range until they were separated by climatic and vegetation changes during the last ice age. However, one piece of evidence suggests a more recent split. Meave Leakey found fossil evidence of white rhino near Lake Nakuru, dated at 1,500 years ago. It is not known whether the fossils belonged to the northern or the southern subspecies.[67]

Current northern white rhino status
Extinct: Uganda, DRC, CAR, Chad, Sudan, South Sudan, north-western Tanzania, Rwanda.
Extant: two surviving females at Ol Pejeta Conservancy in Laikipia, Kenya.

Southern white rhino distribution and status
The historical range was southern Africa south of the Zambezi, with some on the north bank of the Zambezi in southern Zambia. They were found in southern Angola, Namibia, northern Botswana, Zimbabwe, South Africa, grassland/open woodland areas of Mozambique adjacent to Zimbabwe, South Africa and eSwatini.[68] In the nineteenth and early twentieth centuries, the white rhino was exterminated in all of southern Africa except for a very small population in the Hluhluwe-iMfolozi region of what was then Zululand in South Africa. There were said to be just 20 rhinos left in 1885, and under 100 remaining there in 1900. Careful protection and management, including regulated use of limited trophy hunting of individuals, enabled the recovery of the species in Zululand and the use of the population there to restock other regions of South Africa, Zimbabwe, eSwatini, Botswana, Namibia, Zambia and Mozambique – and even to translocate some to Kenya, Uganda, Rwanda and the DRC.[69] The white rhino is now the most numerous of living rhino species, with an estimated 16,830 in the wild. It is listed by the IUCN as Near Threatened.[70]

Breeding and natural mortality
White rhinos mate and calve throughout the year. When a female comes into oestrus a bull does everything it can to keep the cow in its territory. If another

bull tries to mate with the cow or take over the territory, fighting can be intense and lead to serious injury or death.[71] Subordinate bulls only get the chance to mate if they evade dominant bulls when a cow is in season or is outside the territory of another bull.[72] Cows may also be injured or even killed during mating contests, or if a cow is unreceptive and the bull becomes violent. Cows may charge bulls to keep them away from calves or subadult offspring. Bulls can be killed in these encounters.[73] Females generally give birth for the first time at 6.5–7 years old, whereas bulls become fully mature at 10–12 years. Old bulls past breeding age or bulls deposed from their territories behave more like subordinate/young bulls, but may interfere with breeding by other bulls.[74] After mating the gestation period is 16–18 months (about 480 days).

The natural mortality among wild white rhino has been estimated at 1.7% p.a., and that of calves at 8.3%.[75] Mortality results from a number of causes, poaching being the major one in badly affected areas such as KNP and Hluhluwe-iMfolozi. Natural causes include shortage of food during prolonged dry periods. Ferreira et al. note that 'Increased animal mortalities during a drought are primarily associated with limited food availability rather than a lack of drinking water.'[76] White rhinos can also contract TB (found among rhinos in KNP in 2016–17, from analysis of material from carcasses),[77] anthrax and tick-borne diseases. Diseases may be more common among quite densely populated rhino populations in reserves or game ranches because of ease of transmission between animals. Stress-related diseases have been found among white rhino. One such disease is characterised by septicaemia caused by the commensal bacterium *Streptococcus equisimilis*. Four out of five free-ranging, two- to three-year-old white rhino died as a consequence of this bacterium on a densely populated game farm in northern Transvaal in 1992.[78] Fighting between males for territory or the opportunity to mate with females also figures as a cause of death, with cows and calves often being injured or dying as collateral damage during mating or territory disputes. Predation is low and concentrated mainly among calves, sick or injured animals or old, ailing ones. Hyenas and lions are the only real predators of young, sick or old rhinos, though other carnivores may feed on rhino carcasses.

Social behaviour and territoriality

Although more sociable than black rhino, adult males are effectively solitary for most of their lives, whereas adult females may be accompanied by young offspring but also subadults or young adults to which they gave birth. Males occupy territories from which other adult males are excluded, by force if necessary. In Hluhluwe-iMfolozi Game Reserve (GR), males had 'mutually exclusive home ranges 0.7–2.6 km². Female home ranges were much larger and averaged 10–15 km² but with a core of 6–8 km² during optimum grazing conditions'.[79] Rhinos are thought to use what Owen-Smith refers to as 'safe corridors' to access water sources, and resident bulls use warning noises or confront intruding males horn to horn in ritualised confrontations before resorting to violence. Subordinate males use 'defensive snarl displays' in confrontations with more dominant animals.[80]

Female ranges often overlap and are not defended. Males without territories are nomadic until they establish a territory, and they move through the peripheries of occupied territories and do not engage in the scent marking at dung middens or dung scrapes that territorial males engage in to signal their control of territory.[81] Territory sizes may expand or contract according to the density of males, with young males able to control areas that were parts of the territory of older males as population pressures increase.[82] In KNP, Pienaar et al. found that the average male territory was 9.86 km², compared with 22.83 km² for adult females, with the female ranges overlapping with those of a number of males.[83] Wet-season ranges tended to be larger than those of the dry season, as rhino forage further when water is available some distance away from permanent sources.

Neighbouring territorial bulls have been observed to engage in ritualised encounters when they meet at a common boundary, according to Pienaar et al.[84] When rhinos of either sex encounter one another outside territorial areas, combative behaviour can result in order to keep a reasonable distance, protect a food resource or get access to water or shade. If fights develop, rhinos thrust their horns and forcefully press their bodies against one another, which can result in minor injury or in extreme cases in serious injury or death.[85] Such encounters may lead to dominance relationships, which have been observed at feeding places and waterholes.[86] Snarling or grunting noises are associated with encounters, suggesting strategies to avoid physical contact. Mock charges and horn fencing are also used.[87] Rhinos are active for about 50% of the night and day, with no great difference in activity between the two periods.[88]

Diet and foraging

White rhinos are exclusively grazers and prefer short grass areas – often referred to as rhino lawns when they have mown them right down to a level any greenkeeper on a golf course would be proud of.[89] They prefer short grass species such as those in the genera *Cynodon*, *Digitaria*, *Heteropogon* and *Chloris*. Grasses grazed in this fashion have evolved to regrow quickly, providing rapid replenishment of the food resource. Rhinos shift to other areas and even other grass species, such as *Themeda triandra*, which are less palatable, while their favoured plants regrow.[90] As large herbivores with a huge gut capacity, they can consume taller, coarse grasses in large quantities in areas where these have grown up over more nutritious grasses, and so widen the range of grasses consumed during droughts or annual dry seasons.[91]

Rhino interactions with predators

Rhinos have little to fear from predators once they are adult, unless they are weakened by sickness, injury or age. Lions have been known to kill rhinos, as have hyenas. Brain et al. looked at lion predation on black rhino, and came to the conclusion that rhinos less than two years old were at the greatest risk of predation by lions, and that there was an increased danger when the mother of a calf had been dehorned, although no one had observed a rhino calf being killed in this situation. Cows defend their calves vigorously, and examples have been recorded of rhinos killing lions to protect calves.[92] Despite the threat posed to young and subadult rhinos by lions, over a 13-year period of monitoring in the corridor linking Hluhluwe and iMfolozi reserves there were no records of lion predation on rhinos, though there was evidence of hyena predation on small calves. However, the authors found evidence of three subadult rhinos having been killed by lions in Etosha in 1995.[93] All were between three and four years old and killed by strangulation.

Sick rhinos or calves not well protected by their mothers may be attacked and injured or killed by lions or hyenas. Hyenas tear at the tail, anal region, scrotum or underside of a weakened rhino, which may not recover from the injuries or may succumb to blood loss or shock. Young rhinos not successfully defended by their mothers are at much greater risk and may succumb to injuries even where a mother is persistent in defence.[94] Martin notes that young black rhino can be vulnerable to both lions and hyenas, as these young rhinos follow their mothers, so may be attacked and injured before the mother has seen the danger and can defend her offspring. White rhino calves on the other hand tend to walk ahead of the mother, giving her a view of any other animal approaching the calf.[95]

On one occasion, the Namibian conservationist Garth Owen-Smith reported seeing a small rhino calf alone, with no sign of the cow, being attacked by three jackals, which bit at its ears and groin. He wrote 'They were unlikely to pull the calf down and kill it, it would only be a matter of time before a hyena heard the commotion and came to investigate.'[96] A hyena would be perfectly capable of killing an unprotected calf. Corroboration of the ability of hyenas to kill rhinos has come from the South African Wildlife College. In February 2018 the college reported on its website, with accompanying photographs, the killing of a large white rhino cow with an injured leg by a group of hyenas in Manyaleti Reserve. The college's pilot tracked the rhino, which was known to be injured and in need of treatment,[97] and found it near a dam being attacked by five hyenas as it sought refuge in the water.

After a prolonged period of biting at the base of the rhino's tail and grabbing her ears, the hyenas were able to push the rhino's head underwater and drown it. A postmortem by Dr Kyle Piers, who is attached to the Hluvukani Animal Clinic, indicated that the rhino's left front leg was badly injured and that there was trauma to its left lung.[98]

Interactions with elephants and other herbivores

Rhinos, black or white, take little notice of elephants unless a cow with a calf feels threatened and attempts to deter the elephants from coming too close by using angry snorts and mock charges.[99] Rhinos generally give way, at waterholes for example, to adult bull elephants. In July 2009 I witnessed conflict between a white rhino with a calf and a small herd of elephants at the Shamwari Private Game Reserve in the Eastern Cape of South Africa. Vocally, and with short, sharp charges towards the elephants, the rhino cow made plain her desire for the elephants to keep their distance. The rhino calf joined in with squeals and mock charges. However, three adolescent male elephants – which I had already seen engaged in quite rough mock fights among themselves – started harassing the rhino and her calf, leading to much more serious-looking charges. After about ten minutes of increasingly dangerous-looking interactions, several of the adult females in the elephant herd intervened to deter the rhino charges and move the adolescent elephants away. Berger and Cunningham studied interactions between rhinos and elephants at the artificially lit waterholes in Etosha NP, and found that threats by rhinos or elephants only occurred in 18% of encounters, and that in 54% of encounters between female elephants and rhinos, the elephants displaced the rhinos, whereas female rhinos displaced elephants on 14% of occasions. Male rhinos displaced elephants on 65% of occasions.[100] Only once was a rhino seen to strike an elephant with its horn; the seriousness of the injury was unknown.

A modern account of an elephant killing a rhino came from KNP after white rhinos were relocated there from KZN in the early 1960s. The rhino conservationist Ian Player reported that a large white rhino bull had been killed by an elephant at a waterhole near Shingwedzi, after both fell down an embankment and fought, the rhino not giving way to the elephant. The rhino carcass had been gored by tusks on one side of its body and had two deep stab wounds caused by thrusts from the tusks.[101] Elephant attacks on rhinos were prevalent in the mid-1990s in Pilanesburg NP in South Africa. Reintroduced rhinos and elephants coexisted until a group of young male elephants was introduced. In April 1994, they started attacking and killing rhinos. The young elephants were orphans from culls elsewhere and were resettled along with several females. There were no large males for them to associate with once they left the females as they reached about 15 years of age. They formed groups of young males and it was these that killed rhinos; the casualties had serious puncture wounds from tusks, broken ribs and internal injuries from crushing. Greg Stuart-Hill, the region's chief ecologist, said there were few doubts that six dead rhinos found with similar injuries had fallen prey to the same killers.[102] Seventeen rhinos were thought to have been killed in total.[103] Several of the elephants were shot, but the problem disappeared when more adult female and some adult bull elephants were introduced, effectively socialising the remaining young bulls.[104]

Although it is clear that elephants can and will kill rhinos, elephant depletion of browse or grazing during droughts can be more deadly. This happened repeatedly in Tsavo in the late 1960s and early 1970s, when long rain-free periods resulted in less and less browse being available for black rhinos given both the lack of rain and the destruction of bushes and trees by elephants desperate for food. About a third of Tsavo's rhinos died in 1960–1, and more in 1964–6.[105] Daphne Sheldrick, wife of warden David Sheldrick, wrote of seeing rhinos successfully competing with female elephants for access to waterholes but being driven off, and one rhino being attacked by a bull elephant.[106] A study by Landman et al. in Addo Elephant NP in South Africa showed changes in rhino diet in the presence of elephants, with the latter reducing browse availability.[107] There was greater competition, to the detriment of the rhinos, where the two species existed in fenced areas or where there was limited suitable food to begin with. When elephants were present, rhinos had to forage more and consume a wider variety of plants.[108]

Interactions of black and white rhinos with humans

At each stage of evolution from apes and the early australopithecines to *Homo sapiens*, hominins became more upright in posture, more dextrous and more advanced socially and technologically, enabling them to use a greater variety and sophistication of tools and weapons to forage, scavenge from carcasses and hunt. Climatic changes and the corresponding changes in vegetation and habitat increased savanna and open woodland, providing habitat and food for a great variety of herbivores and predators. This also led to greater changes in hominid and then early human societal organisation, food gathering, scavenging and hunting.

One adaptation to savanna living was increased social organisation and expansion of groups, necessary for survival in the more open environment, with the need for defence against predators as well as increased foraging, scavenging and then hunting. As social organisation and tool use advanced, hominins could competitively scavenge from large predators and coopera-tively hunt or trap larger prey.[109] The making of tools and weapons made hunting of large herbivores and competitive scavenging possible with less risk of death or injury.[110] The climatic changes that produced savanna and open woodland provided habitat for a wealth of possible prey and extensive carcass-scavenging opportunities for hominins. In the millennia between the evolution of *Homo sapiens* and the first modern accounts of human observation of rhinos, it is clear that human hunting methods developed the use of spears, bows and arrows and poisoned arrows, as well as pits and other traps for large mammals such as rhinos. Right up until the European penetration of Africa, rhinos would have been hunted for their meat and hides by some hunter-gatherer communities and then combined hunter-farmer communities. However, there was no great effect on rhino numbers or distribution until demand, from outside the rhino ranges in sub-Saharan Africa, for horn or other body parts led to increased killing.

CHAPTER 2

From 20,000 BCE to European penetration

Early human settlement and evidence of wildlife utilisation

Between 20,000 BCE and the first 1,500 years of the Common Era (CE), the distribution of the African species of rhino formed into the ranges across western, central, eastern and southern Africa that were in place when the first European penetration of Africa began, with more detailed recording than before of the presence of wildlife. It is believed that before firearms began to be introduced into Africa around the seventeenth century, there were at least 500,000 black rhino and over 100,000 white rhino on the continent.[1] The two subspecies of white rhino became divided into their northern and southern ranges, the former in a belt across central Africa from southern Sudan to Chad and the latter south of the River Zambezi. Those that had inhabited Rwanda and Tanzania, now known only through fossil evidence and some reports by nineteenth-century hunters and travellers in the Karagwe region, dwindled in number before disappearing altogether.

Black rhinos were present in many areas of West, Central, East and southern Africa – something Norman Owen-Smith says is evidenced by San hunter-gatherer rock paintings in the latter region, with black rhino images found along the Kunene and Zambezi rivers. He noted the similarity in naturalistic representation of animals between the San images and those found elsewhere in Africa, including in Libya and Algeria, where rhinos had once been present,[2] but became extinct owing to the drying of the Sahara and the Mediterranean littoral, and the substantial increase in human populations along the coast and in the northern Nile Valley. The earliest records of rhinos in Africa are these rock paintings. In Niger, rock paintings depicting rhinos have been found that are dated at 2,000 years before the present.[3] The archaeologist Mary Leakey studied painted images on rocks in central Tanzania, where finds of stone and bone tools and other human artefacts were dated at between 50 kya and 1.5 kya.[4] Some early but undated images show humans, crudely drawn antelopes, birds and a large animal with a long upper lip and

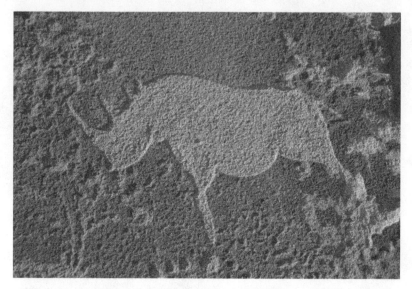

Rhino rock image, Twyfelfontein, Namibia. Courtesy of the Trustees of the British Museum. (Copyright David Coulson/TARA)

what could be two horns, probably a black rhino. One depiction is of tiny humans superimposed on a large white rhino,[5] while another site has images of two rhinos, whose species are not obvious. An image from the Kisese 1 site in central Tanzania clearly shows a white rhino, suggesting their presence in Tanzania. Rhinos are prominent in rock paintings from Twyfelfontein in Namibia (date unknown). The rhino in one picture lacks the large hump at the back of the neck which is characteristic of white rhino, and so could be a black rhino.

As human populations grew in sub-Saharan Africa, settlement (as opposed to shifting hunter-gatherer activity) and agriculture emerged, and the effects of human agency on the environment and wildlife increased. People defended livestock from predators and competed with herbivores for grazing and water. Hunting capability improved, especially with the development of projectile weapons and the advent of ironmaking, enabling humans to kill a wider range of prey animals for food, hides and, latterly, horns or tusks.

From soft to hard boundaries

Around 25,000–20,000 BCE what Reid calls the 'soft boundary' between people and their environment, with humans gathering wild plant foods and hunting/scavenging for meat, was evolving into a 'hard boundary' with domestication of animals, cultivation of plants for food and attempts to exclude wildlife from cultivated areas.[6] As early as 20,000–19,000 BCE, according to evidence from the Nile Valley, humans began the intensive exploitation of wild tubers and fish. Around 15,000 BCE, communities along the northern Nile started harvesting wild cereals,[7] To the west of the Nile, peoples of the Nilo-Saharan and Niger-Kordofan groups, pursuing livelihoods mainly based on foraging, hunting and fishing, developed tool-making, and hunted with spears, bows and arrows.[8]

Domestication of cattle in Africa occurred in Egypt about 10,000 years ago. The climate was wetter there than it is now, providing grasslands for cattle and suitable conditions for early crop production.[9] Archaeological evidence from between 7,000 and 5,000 BCE indicates people were raising animals and growing plants for food in the south-eastern Sahara, the Sudanic regions of southern Egypt and northern Sudan along the course of the Nile, and then spread west and east to West Africa's savannas and the Ethiopian highlands.[10] Ehret says that by the eighth millennium, Saharo-Sudanese people of the eastern Sahara were cultivating grains previously gathered from the wild and building thorn fences to protect livestock from predators.[11] Herding of cattle spread up the Nile into Sudan around 6,000 years ago and then, 1,000–2,000 years later, into Ethiopia and Kenya, with goats and sheep also raised and then cultivation of grains such as tef and finger millet, and later wheat and barley. Population densities were low, and apart from conflict between pastoralists and predators and attempts to exclude animals from cultivated crops, the effects on wildlife, especially megafauna such as rhinos, are likely to have been marginal. Rhinos were once present in the coastal plains of North Africa, but had disappeared or were in the process of disappearing at the start of the period under review. The exact reasons for their disappearance are not clear, but they appear to have declined owing to the effects of climate and vegetational change, with perhaps some influence of human population growth. The human societies there were based on cultivation and livestock husbandry, which, along with the increase in size of human settlements, could have gradually pushed out wildlife.[12]

Farming and livestock cultures spread across North Africa, along the Nile and into areas inland such as the Hoggar Mountains of Algeria and Lake Chad Basin, Sudan and East Africa.[13] Iliffe suggests that metal-working in Africa started in the Nile Valley in the fifth millennium BCE, possibly introduced by migrants from North Africa or the Middle East.[14] The Cushitic and Nilotic communities that moved south and east into the Horn and Kenya around 5,000–4,000 BCE, and into northern Tanzania after 3,000 BCE, brought pastoralism and crop production.[15] They came into contact with and displaced or assimilated hunter-gatherers, the ancestors of the people who became known as Ndorobo, Hadza and Khoi-San.[16] The growth of cultivation and pastoralism enabled increases in the size of settlements in savanna areas. Food production was more reliable than foraging, though this remained an option when drought or other factors limited food production. This started to change humankind's relationship with wildlife, as protection of crops became a necessity.[17] Transhumance and shifting cultivation was practised because of low population density and a lack of competition for land. This limited the depletion of wildlife in any one area through hunting or the presence of large, permanent human and livestock populations. Rhinos may have been opportunistically hunted but did not constitute a threat to humans, their crops or their livestock. This would have been a dangerous activity, and trapping them in pits an arduous one that might not have been repaid in terms of meat and hides; human activity probably had a negligible effect on rhino numbers or distribution, though evidence is lacking and we can only speculate. There is not a great deal of archaeological evidence of the hunting of large mammals such as rhinos or elephants. As Wilmsen observes, 'Megafauna, including … elephant, hippo and rhino, are normally butchered at the place where killed or the carcase found, hence their bones, tusks, teeth and horns are very rare in archaeological sites, and any intercontinental ivory or rhino-horn trade would leave few traces in the locales of production.'[18]

As human societies expanded and developed, trade became part of their economies. This would have included commerce in wildlife products, including at some stage rhino horn and hide. Initially between neighbouring communities, it would have spread to include trade between inland and coastal communities or those who could engage in trade along major waterways, such as the Nile, Niger or Zambezi. Trade grew within sub-Saharan Africa and most importantly between inland communities and the growing communities, kingdoms and then empires along the Nile Valley. Coastal communities began trading across the Red Sea and Indian Ocean.

Egyptian, Greek, Roman and medieval accounts of the rhino

In Dynastic Egypt, animals were frequently represented in wall paintings, statues and carvings; most were species found in Egypt, but some may have been animals presented to rulers by emissaries from other kingdoms. Depictions of the rhino would have been drawn from observations or descriptions of northern white and black rhino in the far south of Lower Egypt and in Sudan, though some may have been taken from ancient rock paintings in the Libyan desert that included rhino.[19] Rock drawings in Egypt itself, such as those near Hosh and Wadi Sab'er on the west bank of the Nile between Aswan and Luxor, depict what is clearly a black rhino, which has been put at the same period as a pair of models of rhino horns from the time of the second pharaoh of the First Dynasty, Hor Aha, dating to around 3,100 BCE. There is also a depiction of a white rhino on a sunken relief from the Eighteenth Dynasty and Tutmosis III (1,479–1,425 BCE) on the west bank of the Nile 16 km south of Luxor.[20] One account tells of Tutmosis of the Eighteenth Dynasty hunting in southern Nubia and killing a rhino (species not given),[21] In keeping with the geographically widespread beliefs over millennia in Africa, Europe and Asia that rhino horn counters or reveals poisons, in Dynastic Egypt rhino horns were believed to have magical powers, and the Bedouin of the eastern Egyptian desert used powdered rhino horn as an antidote for scorpion stings and snake bites. There are also records of rhinos being captured in Nubia and transported down the Nile to Alexandria to be sent to Rome.[22]

The Greek writer and medical practitioner Ctesias, in the fifth century BCE, wrote about a creature with a 'purple head' and a horn, which could be removed when the animal was killed, and used to detect poison if a drinking cup was made from the horn.[23] Ctesias's description makes this animal sound like a cross between an ass and an Indian rhino. The early, totally inaccurate descriptions of the rhino are clearly the origin of the linking in the European mind of the rhino with the mythical unicorn; the Indian one-horned rhino is now the bearer of the Latin name *Rhinoceros unicornis*. In *The Book of Beasts*, a Latin compilation from the twelfth century CE, it is said that the Greeks called the rhino a unicorn and described it bizarrely as a 'very small animal like a kid, excessively swift with one horn in the middle of his forehead, and no hunter can catch him'. The account, clearly based on the myths about the unicorn, goes on to say that the unicorn/rhino can only be caught by leading a virgin girl to where the animal resides; it will then leap into her lap, where it can be caught. It is also, despite its seemingly small size, credited with being able to fight elephants.[24] Later in the work there is a slightly more believable description of the rhino/unicorn as the monoceros, said to have 'a horse-like body, with the feet of an elephant' and a horn in the middle of its forehead.

Along the Mediterranean coast, the rise of the Carthaginian Empire, dating from 814 BCE, introduced a powerful state with large towns, more developed agriculture and a strong army. When Carthage and Rome weren't at war with each other there was a regular trade in live animals and wildlife products from Carthage to Rome. Carthage sent trading caravans into the interior of Africa, exchanging its manufactured and agricultural goods for salt, gold, live animals, ivory, skins, horn and hides.[25] By the fifth century BCE, the trans-Saharan traffic in live animals had become so lucrative that the Carthaginians sent military major expeditions across the Sahara to ensure the control of commerce in wild animals.[26] Trade was disrupted by the defeat of Carthage in the Third Punic War (149–146 BCE), and it became a province of Rome. Romans would then have come into direct contact with African wildlife, including rhinos, in the parts of their growing empire from Mauritania through Numidia, Cyrenaica and into Egypt. Rome controlled the Mediterranean end of trans-Saharan trade and continued the commerce in animals bound for the arenas of the empire.[27] Martin believes that northern white rhino were captured around Lake Chad and sold for display or combat at the Roman games. He writes that, in the first century BCE, Romans called Julius Maternus and Septimus Flaccus crossed the Sahara to Lake Chad, crossing the Fezzan region to reach the lake, and also journeyed eastwards to Ethiopia. They recorded seeing rhinos, though they did not describe them in detail.[28] It is clear that some of the rhinos exhibited in the arena or kept in menageries came from Africa, and there were also Asian one-horned rhinos killed in the arena.[29]

One account suggests that the Emperor Trajan exhibited over 11,000 wild animals in games in 106 CE, including rhinos and elephants from Chad or Bahr al Ghazal in South Sudan.[30] Most of the animals died in combat with gladiators or during an ancient version of canned hunts with the animals trapped in the arena. Auguet researched the role of animals in the arena, noting that rhinos were frequently chosen to fight each other or to fight elephants, but of all the animals forced to fight, the rhino 'was the one whose reluctance to fight showed the most character'.[31] The rhinos were jabbed with lances until they became enraged enough to charge, but they often only charged once against the animal with which they had been matched, and then retreated to the edge of the arena to start butting the podium surrounding the arena. When they were matched against bulls and bears and could be forced into action, the bulls were 'eviscerated like straw dummies … [and] the bears it threw into the air like puppies'.[32] The Roman writer Pliny in his *Natural History* makes a short reference to a rhino appearing in games organised by Domitian, prior to his becoming emperor in 81 CE.[33] Pliny described it as a one-horned rhino, according to Bostock's commentary on Pliny's writings, while other chroniclers of the time said it was two-horned. As Bostock wrote, citing the nineteenth-century naturalist Georges Cuvier:

> At the same games the rhinoceros was also exhibited, an
> animal which has a single horn projecting from the nose …
> This too is another natural-born enemy of the elephant …
> and in fighting directs it chiefly against the belly of its
> adversary, which it knows to be the softest part.

The commentators have been at a loss to reconcile this description with the Epigram of Martial, *Spectam.* Ep. xxii., where he speaks of the rhinoceros exhibited by Domitian as having two horns. It has been proved that this latter was of the two-horned species, by the medals of that emperor, now in existence.[34]

Martial's major work on the Roman games was the *Liber de Spectam*, dated 80 CE. It was written to celebrate the 100 days of games held by Titus to inaugurate the Flavian Amphitheatre.[35] Titus ruled until his death in September 81 CE, when he was succeeded by Domitian. Martial wrote of the rhino:

> it is enormous, the largest of all land animals save the elephant. The adult rhinoceros can weigh as much as three or four tons … it is funny-looking. Apparently clothed in metal plates, it looks as if it has sprung fully-armed from the brow of Nature. It has two horns … when undisturbed it is naturally diffident … but the unexpected aspect of this preposterous quadruped is its explosive anger and incredible power when annoyed.[36]

The rhino that Martial describes appears to have the armour-plated appearance of the Indian one-horned rhino and the two horns of either the black or white rhino of Africa, a common mixing of attributes of the different species. Domitian had a coin minted depicting a rhino. It was not just a way of declaring ownership of this powerful animal, but had everything 'to do with imperial advertisement … "This is my rhinoceros". Domitian's rhino, in its supremacy in the arena. might well stand as a metaphor for the invincible success of the emperor himself.'[37] The coins clearly show a two-horned African rhino.

The Greco-Roman poet Oppian, who lived in the second century CE during the reigns of emperors Marcus Aurelius and Commodus, composed the *Cynegetica* (trans. *The Chase*), in which he gave descriptions of a variety of wild animals, including the rhino. In one passage he sets out some rather fanciful characteristics of a range of creatures:

> The Rhinoceros is not much larger than the bounding Oryx. A little above the tip of the nose rises a horn dread and sharp, a cruel sword. Charging therewith he could pierce through bronze … he attacks the Elephant strong though it be and many a time lays so mighty a beast dead in the dust.[38]

Claudius Aelianus (*c*.175–*c*.235 CE), commonly called Aelian, described a rhino as having:

> a horn at the end of its nose, hence its name. The tip of
> the horn is exceedingly sharp and its strength has been
> compared to iron. Moreover it whets it on rocks and will
> then attack an elephant in close combat, although in other
> respects it is no match for it because of the elephant's height
> and immense strength. And so the rhinoceros gets under
> its legs and gashes and rips up its belly from below with its
> horn, and in a short space the elephant collapses from loss
> of blood ... If however the rhinoceros is not quick enough
> to do as described but is crushed as it runs underneath,
> the elephant slings its trunk round it, holds it fast, drags
> it towards itself, falls upon it, and with its tusks hacks it to
> pieces as with axes.[39]

Romans, and the Greeks before them, were part of global trading networks, and bought rhino horn, ivory, tortoiseshell and other wildlife commodities. In the *Periplus of the Erythraean Sea*, an account of the geography, empires, kingdoms and trade routes along the Red Sea and Indian Ocean coast of Africa in the first century CE, its author referred to the port of Adulis, situated in the region that is now part of Eritrea. From Adulis, rhino horns and ivory were exported, the animals having been hunted inland or along the coast. Rhino horns were shipped along the Red Sea to Egypt and across to Arabia.[40] Some of them went to rich buyers in Greece and Rome. In Schoff's notes on the *Periplus* he records that, under the Emperor Marcus Aurelius Antoninus (about 166 CE), trade emissaries were sent to China offering ivory, rhino horn and tortoiseshell.[41] The trade routes across the Sahara that had supplied Egypt and Rome still operated, though at a lesser volume, after the disintegration of the Western Roman Empire in the fifth century, with European or North African goods being traded by Sahelian communities for ivory, rhino horn, slaves and gold from West Africa.[42]

There is a gap in the written accounts available to researchers of wildlife and trade in animals or animal parts between Africa and Europe from the end of the Roman Empire until around the thirteenth century, when Marco Polo saw Sumatran rhinos in the Far East. and there is evidence of trade between Asian countries and Arab merchants in rhino horns, most of which are likely to have come from Asia and were sold as unicorn horns. These found a ready market in Europe, because of the myths about the curative properties of the horns. Some horn was ground into powder to treat a variety of ailments and to detect poison.[43] Archduke Ferdinand of the Tyrol and the Holy Roman Emperor Rudolf II in the sixteenth and early seventeenth centuries were collectors of rhino carvings. The Museum voor Land en Volkenkunde in Rotterdam has 77 rhino-horn cups, and the Beatty Library in Dublin has 180 rhino carvings from China.[44] What was termed unicorn horn was an important addition to European, Arab, Indian and Chinese pharmacopoeias from the late medieval period right up to the eighteenth century. Unicorn horn was listed as a medicine in England as late as 1789, and the French court, prior to the revolution, used supposed unicorn horn to test the king's food for poison.[45]

East Africa, Sudan and the Horn of Africa

By 2,000 BCE, pastoralist communities keeping sheep, goats and cattle were widespread along the Rift Valley through Kenya and into northern, western and central Tanzania.[46] The area was rich in wildlife, including rhinos, elephants, buffalos, a wide range of antelopes, zebras, wild pigs, and predators such as lions, hyenas, leopards, cheetahs and wild dogs. Archaeological excavations in south Nyanza in Kenya indicate that although pastoralism was practised by the second millennium BCE, remains of wild animals from Neolithic sites show that hunting persisted.[47]

The climate of much of Africa became more arid from the fourth millennium BCE. Savanna and open woodland expanded at the expense of denser forest; this was good for pastoralism and provided suitable conditions for large ungulates, favouring grazers such as white rhino and browsers in open woodland and thornbush areas such as black rhino. There is evidence from the archaeological record that pastoralists, hunter-gatherers and arable farmers lived alongside a diversity of wildlife in the Kenya and Tanzania region.[48] Gifford et al. provide evidence of wild ungulates around Nakuru-Naivasha in the last two millennia BCE from bone assemblages showing 'a diverse range of herbivores living in close proximity to the neolithic pastoralists'.[49] The diversity of wildlife, including, one must surmise, large numbers of black rhino present across suitable habitat in East Africa, was much as it was at the start of the growing penetration of the region in the nineteenth century by European explorers, hunters and traders in the pre-colonial stage of European involvement in Africa.

The fourth century BCE saw the rise of the Axumite Empire in northern Ethiopia and Eritrea. It was the dominant regional power, with strong regional trading links. Hunting was the source of ivory, one of the main commodities traded, but rhino horn was also part of the commerce down the Nile to the Mediterranean. Axumite merchants traded as far north as

Alexandria in Egypt and via Mediterranean routes with Rome and across the Red Sea to Yemen, the Arabian Peninsula and south to Somalia.[50] Anfrey, in his account of Axum and its trading relations, noted that Axum's control of the port of Adulis (40 km south of Massawa) enabled it to trade with India and the Mediterranean (including the Roman Empire). Rhino horn, ivory, hippo hide and tortoiseshell figured in the overseas trade, as did gold, slaves and spices.[51] Farming and hunting capabilities improved when Stone Age technologies were replaced by ironworking over several thousand years.[52] Iron tools enabled more extensive clearing of forests for cultivation. This was linked, between 3,000 and 1,000 BCE, with the arrival of Bantu migrants from Central Africa around the Great Lakes and Rift Valley.[53] Technological innovations were complemented by a growth in intercommunal trade and commerce along the Indian Ocean coast at the end of the BCE period and during the first millennium of the CE period. During the eighth century CE, commerce between the interior of East Africa and the Indian Ocean coast developed.[54] The extent and exact nature of the seaborne trade is hard to assess, but as Sutton writes, 'there is of course literary evidence of sailing in the very earliest centuries AD to Rhapta and other unidentified East African harbours from the Red Sea, itself in trading contacts both with the Roman Empire and with India'. The trade was not substantial but it did include rhino horn.[55] Sheriff believes Rhapta to have been situated in the centre of Tanzania's coastline, well placed to trade with peoples inland to obtain wildlife commodities, including 'a great quantity of ivory, rhinoceros horns and high quality tortoiseshell'.[56]

Rhino horn appears to have been traded along the Indian Ocean coast in the last centuries of the BCE period, with Arabian merchants buying it along with ivory and tortoiseshell.[57] Trading expeditions to East Africa by Arab merchants were taking place by the first century CE. Ehret notes that merchants from southern Arabia traded regularly with the coast of the Horn of Africa, buying rhino horn to sell on in Asia, where it was used in medicines.[58] Trade increased greatly from the eighth century CE, with Arab traders settling on the coast and traders from the Arabian Peninsula exporting African rhino horn and ivory to India and China, and evidence of significant demand for it there at this time.[59] From around 1,000 CE, the trade networks on the Indian Ocean coast grew, with increasing settlement and intermarriage by Arab traders with the Indian Ocean coastal peoples.[60] The Yemeni port of Aden was one of the trading hubs, even before the Islamic period. The level of demand in the Yemeni region itself for rhino horn at this time is unknown, although 'a rhino horn dagger belonging to one prominent Yemeni family is inscribed with the name of a ruler from the early thirteenth century'.[61]

The Swahili trading communities, which emerged from intermarriage between East African coastal peoples and Persian, Arabic and Indian traders nearly 1,100 years ago, and the growing communities of Arab and Persian merchants (at Lamu, Mombasa, Bagamoyo, Kilwa and Zanzibar) traded in rhino horn, ivory and skins, shipping them to the Arabian Peninsula and India.[62] During the medieval and early modern periods, these evolving coastal communities traded goods such as glass beads, cloth and ceramics for raw materials from Africa, including gold, ivory, rhino horn, animal skins and slaves. By the fifteenth century, the Swahili trading ports along the Indian Ocean coast were trading constantly with Asia, with rhino horn among the commodities in demand.[63] The value of wildlife products stimulated hunting for trade rather than just subsistence among communities such as the ancestors of the Ndorobo, Waliangulu and Wakamba of Kenya and the Nyamwezi of central Tanzania; some became both hunters and commercial middlemen, supplying wildlife products from inland to coastal traders. Arab traders working for the Sultan of Oman came to dominate Zanzibar and the surrounding islands, and the port of Kilwa in southern Tanzania, during the latter half of the seventeenth century.[64]

In the second millennium CE, China imported ivory and probably small amounts of rhino horn from East Africa via Arabia. There is evidence of African ivory reaching China in the tenth and eleventh centuries CE.[65] In 1417 and in the 1420s and 1430s, Chinese trading fleets visited the Juba River in Somalia. They took back ivory, other wildlife products, gold and spices.[66] After the fifteenth century, Chinese imports of commodities from East Africa were obtained via Arab and Indian traders on the East African coast. There is evidence from examination of the history of Chinese art forms, including carved rhino horn pieces, that 'the primary characteristics of ancient Chinese society were those in which rare resources were acquired to show the potential of power' and that 'rhino horn is a valuable, rare material' – and so was a valued and traditional medium for artistic expression.[67] Carved rhino horn art and items such as cups and bowls date back to the Tang Dynasty (618–907 CE),[68] and grew again in popularity in the mid-fifteenth century, their forms and use developing over the next 100 years. The Ming (fourteenth to seventeenth century) and Qing/Ching (seventeenth to twentieth century) Dynasties of China were the most prosperous and the most important for the development of rhino horn carving. Rhino horn products were used to replace hooks, belt buckles and hair clips made from other, less valuable materials.[69] During the Ming Dynasty, China still had rhinos in the central and southern regions, but shortages of horn for carving occurred as numbers declined there, requiring imports from elsewhere in Asia or from Africa. There was at the time significant trade in a variety of products along the famous Silk Road from China through Central Asia to Persia and beyond.[70] In 1850, carving of rhino horn seemed to abruptly stop, presumably owing to a shortage of horn, and rhino horn cups or bowls began to take on a value beyond the artistic merit of the carving or the economic value of the material:

> Carved rhino horn cups were initially used to pour wine as
> an offering to ancestors at celebrations ... [and] the belief
> that cups made of rhino horn neutralize the poison because
> the rhino's metabolic system extracts only the beneficial
> elements of the herb and is stored in the blood it is also
> believed that the horns belonging to the male rhinoceros
> have special properties based on the principle of rarity.
> Rhino horn cups are therefore used in beverages to prevent
> ingested toxins from entering the body.[71]

Emslie and Brooks believe that the poison-detecting belief may not be entirely groundless as 'the alkaloids present in some poisons react strongly with the keratin and gelatine in the horn'.[72] The demand in China would have been a motivating force for the export of horn by the Omani Arab, Afro-Shirazi and Swahili communities that controlled overseas trade on the Indian Ocean coast.

Ethiopia, Somalia and Sudan

In western Ethiopia, excavation of sites at Ajilak (Gambella region) indicated extensive hunting of wild fauna in the first millennium CE.[73] The region's long-grass plains and bush terrain provided ideal habitat for a variety of ungulates, including rhinos, conceivably of both species. Leo Africanus, the Arab traveller and chronicler of the early sixteenth century, described life, the region's institutions, religion and inhabitants and its wildlife and trade in ivory and rhino horn in his account of travels in Africa.[74] He wrote that the people of Abassia, as he called it, paid tributes to the monarch in the form of rhino horns, ivory, hides, salt, gold, silver and corn.[75] An account of the history and fauna of Ethiopia in the early seventeenth century notes the presence of two-horned rhino in Ethiopia. Pedro Páez's *History of Ethiopia* refers to rhinos, calling them *abada* and reporting that they had one horn on the nose and one on the forehead.[76] The horn was black and thick at the root, with the horn on the forehead three fingers in width and three spans (about 67 cm) long. Páez noted that some of the horns from the rhinos had been sent to the Ottoman Sultan from Ethiopia.[77]

There were reports from European travellers to Ethiopia of a rhino, referred to as wide-cloaked, that appeared to have two horns like African rhinos but an armour-like skin more akin to that of the Indian one-horned rhino. Rookmaaker and Kraft investigated the accounts of this animal, and noted: 'The alleged wide-cloaked rhinoceros has the appearance of the armour-plated *Rhinoceros unicornis* endowed with two nasal horns. It was observed in Ethiopia first by James Bruce in 1772 and again by William Cornwallis Harris in 1842.'[78] It seems that there was a mounted example of this curious rhino in the Bavarian State Collection of Zoology in Munich from 1802, but it was destroyed in 1944. Rookmaaker and Kraft write that a 'partial lower jaw taken from the hide is still available, and is identified as one of a juvenile *Diceros bicornis*', adding that 'As no animal of this description has been seen again in Africa, it is presumed Bruce and Harris were led astray in their observations and recollections ... There is compelling evidence that the specimen was shaped like an Indian Rhinoceros by a taxidermist around 1780 following available representations in the literature.'[79] The authors also noted that 'The plate is easily recognized as a copy from Buffon's "Histoire Naturelle" ... a depiction of Clara, the famous Indian rhinoceros which toured Europe between 1741 and 1758 ... Bruce, or his publishers, merely added a second or posterior horn to make the animal double-horned.'[80]

Prior to the decline in Ottoman direct rule, increasing European interventions and then joint British/Egyptian hegemony over Sudan, there had been a centuries-old slave, ivory and wildlife product trade conducted from Khartoum. Rich Arab merchants there sent slaving and trading expeditions up the Nile from Khartoum to what is now South Sudan, but also into northern Congo, and as far west as present-day CAR and Chad. The main motivations of the trading/slaving expeditions were to obtain slaves, ivory and cattle, but rhino horn, hides and other animal skins were also in demand. Merchants sent the commodities to Khartoum, from where they were exported to Egypt, the Arabian Peninsula, Europe and Asia. Baggara and Rizeigat Arabs from Darfur and Kordofan were hired to provide the main armed force in the trading and raiding expeditions.

West and Central Africa

Savanna, thorn bush and dry woodland in West Africa provided suitable habitat for rhinos. Northern white rhino ranged across the northern Congo savanna, the CAR and Chad to the east in north-western Uganda and Sudan. Black rhino were likely to have been found across an area covering northern Cameroon, northern Nigeria, the Lake Chad Basin, Niger, northern Benin, northern Togo, Burkina Faso and possibly northern Ghana, Ivory Coast and as far west as Guinea, Guinea-Bissau, eastern Sierra Leone and Senegal. Proof of the presence of rhinos to the west of northern Nigeria through to the Atlantic coast is lacking, despite maps produced in the sixteenth century that show their presence.[81] The drying out of the region and increased desertification occurred around 2,000 BCE,[82] and the effects of human population expansion combined with deforestation progressively reduced wildlife ranges and numbers up to the present.[83]

From around 500 BCE, better tools and agricultural techniques enabled an expansion of agriculture as well as improved weapons for hunting and protection against dangerous animals. The population groups that coalesced to become the pastoralist Fulbe (called Peuhl in Senegal and Fulani in Nigeria) moved west and south as the Sahara dried out and grazing land was lost. They traversed the whole region of West Africa between the desert and the dense forest with their livestock.[84] By the beginning of the second millennium CE, populations had increased, settled communities had grown and kingdoms or empires had evolved, such as Ghana, Kanem, Songhay and Takrur in the Sahel.[85] The depletion of wildlife would have been a gradual process in savanna and Sahel regions. Hunting was not recorded as being of great economic importance. European state expansion and imperial competition for trade and, much later, territory kickstarted exploration along the coasts of West Africa and inland along river systems. Ivory was sought keenly, but was just one among many commodities that Europe obtained from Africa. European merchants exchanged cloth, cowries, beads, firearms and metal goods for wildlife products, slaves, gold, spices and copper. The Portuguese started the process when they tried to circumvent the hostile Islamic rulers of North Africa who monopolised trans-Saharan trade. The Portuguese were joined and competed with by British and Dutch merchants, who sent expeditions as far as the Kingdom of Benin.[86]

Southern Africa

Looking at the settlement of southern Africa since 12,000 BCE, the initial focus is on San and Khoikhoi peoples (collectively known as the Khoisan) moving south from near Lake Malawi through eastern Namibia and the Okavango. Others migrated from the northern end of Lake Malawi via the Zambezi and Limpopo Rivers. The Khoisan peoples had cultures of foraging, hunting and scavenging. Later contact with other communities, including early Bantu migrants, led to the Khoikhoi adoption of pastoralism in addition to hunting.[87] Cave and rock paintings of the Khoisan peoples recorded the wildlife around them and represented hunts. One large collection of San rock paintings is to be found in the Tsodilo Hills of north-western Botswana, near the Namibian border. The hills retain 'a remarkable record ... of human culture and of a symbiotic nature/human relationship over many thousands of years'.[88] The paintings on the rocks cover a large timescale – from about 2.5 million years ago through to the nineteenth century – and provide 'an insight into early ways of human life, and how people interacted with their environment both through time and space'.[89] They include images of both black and white rhino, denoting their presence in this arid region.[90] There is evidence from rock paintings in Zimbabwe of the presence in that region of both black and white rhino, as Roth notes:

> The early occurrence of both African species over large
> parts of Rhodesia south of the Zambesi is well supported by
> numerous bushman rock paintings ... the black rhinoceros
> at one time inhabited the whole of the densely covered
> Zambesi valley, and still extends north of it, no evidence was
> found in the literature that the southern white rhinoceros
> ... has ever in fact occurred in the low, hot valley below
> the Zambesi escarpment ... in Rhodesia there is abundant
> evidence that the species inhabited and perhaps preferred
> the watershed plateau between the Zambesi and Sabi
> drainages.[91]

Namibia is known for its extensive rock art, providing evidence of human activity and awareness of the wildlife that people coexisted with and hunted. According to Sullivan and Muntifering, 'Images of rhino are a notable component of rock art assemblages, from the Orange to the Kunene Rivers in west Namibia'; a site near Twyfelfontein has images stretching over two millennia, and contains 46 images of rhinos, of which 20 are clearly black rhino and the remainder most likely white.[92] In his research into the presence of rhinos in Namibia, Joubert found rock carvings south of Bethanie, near Keetmanshoop in southern Namibia, south-east of Luderitz, near Om Aar, at Brandburg, Kuiseb River, at Spitzkoppe, in the area south of Gobabais and at numerous other sites, indicating a wide distribution, and leading him to note that the large number and locations of rock carvings and paintings accord with reports from early European travellers of the widespread presence of rhinos in a variety of habitats, including the arid semi-desert regions and the dry Kaokoveld.[93] He concludes that 'It may therefore be assumed that in the era before 1900, the black rhinoceros was distributed from the Cunene River in the north, down to the Orange River in the south, and extended westwards to the eastern boundary of the Namib Desert.'[94]

Image of a rhino carved into rock in the Han Mountains, Namibia.
(Wikimedia Commons, Public Domain)

The original San and Khoikhoi inhabitants of southern Africa hunted a diversity of wildlife for food, but there is no evidence that the rhinos were sought after as food, unlike the hippos, which were highly valued by San hunters for their meat and fat.[95] They hunted with basic weapons such as arrows tipped with poison and spears. With the arrival in southern Africa of Bantu communities with iron weapons and tools, hunting weapons improved, with hunter-gatherers trading meat, skins, ostrich feathers, honey and other wildlife or natural produce for iron artefacts and other commodities.[96] Bantu migration south into Zimbabwe, Namibia, Botswana, South Africa and Mozambique pushed the San and Khoikhoi north-west and into the fringes of the northern Karoo and the Kalahari. The incoming Bantu agro-pastoralists settled in the eastern half of southern Africa about 2,000 years ago.[97]

By around the year 700 CE, Bantu-speaking communities had established settled communities in south-west Zimbabwe, north-eastern Botswana and northern South Africa in what became known as the Leopard's Kopje culture, combining pastoralism and cultivation with ironworking.[98] Trade developed between inland communities and those on the Indian Ocean coast. An important source of ivory, horn, hides and skins for the merchants at the southern Indian Ocean ports was between the Zambezi and Limpopo Rivers, and particularly the region where Botswana, Zimbabwe and South

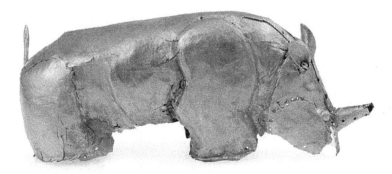

The golden rhinoceros of Mapungubwe.
(Wikimedia Commons – University of Pretoria Museums)

Africa meet. This included the kingdom of Mapungubwe, where hunting and trading took place from the ninth to the early fifteenth centuries.[99] It supplied wildlife products, gold and other commodities. That rhinos were common and in some way important to the Mapungubwe people was demonstrated when in 1934 archaeologists unearthed an intricately made rhino – with gold foil covering a wooden body. It was recovered 'from a royal grave at the site of Mapungubwe in northern South Africa close to the border of Zimbabwe. Its creation in the thirteenth century is a reflection of the wealth of the state of Mapungubwe, southern Africa's earliest known kingdom.'[100]

When the Mapungubwe civilisation went into decline in the late thirteenth century, trade switched to the Great Zimbabwe kingdom between the Limpopo and Zambezi, and later the Mutapa kingdom of Zimbabwe (1430–1760 CE). They both traded with Swahili merchants on the Indian Ocean coast and, from about 1560, with Portuguese trading posts established along the Mozambican coast.[101] The Portuguese competed to obtain ivory and other wildlife products with the Swahili-speaking and Arab traders, who had ports on the Indian Ocean coast from Somalia and Lamu in northern Kenya to the northern Mozambican coast. In these southern African kingdoms such as Mapungubwe and Great Zimbabwe, and also the Manekweni community of Bantu migrants living in the coastal lowlands of southern Mozambique from the twelfth to the sixteenth century, meat from hunting made up a relatively small part of food intake and much hunting was for horn, ivory or skins to trade on the coast.[102] The end of Great Zimbabwe's power led to the growth of the Monomutapa kingdom further north. At the time of Monomutapa rule, rhinos, elephants and lions were believed to be there in large numbers. The sixteenth and seventeenth centuries saw the establishment of Bantu settlements in northern and eastern South Africa that would become Xhosa, Nguni, Sotho and Tswana kingdoms.

Arrival of the Portuguese and competition
for Indian Ocean trade

When the Portuguese rounded the Cape and established trade along the Indian Ocean coast and on to India, they competed with established Arab and Swahili traders, especially those based at Kilwa and in northern Mozambique. The Portuguese established trading posts on the coast and the Zambezi Delta, obtaining ivory and rhino horn from the interior of south-central Africa and exporting much of it to India. The trade helped form a class of Indian merchants, known as Banyans, based in Indian Ocean ports and playing an important role in meeting the increasing demand for wildlife products by financing trading caravans sent into the interior.[103] Gujarat was the main destination for the rhino horn, where, as Emslie and Brooks point out, the idea developed that rhino horn was an aphrodisiac, as for a period it was used for that purpose there – a mistaken and ignorant idea now perpetuated by the world's media as a reason for Chinese use of rhino horn.[104]

The Portuguese took control of the Island of Mozambique and established a trading post at Sofala. This enabled them to take part in the growing trade with peoples based inland, via the Yao people of Mozambique and Malawi, who traded with inland communities. The Portuguese bartered guns with inland communities for slaves, ivory, horn and skins. The introduction of guns, and the steady expansion of human settlements and agriculture, all had the effect of gradually reducing wildlife populations in regions near the coast or main river routes.[105] Communities with guns could hunt more and raid neighbours to seize slaves for sale and tradable commodities, something the Yao undertook with alacrity. The supply of wildlife products, spices and copal (an aromatic tree resin) was extremely profitable for the coastal merchants and increased the commercial exploitation of wildlife in East and southern Africa. The merchants commissioned expeditions to shoot elephants and rhinos or buy horn, ivory and skins from hunters in the communities there. Until the development of trade from the interior to the coast, rhino horn had little value to hunters and the communities from which they came. If rhinos were killed, it was for meat and hides, not their horns. Meat was valued by some communities, and hide was used to make shields for warriors.

In what is now the Mpumalanga province of South Africa, communities developed in the area between Ohrigstad in the north and Carolina in the south, where archaeologists have found densely walled settlements with roughly circular homesteads.[106] The area is believed to have been the site of dispersed communities called the Koni, and has evidence of trade with the Indian Ocean coast: 'glass beads from the seventeenth to the early nineteenth century were found at the site of the Ndzundza capital KwaMaza ... While the key items in demand at the coast were ivory and to some extent rhino and other horns.'[107] The Pedi kingdom, growing in power in the late eighteenth century, established hegemony over the Bokoni, and the area was later largely depopulated after raids by the Ndwandwe, the Swazi and the Ndebele.[108]

European settlement and the growth of rhino hunting for horn, meat and sport, 1600–1800

By the early seventeenth century, European penetration of Africa had begun a new phase that would lead to European colonial occupation across the majority of the continent over the next 300 years. The Dutch East India Company (VOC), established in 1602, had trading ports in Asia to supply spices from Indonesia and needed resupply bases on the way there, which led to a Dutch settlement at Cape Town. This period saw increasingly aggressive trading practices on the part of the competing Portuguese, Dutch, British and French. Their trading and demand for wildlife products, gold and slaves prompted greater exploitation of African wildlife for hides, horns and tusks, and the expansion of the slave trade. The arrival of European traders and hunters, who ventured into the interior for profit and adventure, facilitated the importing and widespread distribution of firearms from the seventeenth century onwards. European traders encouraged increased hunting by indigenous people to supply the European demand for wildlife products. Arab traders sold muskets and gunpowder in North Africa, parts of West Africa and along the Red Sea to the kingdoms of the Horn of Africa and East Africa.[109]

Dutch occupation of South Africa's Cape region

The Dutch occupation of what became Cape Town and the surrounding areas commenced in 1652. In his journal, the first governor of the VOC, Jan van Riebeeck, recorded the occupation of the settlement and establishment of the revictualling station for VOC fleets bound for the East Indies. He wrote of trade and conflicts with the Khoikhoi and San communities, and about the wildlife there. The only early reference to rhinos was in his account of a journal kept by fugitives from Company employ who were captured after escaping from Cape Town in October 1652. One of them, Jan Blanx, recorded that having walked seven miles from the settlement they were confronted by two rhinos 'which threatened to attack us'.[110] The men escaped the rhinos but not the soldiers of the VOC. Cloudsley-Thompson wrote that at the time of the arrival of the first Dutch occupiers at Cape Town, the region abounded with wildlife, including black rhinos, hippos, elephants, buffalos, lions, wild dogs and a diversity of antelopes.[111]

On another occasion, a group of Company soldiers journeyed for three and a half days towards Saldanha Bay (146 km north of Cape Town) to investigate possible opportunities for trade with the local communities. They returned to report seeing 'many elephants, rhinoceroses, elands, harts, hinds and other game'.[112] There is little further mention of rhinos in Van Riebeeck's journals or correspondence with the VOC, but on 4 September 1608 he recorded that the directors of the Company had written to him that they had been informed that large amounts of ivory were available at the Cape. He wrote that some Company employees or free settlers had obtained ivory from Khoikhoi or San hunters and had sold it privately to sailors on passing ships. Van Riebeeck added: 'Whereas it is presumed that the freemen [Dutch released from Company employ], before this allowed to barter for tusks, rhinoceros horns and ostrich feathers on condition that they were to sell them only to the Company at 4 times the value of what they paid – slyly sell what they get to the ships' crews and others', and he resolved to stop the freedom to barter for ivory, horn and feathers other than by the Company, with fines to be introduced for breaking the regulations.[113]

One ship that went along the coast to buy livestock from Khoikhoi communities dwelling there which Van Riebeeck called the Chainouqas and the Gorachouqas (the latter later being known as Griquas) came back with no cattle; however, the crew reported that two or three ivory tusks had been seen, and some rhino horns, but had not been purchased for the Company.[114] In June 1659, Van Riebeeck recorded that 'a few rhinoceros horns' had been obtained from the Khoikhoi community led by Oedasoa.[115] Another exploratory expedition from Cape Town journeyed out past Salt River and Tygerberg, and near a hill they named Riebeeck's Kasteel (80 km north of Cape Town) saw lions, rhinos, wild horses (presumably quaggas or zebras), hartebeests and ostriches.[116]

Dutch observations of rhinos in the Cape led to a series of disagreements among amateur naturalists regarding the nature of the African rhinos and how they compared with Asian ones, especially the famous but fanciful representation by the German artist Albrecht Dürer of an Asian rhino. One problem, as Rookmaaker notes, was that

> The easiest distinction between the rhinoceroses of Africa
> and that of Asia would have been in the number of horns,
> two rather than one, but unfortunately Dürer's rhinoceros
> could be interpreted to show either one horn (large
> and nasal), or two horns (incorporating the one on the
> shoulders). It was necessary, therefore, to show that the two
> horns of the rhinos in Africa were both definitely positioned
> close to each other on the nose.[117]

Accounts produced by Dutch settlers and officials were at best ambiguous about the appearance of both black and white rhinos, and seemed always to refer back implicitly to Dürer's image. Pictures from the time copied Dürer's image and put it in an African context, in a scene with Khoi people, an ostrich and Table Mountain in the background.[118] One German naturalist, Peter Kolb, visited the Cape and, without ever having seen a rhino, said it had a dark-coloured horn on its nose and a small one further back. In 1727, he illustrated the account with a picture of a Dürer-style Asian rhino attacking an elephant; this inaccurate image was later replaced with one drawn after seeing the mounted skin of a two-horned African rhino at the University of Leiden.[119]

The Portuguese had made land on the Namibian coast a couple of times on their voyages to and around the Cape but had not established a foothold there, and neither had the Dutch. In 1760, a Dutch hunter, Jakobus Coetzee, accompanied by 12 Griqua hunters and servants, journeyed from the Cape into southern Namibia to shoot elephants for their ivory. In the Warmbad area of Karas in southern Namibia they reported seeing 'a multitude' of rhinos, lions and giraffes, animals that were already becoming rarer in the settled areas of the Cape.[120] In June 1792, a Dutch settler, William van Reenen, wrote a report for the Cape administration on his expedition into Namibia to search for communities of Hereros. He failed to find them, but did come across a settlement of Bergdamas (now more usually known as Damara) on the Namibian central plateau. He lost 140 oxen on the journey, but made up for the loss by shooting 65 rhinos (obtaining meat, hides and horn) and large numbers of other game.[121] William van Reenen's two brothers and a man named Peter Pienaar set out on a new expedition into Namibia, going by ship to Walvis Bay. They intended to buy cattle but failed. On a journey along the Swakop River valley, Pienaar reported seeing 300 rhinos (species not stated), a greater number of elephants and many gemsbok, springbok and lions. He said that he shot 20 rhinos and three elephants.[122] The meat from them was used to feed the local guides and servants whom he recruited to go with him.

Dürer's rhino: the image of rhinos for Europeans for nearly 300 years

The woodcut by Albrecht Dürer of an Indian one-horned rhino is one of the most famous and misleading representations of a rhino. Yet it became the go-to rhino image for Europeans for centuries and was often taken to be representative of African as well as Indian rhinos, causing confusion about the appearance of African rhinos.

Dürer had never seen a rhino, and relied on second-hand accounts. A sketch of the animal had been sent to Pope Leo X in Rome by King Manuel of Portugal, who also sent the rhino by sea as a gift to him, but it died in a

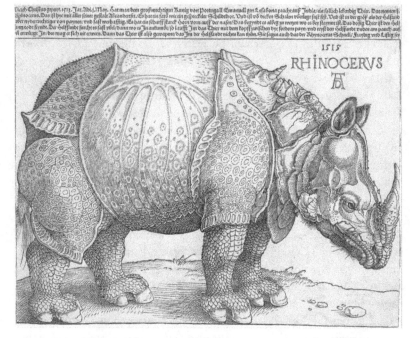

The famous woodcut of 1515 by Albrecht Dürer. (Public Domain,
Wikimedia Commons)

shipwreck so was never seen by the Pope. Dürer probably saw a Portuguese
newsletter that was sent to Nuremberg (where he lived) describing the
supposedly bitter rivalry between the elephant and rhino, taken from Pliny
the Elder.[123] These elements were used to build up Dürer's depiction, and the
art historian Ernst Gombrich said they were supplemented by elements of
Dürer's own imagination, especially ideas used in his depictions of exotic
beasts and dragons.[124]

CHAPTER 3

European penetration and occupation

B y the end of the eighteenth century, substantial numbers of guns were entering Africa, carried or traded by European hunters or traded with African hunters for slaves, horn, ivory and skins by Swahili, Omani and Indian merchants. They had a major effect on the ability to hunt for food, horn, hide, skins and ivory, and to raid for slaves. By 1888, around 100,000 guns a year were coming in through Zanzibar alone.[1] As Martin notes, 'Technical improvements to fire-arms in the nineteenth century made hunting easy. The new guns were more accurate and had a fire-power that made shooting big game a possibility for many who had never before hunted.'[2] The greater range meant that rhinos were increasingly vulnerable as hunters didn't need to get close to kill them. One Boer trader in South Africa is said to have supplied guns to 400 local hunters for the sole purpose of getting them to kill rhinos for their horns and hides.[3]

The extent of killing was staggeringly high, with rhinos and elephants suffering greatly from hunting for horn, ivory and sport. Between 1849 and 1895, about 11,000 kg of horn were exported from Africa annually, amounting to about 170,000 rhinos, with tens of thousands more killed in the first 60 years of the twentieth century.[4] Ivory was the main preoccupation of European hunters in Africa in the nineteenth century, but rhino horn was a very saleable commodity and hunters did not pass up opportunities to shoot rhinos. They used the meat to feed their retinues, and sold the hides and horns. Thousands of both black and white rhinos were killed, and the latter were brought near to extinction in southern Africa. By the late nineteenth and early twentieth centuries some hunters, colonial officials and naturalists were aware that the massive scale of hunting was depleting populations of rhinos and other species, and they started setting up reserves, tightening hunting laws and levying taxes on ivory and horn exports.

One move, more symbolic than practical, was the drawing up of the Convention for the Preservation of Wild Animals, Birds and Fish in Africa, which was signed at a conference in London in May 1900 that was attended by leaders from Britain, Germany, Italy, Spain, France, Portugal and Germany. It aimed to save species from 'indiscriminate slaughter' and to ensure the 'preservation throughout their possessions in Africa of the various forms of animal life existing in a wild state which are either useful to man or are harmless'.[5] It set out to impose export duties 'on the hides and skins of giraffes, antelopes, zebras, rhinoceroses, and hippopotami, on rhinoceros and antelope horns, and on hippopotamus tusks', to generate income from the hunting of these species.[6] In Britain, the conference and convention led to the formation of the Society for the Preservation of the Fauna of the Empire, which began to publish the *Journal*. However, little was done in practice to reduce the scale of legal hunting or seriously assess the effects on rhino populations of hunting for rhino horn or sport.

African rhino in the nineteenth century: status and species identification

Many European hunters who went on long expeditions sold specimens to collectors and museums in Europe and North America. They were interested in the wildlife beyond just shooting it, and engaged in arguments about how many species of rhino were found in Africa and how they differed. A more informed debate than the previous ones, which had been obsessed with Dürer's rhino image as a representation of all rhinos, developed based on first-hand evidence. At first they labelled what are now categorised as subspecies of rhino as distinct species. Minor differences between widespread populations were interpreted as indications of a diversity of species rather than subspecies or geographical variations in colour or horn configuration.

In July 1846, William Cotton Oswell saw a white rhino which he thought to be a different species from those so far identified, having a long anterior horn curving forward, and he used the local name *quebaaba* for it. It became known as *Rhinoceros oswelli*.[7] He saw five and shot two. On the same trip he shot two black rhino of the type referred to as *Rhinoceros keitloa*, known as the black rhino of the west. The proliferation of attributions by hunters fed into the taxonomic system. The naturalist Gray in 1869 divided African specimens in the British Museum into *Rhinaster bicornis* (*bovili*), *R. keitloa* (*keitloa*), *Ceratotherium simum* (*mahooboo*) and *Ceratotherium oswellii* (*kobaaba* or *quebaaba*). Drummond added a fifth species, based on direct observations in Natal, by dividing *Rhinoceros bicornis* into two sympatric subspecies, named *Rhinoceros bicornis minor* (*borele*) and *Rhinoceros bicornis major* (*kulumane*).[8] He wrote in 1876 in a paper for the ZSL that 'at present naturalists have arrived at no decided conclusions as to the number of species of Rhinoceros inhabiting Africa'. He believed there were four species, or five if you included *Rhinoceros oswelli*.[9] In 1881, Selous wrote to the Zoological Society that:

> In those portions of Southern and South Central Africa
> in which I have hunted I have only met with two true
> species of Rhinoceros – namely the large, square-mouthed,
> grass-eating species (*Rhinoceros simus*), and the smaller
> prehensile-lipped rhinoceros, which feeds exclusively upon
> bush (*R. bicornis*) … there are only two species in South
> Africa, or, indeed, in all Africa.[10]

We now categorise all the black rhinos as one species, *Diceros bicornis*, with four subspecies: western (*Diceros bicornis longipes*, declared extinct by the IUCN in 2011);[11] eastern (*Diceros bicornis michaeli*); south-western (*Diceros bicornis bicornis*); and south-central (*Diceros bicornis minor*). The white rhino is categorised as having northern and southern subspecies.[12]

In the nineteenth century, black rhinos of the various subspecies were found in the Horn of Africa, East Africa, across suitable habitat in Central Africa and into the grassland and less arid Sahel regions of West Africa as far west as northern Nigeria. In southern Africa they were found from Zambia and Malawi south through Mozambique, Zimbabwe, northern Botswana, Angola, Namibia, South Africa and Swaziland. The northern white rhino was still to be found across the belt from southern Sudan, Uganda, northern Congo into the CAR and Chad, and possibly in eastern Rwanda and north-western Tanzania. The southern white rhino was found south of the Zambezi in parts of Zimbabwe, Mozambique, South Africa, Swaziland, Botswana and Namibia, but disappeared from most of its range by the end of the century. The last white rhino in Zimbabwe was shot in 1895.[13]

In the early part of the nineteenth century, the black rhino was the most numerous of the world's five rhino species, with several hundred thousand animals across the range from west-central Africa to eastern and southern Africa.[14] Relentless hunting for their horns, hides and meat severely reduced numbers, and by 1893 only two breeding populations of about 110 animals were found in southern Africa.[15] They remained more widespread and numerous in East Africa, and were even considered an obstacle to agricultural expansion plans in Kenya in the mid-twentieth century – with about 1,000 animals killed in 1946–8 by game wardens. Northern white rhinos were progressively depleted by legal hunting and poaching for horn, being reduced to very small numbers in Uganda, Chad and the CAR and southern

Sudan, and with only a viable population of 1,200 surviving in Garamba NP in the north of the Belgian Congo by the early 1960s.

East Africa

During the eighteenth and nineteenth centuries, Omani-Arab-controlled Zanzibar was 'the seat of a vast commercial empire that in some ways resembled the mercantile empires of Europe of the preceding centuries'.[16] It was only challenged and then subordinated by the onward march of European colonialism in Africa in the late nineteenth century. Its rulers and merchants, and then the British, profited from supplying European, Asian and North American markets with ivory, rhino horn, hides, leopard skins, copal and spices – and, prior to British suppression of it, the slave trade. Zanzibar's merchants operated from the periphery of the developing global capitalist system, meeting the demands of the capitalist industrialising West for luxury goods, such as ivory and horn, and raw materials from Africa, as well as trading with India and China.[17] The Omani, Afro-Shirazi, Swahili and Indian merchant communities continued to trade under European rule, but the Europeans became the gatekeepers and tax collectors for wildlife commodity trading.

Reports from European explorers of vast, untapped herds of elephants, rhinos and other wildlife led to an influx of hunters in the nineteenth century. This was encouraged by the Imperial British East Africa Company (IBEAC), which had been chartered in 1888 to economically exploit Kenya. Along with Uganda, this had been designated as a British area for colonisation at the 1884–5 Berlin Conference, at which Africa was arbitrarily divided up between the colonial powers.[18] The IBEAC saw major advantage in using wildlife commodities to finance the development of the colonies through income from exports, taxes and hunting licences, with hunting for meat being a way in which settlers could feed themselves until cash crops could be harvested. Wildlife numbers in British-controlled Kenya and Uganda were so substantial and widely distributed that it was presumed there would not be serious depletion, as had happened in southern Africa following extensive settlement and hunting by Europeans.

The Ndorobo, Waliangulu, Wakamba and other communities who had for centuries hunted for meat, ivory, rhino horns and skins were still engaged in and partly dependent on meat from hunting, and were trading wildlife products with other communities and coastal merchants at the time of the British occupation. Their hunting was banned and deemed to be poaching under the IBEAC and then the British colonial administration. Despite this, hunting by Kenyan communities continued and increased when cattle diseases, such as rinderpest in the early 1890s and then drought in 1897–8, reduced other food availability. Access to game meat and wildlife products to trade was a means of survival as livestock died. Rhinos, along with elephants, hippos and zebras, were unaffected by rinderpest, and were hunted more following the loss of huge numbers of cattle, buffalo and other ungulate species.[19] Ivory and horns were sold illegally by hunters to the coastal traders or to Somalis, who smuggled them into Italian-controlled Somaliland.

The IBEAC raised income by selling game licences to settlers, officials and visiting hunters, and taxing exports of ivory, horn and skins. There was a massive growth in hunting by Europeans, Afrikaner migrants and, later, visiting Americans. In 1893, one British hunter, Garner Muir, killed more than 80 rhinos in under three months in the Machakos region, without let or hindrance from the authorities.[20] Hunters and naturalists journeying through Kenya reported an abundance of black rhinos. The establishment of the British East Africa Protectorate in 1895 led to the imposition of new hunting and game regulations in 1899, allowing settlers to shoot any game on their land, while licences were required for visiting sportsmen and resident officials. Sir Charles Eliot, Commissioner for the Protectorate in 1900–4, wrote that the regulations were hard to enforce as 'Great difficulty is experienced in inducing sportsmen to comply with the prescribed formalities, but the main object of preserving from destruction the enormous quantity and variety of game which inhabits this Protectorate has been attained.'[21] Settlers, in particular, objected to limits on the numbers they could shoot, arguing, as Elspeth Huxley wrote, that rhinos, elephants and buffalos would destroy fences, and that predators would take their stock.[22] Eliot reported that in 1902, 13 male rhinos and five females were shot on licence.[23] The basic sportsman's or public officer's game licence allowed the holder to shoot two rhinos a year.[24] In 1904–5, the annual game returns for the Protectorate showed that 98 male rhinos and 27 females were killed on sporting licences, two rhinos were shot on settlers' licences and 38 on public officers' licences.[25]

In 1906, Hayes Sadler, Commissioner for the East Africa Protectorate, warned in a letter to the government minister Lord Elgin of the need to protect the rhino in Kenya before they became extinct, noting their decline in recent years, though with no mention that they were still being killed legally on hunting licences.[26] Sadler called for the employment of game rangers to police the hunting regulations and ensure the preservation of endangered species. He said this was needed 'if the question is to be taken up seriously with a view to preserving the game from extinction within the next decade or two, more particularly the rhinoceros, greater kudu, roan and sable antelopes … all of which have sadly decreased in numbers within the last 16 years'.[27] The Game Ranger's Department was established in 1907, with

Lieutenant-Colonel John Henry Patterson of *The Man-Eaters of Tsavo* fame as head warden,[28] and with a total of four white officers, which rose to six by 1927 and 17 by 1958.[29]

On safari in the East Africa Protectorate at the start of the twentieth century, naturalist and hunter Abel Chapman noted that near Lake Baringo his hunting expedition trekked through a landscape 'horrid with bush and thorn – bad going for the heavily-laden safari, especially when rhinos filled their breasts with frequent alarm.'[30] North of Baringo, Chapman encountered a large male black rhino. He and his companion shot five bullets into the animal but failed to kill it, and were unable to find it dead or alive the next day.[31] He wrote that another British hunter, Benjamin Eastwood, encountered seven rhinos together on the Laikipia Plateau. He shot one and then went off in pursuit of an oryx, encountering another rhino which he shot but failed to kill. It charged him twice, and injured him so badly he lost his right arm.[32]

It is worth emphasising that many settlers justified the killing of a large number of rhinos, claiming they were potential pests and likely to tear down fences, but generally the view was that they were no threat to crops. The Hungarian naturalist, Kálmán Kittenberger, who travelled through Kenya, German East Africa and Uganda, said there was little evidence that rhinos were crop pests or that they competed with domestic stock for grazing or water.[33] This view was supported by the experienced British hunter, Robert Foran, who said that he would not rank 'the rhino as really a danger in normal circumstances, or a serious menace to either human life or property'.[34] Foran believed that the combination of poor eye sight, a strong sense of smell and good hearing got rhinos into trouble, as they would move towards a scent or sound to investigate it, and that they were stupid and aggressive, preferring to start a fight rather than have one 'forced upon him'.[35] Foran, showing a callous disregard for animal suffering, recounted the story of a hunter, accompanied by Wakamba guides, who shot at a black rhino but failed to kill it. The Wakamba fired arrows and threw spears at it until it 'resembled a gigantic, bounding porcupine', escaping with spears and arrows embedded in it.[36]

Tales of big game hunting were lapped up by Victorian and Edwardian Britons. Hunting was viewed as the exercise of imperial power, and ideal training for those running the empire. Mackenzie wrote that 'The importance of the Hunt can be identified at every level of the theory and practice of the imperial ethos.'[37] In the late Victorian period and the early twentieth century there was substantial European settlement on land taken from the original inhabitants, and extensive sport and commercial hunting by settlers and visiting hunters. Just before the First World War, the commercial exploitation of big game became a lucrative enterprise, with rhinos, elephants, buffalos, lions and leopards, known as the Big Five, being highly prized trophies, and these and other game being shot in unlimited numbers. However, settlers, colonial officials and white hunters blamed Africans for the depletion of game, and described 'the transformation of African hunters into criminal poachers'. They decried their hunting methods as cruel, wasteful and barbaric.[38] In 1902, a government ordinance created two game reserves (GRs), the Northern (from the Ugandan border to Marsabit and Lake Baringo) and the Southern (from the Uganda Railway in the north to the border with German-ruled Tanganyika, and from Tsavo in the east to Uganda in the west),[39] in which hunting was supposedly regulated.

Former US president Theodore Roosevelt undertook a long safari in East Africa in 1909. His own account claimed that it was in the cause of conservation, with specimens sent back to the Smithsonian Institute.[40] He and his son Kermit killed 512 mammals; in total, 5,013 mammals, 4,453 birds and 2,322 reptiles or amphibians were shot by the whole entourage. Kenyan governor Frederick Jackson, an experienced hunter, wrote that 'he so unduly exceeded reasonable limits, in certain species, and particularly the white rhinoceros, of which he and Kermit killed nine'.[41] The hunting expedition covered Kenya, Uganda, northern Congo and Sudan. Roosevelt recorded that he and his son killed nine northern white and eleven black rhinos; Roosevelt himself shot eight black and five white rhinos.[42] Like many hunters, he referred to rhinos as dumb brutes, a 'monster surviving over from the world's past'.[43] Roosevelt's expedition was obsessed with shooting sets of animals. He wrote that his colleague Heller had shot a bull rhino and 'needed a cow and calf to complete the group'. They found a large cow and calf and proceeded to kill them both, noting rather gleefully that the calf 'when dying uttered a screaming whistle, almost like that of a small steam-engine'.[44]

Hunters often ascribed sheer hate, malevolence or other evil motives to rhinos they were attempting to kill, rather than an instinct for survival or in the case of cows for protecting their calves. Roosevelt was obsessed with the idea that black rhinos were by their nature aggressive and a threat to humans, rather than being short-sighted, bush-dwelling animals that only charged when cornered or threatened:

> [Rhino] cows with calves often attack men without
> provocation, and old bulls are at any time likely to become
> infected by a spirit of wanton and ferocious mischief and
> apt to become man-killers … Where we were by Nyeri …
> the rhinos had become so dangerous, killing one white man
> and several natives, that the District Commissioner who
> preceded Mr. Browne was forced to undertake a crusade
> against them, killing fifteen.[45]

The British hunter Chauncey Stigand painted a very different picture of their temper, noting that if not disturbed they would ignore passing hunting parties if they felt no threat, and stating, 'Sometimes I have passed within one hundred yards ... I have never met with any misadventure in so doing, and generally the rhino has not taken the slightest notice of us.'[46] Stigand noted that the Ogiek hunter-gatherers of the southern slopes of Mount Kenya and the Mau Forest hunted rhinos and elephants, and would camp by a rhino carcass for days until the meat was consumed. They would use deep pits to trap and then kill the rhino.[47] Stigand wrote that as a result of heavy European hunting, 'In the Athi Plains ... and in the Rift Valley, the rhino is now practically non-extant ... Fortunately a large portion of the country is thick bush and grass intermingled, so the rhino inhabiting such strongholds will defy the sportsman for many years to come.'[48]

Accounts of the sort of celebrity safaris that were carried out by Roosevelt and members of the British aristocracy, with photographs showing the lions, elephants, rhinos and buffalos killed, made hunting tourism popular with the wealthy and a source of income for the colony.[49] They flocked to hunt in Kenya on safaris run by people such as Denys Finch Hatton, Bror Blixen, and later Arthur Hoey, R.J. Cunninghame, Bill Judd and Leslie Tarleton. A Kenyan government report printed by the *Journal* in 1913 recorded a rise in game licences issued from 829 in 1910–11 to 973 in 1911–12, with a particular increase in travellers' licences. Stigand was in favour of having limited quotas for some species, including the rhino, and took a dim view of the honesty of many hunters when it came to observing hunting limits:

> Supposing, for instance, you have already shot two rhinos, and a third charges the safari and is killed in so doing, the trophy must be handed in to the Government ... rules are framed for weak mortals, and game-rules are no exception ... If a sportsman was allowed to keep the trophy of any rhino that charged him, just think of the number of men who would annually be charged several times by rhinos.[50]

The Protectorate game report for 1911–12 said rhinos were still common in some areas, but had 'undoubtedly decreased considerably of late years in the more accessible parts of the country', which led to more restrictions on how many could be hunted. Even so, in the years covered, 157 male and 39 female rhinos were killed on licence.[51] The Game Ordinance of 1921 for Kenya detailed that 'Nothing in this section contained shall be deemed to prevent the export for sale of elephant ivory, rhinoceros horns, hippopotamus tusks or zebra hides which have been lawfully obtained', and that special licences could be purchased to hunt large game, at the following costs:

> For a Licence to hunt, kill or capture one elephant . . . £15
> For a Licence to hunt, kill or capture two elephants . . . £45
> For a Licence to hunt, kill or capture one rhinoceros . . . £5
> For a Licence to hunt, kill or capture two rhinoceros . . . £15
> Female rhinoceros were not to be hunted without special permission being granted.[52]

In 1924, Captain Keith Caldwell, assistant warden in the Kenya Game Department, wrote that:

> Until three years ago rhino were included on the ordinary licence. One could be shot on Crown land, and as many as you liked on private land; but the slaughter became so great that, to keep a check on rhino killing, these animals were put on a special licence ... The supply of rhino horns did not diminish; and when they were produced they were always supposed to have been killed on private land previous to 1921. A notice was then inserted in the official *Gazette,* and in the Press, stating that all rhino horns in the possession of private individuals had to be registered; and that failing production of the registration certificate, or the special licence under which the animal was shot, such horns were liable to confiscation. The result has been to help to shut down the rhino horn trade, and to save the lives of a number of animals which were being killed for profit.[53]

In his survey of Kenyan wildlife published by the ZSL in 1922, Hobley blamed overhunting for the decline in rhino numbers.[54] He recorded that during the First World War, 'troops of both forces' shot many rhinos in the Southern GR and along the border with German-controlled Tanganyika.[55] Another cause of the depletion of rhinos, he added, was the increase in the price paid for rhino horn, which had encouraged 'native hunters to slaughter these animals' and sell the horns to 'Indian traders, who surreptitiously shipped them out in considerable quantities' to meet demand in Arabia and East Asia.[56] Hobley's view that rhinos were falling in numbers was partially corroborated by the 1923 game report, in which chief game warden Blayney Percival said black rhinos were not thriving in the Northern reserve because of poaching by Ethiopian gangs and by the Turkana people.[57] Caldwell blamed the depletion of rhinos not on the obvious overhunting by settlers and visiting hunters, but on

> the native hunter ... In the old days they lived on the game
> for meat, and the game and they got on very well together.
> When you have the money-making and profit element
> introduced you get a totally different state of affairs ... Then
> came the Somalis from the north ... he saw at once that
> the easiest things to be taken out of the country were ivory
> and rhino horn. He used the tribes as hunters and exported
> ivory – our ivory – and rhino horn. The great obstacle
> to putting down this trade is the existence of a 'fence' or
> receiver – to wit, Italian Somaliland.[58]

Caldwell said that it was demand in China that was encouraging the trade among Somali and other coastal merchants, who paid indigenous hunters to poach. Some poached horn was taken to Zanzibar by dhow, where it was laundered as legal horn. He said that on checking the Zanzibar customs ledgers he found that '40,000 rupees worth of horn was imported, and 100,000 rupees worth was exported', with the difference made up by illegally smuggled horn.[59] In his game report for 1926, Game Warden Ritchie lamented the poaching of rhinos for their horn, particularly in areas near the border with Italian Somaliland. He said the price for ivory and horn meant that 'Elephants and rhinoceros are thus, at the present time, regarded as potential wealth alive; and ... a certain fortune dead'. Rhino horn, he said, was selling for 36 shillings (about £140 in 2022 values), almost double the price of ivory.[60]

To combat poaching, the Kenyan authorities banned all trading in ivory or horn unless the horn or tusk was accompanied by a government licence allowing the killing of the animal. Ritchie employed a game ranger, C.G. MacArthur, to deal with what was termed native poaching. MacArthur was strenuous in his pursuit of poachers selling ivory, rhino horn and leopard skins to coastal merchants.[61] In one operation in Malindi on the Kenyan coast, MacArthur seized 218 rhino horns weighing 955 lb, resulting in a fine of £560 (equivalent to £40,684 today). In another two investigations in Nairobi, 72 rhino horns were seized.[62] In 1930, MacArthur arrested 495 African poachers, and large fines and terms of imprisonment were imposed. In 1931, there were 477 poachers arrested.[63] However, poaching and smuggling continued. In 1937, Keith Caldwell pointed out that the area of Jubaland, which the British had ceded to Italy, 'used to teem with Rhino and Elephant. Both have now decreased greatly as the result of wholesale killing by Natives, who have sold the tusks and horns to traders in Italian Somaliland.' He added, 'In 1920 the

Northern Game Reserve of Kenya was full of Rhino. I remember … seeing three or four a day. I have just completed another long trip there, and saw only one Rhino … The wholesale destruction there is the work of Native poachers, who have killed the Rhino for his horns.'[64]

There was significant Somali involvement in poaching and cross-border smuggling in the period of the Italian occupation of Somaliland. Dalleo chronicled the situation in the first four decades of the twentieth century:

> For centuries the peoples of the Horn of Africa have
> supplied foreign consumers with game products and
> trophies. Due to the introduction of colonial governments,
> however, these peoples, such as the Somali of Kenya, faced
> restrictions on such activities. Organized poaching played
> an important part in the economic life of the Somali of the
> Northern Frontier District (NFD) … Cooperating with
> many other ethnic groups, they dealt in game trophies such
> as ivory and rhino horn.[65]

He estimated that from 1 July 1891 to 31 December 1893, the value of exports from the Somali port of Kismayu amounted to 248,713 rupees (about £2.51m in current values), including ivory and rhino horn worth 100,683 rupees (about £1m in current values).[66] Often the rhino and ivory horn trafficked from Kenya to Somaliland was legalised by the Italian colonial administration and exported back to Kenya. Ritchie wrote in 1936: 'A very large quantity of Kenyan poached ivory and rhino horn found its way back to Mombasa accompanied by papers showing "legality" of export from Kismayu. In seven months in 1938, 1,408 lb of rhino horn from Kismayu was landed at Mombasa, being the trophies of some 200 animals.'[67] Ritchie noted in his game report for 1926 that the employment of investigating agents had helped to reduce the poaching, smuggling and sale of illegal ivory and rhino horn; this had been helped by paying 'native' informers for information on poachers and traders.[68]

An indication of the growing concern about the poaching of black rhinos was the article published by the *East African Standard* on 7 March 1930 reporting that 'poaching and smuggling of rhinoceros horn has become a serious problem for the Game Department in spite of the exercise of every possible precaution and vigilance by the Department … Occasionally the culprits are brought to justice and the punishment is heavy.'[69] It said that five Barawa men from southern Somalia had been convicted of illegal possession of 187 rhino horns, estimated to come from 94 rhinos. The report went on to say that the rhino horns had been bought by the men through intermediaries, transported to Malindi and from there would be sent abroad by dhow. It was thought that the rhinos had been killed in the Ukamba GR by Wakamba poachers. The newspaper said that the recent increase in poaching was 'due to the fact that recent droughts resulting in food shortages, and the increasing inducements offered by dishonest traders … have tempted the native more and more to take part in these profitable ventures'.[70] The Wakamba used poisoned arrows to kill the rhinos, and bought the poison from the Giriama, the report said.[71] The police and game authorities sought out makers of the poison, and in the Kitui region had seized 400 lb, which could have been used on 25,600 arrows.[72] While the Game Department was fighting poaching and smuggling, the colonial administration was still benefiting from rhino horn sales. The game report for the years 1932–4 stated that the total weight of rhino horn sold was 578 lb, in 1933 it was 546 lb, in 1934 it was 2,418 lb,[73] and in 1935 1,353 lb of rhino horn were sold from government stocks. The horn was sold for a higher price than previously, the average price for best-quality horn being Sh. 15/50 per lb, compared with Sh. 12/50 per lb in 1934.[74] At this time, there was pressure on the Game Department from settlers around Nyeri in the Central Highlands to reduce the number of rhinos, which they said were a threat to life – one settler having been killed by a rhino. The honorary game warden, Mr Cunninghame, killed 20 rhinos at Ngobit and 27 at Nyeri. Another honorary warden, Major Kingdon, shot 14 rhinos. The problems around Nyeri were human-created, as scrub burning to clear land in the Abderare Range had driven many rhinos from their normal ranges.[75]

In 1930, the naturalist R.W.G. Hingston produced a report published by the Society for the Preservation of the Fauna of the Empire recommending that the Southern reserve in Kenya be converted fully into an NP, providing protection in particular for rhinos, buffalos, elands, giraffes, elephants, lions and cheetahs.[76] Caldwell didn't think this was likely, and said in an interview broadcast by the BBC in 1938 that although ivory and leopard skin smuggling was continuing, 'it's chiefly the rhino horn that people try to smuggle nowadays'.[77] He said that rhinos were easy to kill and were 'money for jam' for 'native' poachers. Mervyn Cowie, famed for his role in founding Kenya's first NPs, hunted rhinos for profit in the 1930s. He wrote that if he shot a rhino with a front horn weighing at least 15 lb he could sell it at 40 shillings per pound and make a considerable sum once he had paid for the licence.[78] This was at a time when Cowie was publicly lamenting the loss of wildlife through overhunting. He admitted to pangs of conscience about the hunt, but said: 'In any case I argued, what does one rhino matter? If I don't bag him, somebody else surely will.' Cowie also expressed anger that poachers in Tsavo trapped and killed rhinos 'for a few pounds of horn on its nose' – exactly what he had been doing 20 years earlier.[79] Cowie and other hunters could get licences to kill rhinos for profit, but when it came indigenous peoples hunting, as they had for centuries, he called it a 'racket'.[80]

The professional hunter Sydney Downey noted that the number of licences granted by the government to shoot rhinos had increased from 33 in 1937 to 98 in 1951 and 204 in 1957, at a time when numbers were thought to be falling.[81] Downey reported that during the Emergency period, usually called the Mau Mau Emergency, 13 rhinos, 18 elephants and 40 giraffes were killed by poachers around Isiolo in 12 months, when game wardens had been drafted into the security forces to carry out counter-insurgency operations.[82] While wardens were fighting poaching and smuggling, and lamenting the decline in rhino numbers, in one operation between 1946 and 1948 the professional hunter J.A. Hunter killed 1,000 black rhinos in the Makueni district south-east of Nairobi to clear land for settlement by Europeans who were being encouraged to move there after the Second World War.[83] Rhinos were also killed by the Game Department if they were deemed to be problem animals, or in areas that were to be cleared for commercial farming by white settlers, with 45 rhinos killed in 1954 and 17 in 1957.

The game ranger and anti-poaching investigator Rodney Elliott was assiduous in combating African poaching, and tried to prosecute white hunters who broke the rules. He believed that ways should be found to enable African communities to benefit from the utilisation of wildlife, writing that 'African support for conserving could only benefit if game was profitable to the people. Consequently I was ever on the look-out for means to obtain money from wildlife into the Local District Council's coffers.'[84] However, prosecutions did not succeed in stemming the decline in rhino numbers caused by poaching and overhunting. Some survived in the Amboseli region, Tsavo, Makueni-Makinda, Lake Nakuru, Mount Kenya and Maasai Mara, but numbers were massively reduced. In the late 1950s, in Amboseli there was the first outbreak of spearing of rhinos by Maasai in protest at the establishment of the protected conservation areas in the land where they had traditionally grazed their cattle; two rhino cows and their calves, two elephants and two lions were killed. At the Marsabit Reserve, Samburu and Boran pastoralists killed four rhinos, 11 lions and 26 leopards in a conflict over lost grazing lands.[85]

On 28 January 1945, the ordinance for the establishment of NPs in Kenya was promulgated, having been given assent by the colony's governor.[86] Nairobi NP (NNP, which still houses black and imported white rhinos) was the first to be created, with Tsavo East and West NPs following.

The Tsavo anti-poaching campaign, drought and rhino die-off

Tsavo became an NP in 1948 and was seen by the colonial administration as uninhabited 'pristine wilderness … raw, man-free nature'.[87] This was not so.

In the north-west of the park, the Wakamba and Waliangulu lived on the fringes of the park and had traditionally hunted in it and collected firewood and honey. Tsavo was huge, and there were substantial populations of elephants and black rhinos, with a diversity but not huge numbers of other game – but it was prone to droughts.[88] Few resources were put into development, patrolling, surveying animal numbers or assessing the level and effect of hunting by communities bordering the park. It was soon realised that it was unmanageable as one unit, and it was split into east and west parks. David Sheldrick was appointed warden of Tsavo East, assisted by Bill Woodley. On taking up the post, Sheldrick became aware of illegal hunting by Wakamba on the northern edge of the park, and the continuing hunting by Waliangulu.[89] The former head of the Game Department and experienced conservationist Keith Caldwell said, after a visit to Tsavo in 1951, that he was aware that Wakamba and Waliangulu hunted in the reserve but concluded, 'I doubt if they do serious damage.'[90] This did not halt anti-poaching operations.

In the mid- to late 1950s, Sheldrick and Woodley led anti-poaching operations in Tsavo, combating the continuing hunting of elephants and rhinos. Using arrows tipped with *Acokanthera schimperi* poison and fired from large bows, Wakamba and Waliangulu were able to kill elephants and rhinos silently. About 2,000 Waliangulu lived around the newly established park boundaries and they had a small but experienced group of hunters, who became the chief targets of the anti-poaching campaign. The hunters travelled in small groups on foot, using poisoned arrows rather than guns, making them hard to trace and catch. With relatively few hunters involved and similarly low numbers of Wakamba hunters in northern areas of the park, there was a regular but small off-take of elephants and rhinos,[91] although one Waliangulu hunter, Wambua Ngula, is estimated to have killed 240 rhinos, 300 elephants and 31 buffalos in seven years.[92] At first, little real success was achieved and little or no effort was made to catch the traders who bought the horn and ivory from the poachers. The smugglers were adept at covering their tracks, and had the money to hire lawyers and even bribe officials to ensure that the smugglers continued in business.[93]

Following the suppression of the Kenya Land and Freedom Army (Mau Mau) uprising, in which both wardens were involved, Woodley and Sheldrick used militarised tactics, intelligence gathering and raids on the villages of suspected poachers outside the park.[94] The killing of a Kenyan game ranger by a Waliangulu hunter on 15 January 1955 was the turning point in operations against poachers, and a much more strenuous approach was taken; this was successful in catching poachers but not in breaking up the smuggling networks they supplied. Sheldrick's approach was to be copied in many countries, with a stress on fast, tough responses by well-armed units of rangers acting like counter-insurgency forces. During the anti-poaching operations, from November 1956 to August 1957, 8,684 lb of ivory and 264 lb of rhino horn were recovered.[95] North of Tsavo, in the region between the Galana and Tana rivers, there had been significant poaching, with 100–200 rhinos killed there annually between 1953 and 1957.[96] Ian Parker was involved in anti-poaching operations in the Tsavo and Galana areas. He wrote that 429 mainly Waliangulu and Wakamba poachers were arrested and 35 lb of rhino horn worth £25,000 were seized (£729,000 in today's money).[97]

Even when poaching declined as a result of the anti-poaching efforts, threats to the rhino remained, chiefly drought and elephant damage during droughts to the browse on which rhinos depended. In 1959, increasing elephant numbers were leading to substantial vegetation loss, changing woodland to dry savannah. Daphne Sheldrick, wife of David and later head of the famous Kenyan elephant orphanage, said that areas of Tsavo resembled 'a battlefield or lunar landscape'.[98] The loss of woodland vegetation reduced numbers of browsers such as black rhinos, gerenuk and other antelopes. The increase in elephant numbers, without any poaching or licensed hunting within the park, was because birth rates exceeded natural mortality. When drought hit, the elephants depleted the food sources and destroyed trees. It was a cruel irony that the protection of the elephants from local hunters led to a situation where a cull well into the thousands might be needed to protect the black rhino from starvation.[99] By 1960, Tsavo was in crisis as elephants tore trees apart and the shrub/bush layer beneath them was stripped away. Hundreds of black rhinos died as the drought continued. Veterinary examinations of some of the rhino carcasses showed that they had died of starvation rather than thirst – some were even found in rivers such as the Athi, Tsavo and Sabaki, and at Lugard's Falls, where 300 died in close proximity to water.[100]

German East Africa/Tanganyika

In the mid-nineteenth century, the explorer Richard Burton travelled widely in East and Central Africa and the Horn of Africa, often noting the presence of wildlife and its utilisation by local communities or commercial hunters.[101] He reported that black rhinos were common in the wooded areas and plains between the Indian Ocean coast and Lake Tanganyika.[102] In some districts, local people killed rhinos for meat and believed that the marrow from rhino leg bones would cure epilepsy, one of the few African examples of medicinal use of rhino body parts.[103] Rhino horns were a valuable commodity and hunters traded them to merchants from Zanzibar, while hide was used to make whips and other items requiring durable leather. Burton wrote that the 'black rhinoceros with a double horn is as common as the elephant in the interior. The price of the horn is regulated by its size. Upon the coast a lot fetches

from 6 to 9 dollars per frasilah [about 35 lb] which at Zanzibar increases to
from 8 to 12 dollars.'[104] Horns were exported to Muscat and Yemen or to Asia.
Burton said that white rhino were not found in the regions over which he
travelled in East and Central Africa.

Henry Stanley, in his account of his journey in 1874–6 to seek the source
of the Nile, noted that between Bagomoyo and Lake Victoria, rhino hide
was used to make shields. He made no mention of use of the horn.[105] The
Arab sultans of Zanzibar at one stage had bodyguards from Baluchistan, who
carried rhino hide shields.[106] Stanley wrote that while in the Akagera region
of Rwanda in March 1876, his guides took him to an area where he saw three
white rhinos and four that he described as 'black brown' (not making it clear
whether he meant black rhinos). His guides wanted him to shoot one for meat,
but as he could not get 'a certain shot, I was loath to wound unnecessarily, or
throw away a cartridge'.[107] A few days later, near Chief Rumanika's territory in
the Karagwe kingdom of north-western Tanzania between Rwanda and the
western shores of Lake Victoria, he shot what he described as a white rhino,
and his retainers carried the meat back to camp.[108]

Writing of his 1860–3 expedition to find the source of the Nile, Speke
wrote of the 'morose rhinoceros' being less numerous than elephants and
found in 'very thick jungle' in East and Central Africa,[109] suggesting that
it was the black rhino. In the region he called Kanyenye (near Tabora),
Speke was told by local people that they could guide him to where black
rhinos could be found. Speke accompanied them and shot a black rhino,
which was cut up to supply meat for his porters and trade for grain in the
local village.[110] From his accounts of his journey one can deduce that much
of central Tanzania, including quite thick forest, was inhabited by black
rhinos. In the territory of Chief Rumanika in Karagwe, Speke described the
habitat as:

> wild, and very thinly inhabited, this was greened over
> with grass, and dotted here and there on the higher slopes
> with thick bush of acacias, the haunts of rhinoceros,
> both white and black … and as I had never before seen
> white rhinoceros, killed one now; though, as no one
> would eat him, I felt sorry rather than otherwise for what
> I had done.[111]

In the last years of the nineteenth century, the Society for German Colonization was established and became the vehicle for establishing a German
presence in Africa. It concluded treaties with several chiefs on the Tanganyika
coast in the early 1880s. On 3 March 1885, the German government granted

it an imperial charter to establish a protectorate there. German ambitions conflicted with the hegemony of the Sultan of Zanzibar, and the Germans dispatched five warships to force him to concede in August 1885. The British and Germans agreed to divide the mainland between them, and confirmed the borders between Kenya and Tanganyika in a treaty of 1890. German rule was established over Bagamoyo, Dar es Salaam and Kilwa. Occupation and military conquest of mainland communities followed. The sultan was forced to accept subordination to the British.

Even under the hegemony of the powerful Zanzibari traders such as Muhammed Bin Hamid (also known as Tippu Tip), the mainland African communities were able to hunt as they chose, and the lack of territorial borders gave the ivory and rhino horn trade fluidity.[112] The imposition of German rule changed everything. Locals were not banned from hunting, but licences were introduced and the Germans sought to control commerce. They established protected areas to prevent the loss of some species to overhunting, and by 1919, when the British were given control of the territory under a League of Nations mandate, there were 20 reserves in Tanganyika. The British retained these and exerted closer control over hunting than had the Germans, introducing a Game Ordinance in 1921; this limited indigenous hunting to encourage the generation of income from visiting hunters. A Game Department was set up with Charles Swynnerton as its director. Its report in 1922 said that black rhinos were abundant in northern districts, became scarcer towards central Tanganyika around Tabora, but were present in small numbers at Mahenge (adjacent to what became Selous GR and later Nyerere NP) and further south at Tunduru.[113] In north-western Tanzania and on the fringes of what is now the Serengeti NP (SNP), there was a history of hunting by the Sukuma and Asi people, the latter hunting rhinos and trading the horn to the Sukuma, who sold it to coastal merchants.[114] In 1929, all hunting was banned in Ngorongoro Crater to preserve the rhinos, lions, cheetahs and buffalos there.[115] European and American hunters included rhinos in their wish lists when hunting in Tanzania. One party of Italian hunters in 1929 killed at least six and took their skulls back to Italy.[116] Poaching was a problem, with Indian traders on the coast, as Swynnerton put it, 'acting as inciters' of poaching and smuggling illegally obtained rhino horns out of the territory. He noted that the Zanzibar colonial authorities had instituted regulations to try to ensure that only documented horn or ivory was exported from there.[117]

In the Tanganyika Territory annual game report of 1932 it was noted that rhinos were still found in reasonable numbers in Northern, Tanga, Central, Eastern and Western provinces, but that 16 people had been convicted during the year of illegal possession of ivory or rhino horn and that in areas of Central Province, 'rhinoceros are killed annually by natives, who either bring them in as Found ... Rhinoceros Horn for a reward, or dispose of them to some receiver for a small percentage of their value'.[118] In Tanga, game officials had shot several rhinos that had been damaging cotton crops.[119] In the mid-1930s, there were increasing reports of poaching by local hunters, and the Game Department blamed the poaching on game dealers who encouraged

> this illegal traffic in ivory and horn by holding out monetary
> gain ... Poisoned arrows are responsible for most of the
> killing, but a muzzle-loader in a native's hands over a
> water-hole at night at a few yards' range is a deadly weapon.
> Rhinoceros horn is bought ... at a price varying from 1s.
> 50c. to 2s. per lb. As its market value at the moment is 12s.
> per lb. even if it is disposed of by the receiver for only half its
> market value a very handsome profit is still being made.[120]

Game Department records showed that 55 rhinos and 52 elephants were killed by poachers in northern Tanzania in 1936. The Game Department was involved in the legal rhino horn trade. Between 1947 and 1956, the Department recorded that revenue totalling £26,369 (about £850,772 in today's money) had been collected from the sale of rhino horn.[121] There was evidence of continued rhino poaching in many areas of Tanzania during the Second World War. The Game Department report for 1941 noted that in the Dodoma district

> although 1 Elephant only and 4 Rhinoceros were killed
> departmentally no less than 80 Elephant tusks and 35
> Rhinoceros horns were brought in as 'Found' ... It is
> difficult to believe that 40 Elephant and 18 Rhinoceros died
> natural deaths in the one district ... The real reason can be
> attributed to native hunters using various methods such as
> poisoned arrows, heavy down-pointing spears set high up
> in a tree with a trip rope for an animal to knock against,
> muzzle-loaders, etc.[122]

The Tanganyika colonial administration was paying for 'found' rhino horn, so poachers could sell horn that they had poached to the authorities, claiming they had found it on the carcasses of already dead animals.

Even though poaching was occurring in many areas, black rhinos were seen regularly across the country. During his African tour of 1957, the secretary of the Fauna Preservation Society noted the presence of rhinos in

Serengeti and Ngorongoro, adding that in the Ngorongoro Crater rhinos and lions had been speared by the Maasai when they thought that they threatened the well-being of their cattle, though there were suspicions that the welfare of the cattle was a pretext for engaging in hunts.[123] As the protected areas of Serengeti and the Ngorongoro Conservation Area (NCA) expanded and people were excluded or limited in their access, the level of 'conservancy-pastoralist confrontation' led to 31 rhinos being speared in the NCA in 18 months in 1959–60 and another 12 in 1961, compared with only 17 killed between 1952 and 1959.[124] In Serengeti, there was conflict with Maasai pastoralists as a decision was made to expand the NP to link it up with the Mara area of Kenya. The warden of Western Serengeti, Myles Turner, said that the expansion would join an area with substantial populations of rhinos, elephants, buffalos and lions to the Serengeti and create a protected area that would safeguard much of the wildebeest migration route.[125]

Uganda

In Uganda, under first the IBEAC and then the British protectorate, ivory and rhino horn were an 'important subsidy to European imperial activities', and hunting could be an attraction for settlers, officials and visiting European hunters, with lions, buffalos, elephants and rhinos available to hunt.[126] Frederick Jackson, who became Lieutenant-Governor of Kenya and then Governor of Uganda, was active in the 1880s running ivory, horn and hide gathering expeditions for the IBEAC.[127] Under British rule, taxes and licences were put in place to garner income from commercial and sport hunting and exports of wildlife products. Licences allowed a hunter to shoot three rhinos, 12 elephants, six hippos, one giraffe and one buffalo.[128] In 1905, Colonel Delme Radcliffe, a member of the Society for the Preservation of the Fauna of the Empire, wrote a report on wildlife in the Nile Province of Uganda and south-west Uganda, noting the occurrence of rhinos, but without identifying which species, though it is likely they were northern white rhinos:

> In Uganda Proper they are nearly, if not quite, extinct.
> On the east of the Nile and in the Nile Province there are
> still a good number. North of the Aswa River they are still
> more common. They seem sufficiently protected by the
> Game Laws; natives do not interfere with them … and
> there should be little risk of their extinction … south of the
> Kagera River they exist in extraordinary numbers, literally
> in places almost in herds; whereas to the north of the Kagera
> River not a single one is to be seen.[129]

Kittenberger hunted in Uganda before the First World War and bought a white rhino licence for £25, noting that only two or three were issued each year because of the scarcity of the rhinos. He shot a white rhino in West Nile and donated the hide to the Hungarian National Museum.[130] R.T. Coryndon, the Commissioner for the Uganda Protectorate, wrote a report on elephants in Uganda in 1921, but also noted the comparative scarcity of black rhinos and northern white rhinos, recording that:

> The case of the common black rhinoceros need hardly be considered at present, for there are few in the country … It is probable that, speaking generally, not more than ten are killed in a year. I do not believe that the case of the great white, or square mouthed rhinoceros (*R. simus*) need cause any greater apprehension. Their range is very limited. It does not extend more than probably forty miles along the left or west bank of the Nile above Nimule, and, say, fifteen miles inland.[131]

In 1926, the Uganda Game Ordinance allowed visiting hunters and residents to buy a special licence to shoot one white rhino. Licences could be obtained to shoot black rhinos in West Nile and Gulu districts.[132] The Game Department report estimated that there were about 130 white rhinos in Uganda.[133] This was down from the 1925 census figure of 150. The Game Department records showed that 15 white rhinos were known to have been killed in 1925, 18 in 1926, 17 in 1927 and five in 1928 – while 28 had been born between 1925 and 1928. This calculation put the 1928 population at 123.[134] In 1929, the sale of ivory, rhino horn and hippo tooth brought in £16,159 (the equivalent of £1,140,984 in today's money). The report added that 73 horns from 47 black rhinos had been recorded as legal trophies; some of the horns were from previous years, with 50 taken on licence in 1929. Aware that black rhinos were being killed or horns recovered by local people from rhino carcasses, the Game Department offered a five shilling reward for any horns handed in to them – this led to 85 black rhino horns being handed in at Moroto, totalling 377.75 lb in weight. At the customs post at Mbarara, in western Uganda, it was recorded that 335 lb of rhino horn had been imported from the Belgian Congo. In addition, six white rhino horns, weighing 37 lb, were recorded as having been found by the Game Department, two from a rhino illegally killed by villagers.[135]

In his survey of wildlife conservation in East and Central Africa, Hingston said that strict protection of the remaining white rhinos in Uganda was absolutely vital, but that a 'A special sanctuary for it is out of the question owing to the presence in its area of a large native population.'[136] Despite the recommendation of strict protection, white and black rhinos were being killed and the horns exported as trophies or sold commercially. The game report for 1931 recorded that rhino horn, hippo tooth and ivory sales brought in £15,266 12s 58d (£1,303,253.15 in today's money) – the majority will have been from ivory, although rhino horn fetched a higher price per pound than ivory. The annual game report for 1932 recorded that the white rhino was increasingly threatened by illegal killing.[137] The next year's reports struck a more hopeful note, stating that in the Murchison Falls GR 'black rhinoceros and buffaloes, in the past considered rarities, are now frequently seen'.[138] The black rhino was said to be 'not uncommon' in the Northern and Eastern provinces and in the south-west of Lango district.[139]

In 1935, the annual game report recorded that rhino horn sold for between £12 (£1,091 in today's money) and £15 per pound.[140] The following year's game report stated that black rhino numbers were increasing in Bunyoro and Gulu regions, though no figures were given, but in parts of Gulu and in Lango they were blamed for crop destruction. The report added: 'It may be necessary in the future to take energetic action against these creatures if they persist in their marauding tendencies.'[141] In the Game Department report for 1939, a report by Captain Salmon to the department suggested a 50% increase in white rhino numbers over the previous decade, with numbers estimated at 220, including many young rhinos, although in the southern part of the range, where 'innumerable spiked and leather thong foot traps were found, scarcely a juvenile was seen'.[142] By 1949, it was believed that there were about 24 white rhinos in the Mount Kei Crown Forest Reserve, west of the Nile, with more at the Mount Otze sanctuary in West Madi, and the largest number were to be found at Era Crown Forest Reserve in West Madi, which was not a sanctuary.[143] It was estimated by the Game Department that there were 150–200 white rhinos in West Madi and the West Nile district as far south as the Arua–Rhino Camp road.[144] By 1955, the white rhino population was put at 350. The tragedy for them began in 1956 when poaching by hunters from southern Sudan became widespread.[145] Edroma wrote that in the Ajai-Inde Swamps and the Otze and Kei sanctuaries in West Nile, poaching was so heavy that the population dropped from over 335 in 1958 to 80 in 1962.[146]

Horn of Africa, Sudan and Central Africa

The distribution of black rhinos in the Horn of Africa is not well documented, with few estimates of numbers and locations. Much of the information that is available comes from hunters in the late nineteenth and early twentieth centuries. In 1892, a British hunter, Harald Swayne, undertook a series of hunting expeditions in British Somaliland. He noted that the main port, Berbera, was an important trading hub for exports of hides, ivory, and rhino and antelope horns brought from the interior, including the Ogaden and Harar regions bordering Ethiopia.[147] Rhino hide, he said, was sought after by some Somali clans, such as the Dolbahunte and the Isaaq, to make shields.[148] In the district of Dagahbur, now within the Somali region of Ethiopia, he tracked and shot his first two rhinos, having to fire four bullets into each of them.[149] When he wanted to get the assent of the Ethiopian ruler Ras Makonnen for his charting of trade routes from Aden that would cross Ethiopian territory, he sent him as a gift two of the rhino horns he had obtained from the animals that he shot. Ras Makonnen sent rhino horn cups to Swayne as a present.[150] Swayne said that rhinos were present in the largest numbers along the Somali/Ethiopian border and south-eastern Haud in Ethiopia; others were in areas south of Berbera, near the coast.[151] The naturalist Drake-Brockman wrote a short paper on the wildlife of Ethiopia in 1908, noting that only a few rhinos were still to be found in Western Ogaden and southern Borana.[152] In his survey of the mammals of Somaliland in 1910, he wrote that black rhinos were never seen north of Burao (east of Hargeisa in British Somaliland), but were found near the Ethiopian border in Haud and the Nogal Valley, and were plentiful in the Ogaden.[153] When demand for horn grew, hunting in Ethiopia and parts of Somalia increased. British diplomats based in Ethiopia reported that large numbers of rhino horns were exported from Ethiopia and Somalia in the early years of the twentieth century; Mogadishu, Merca and Brava in Somalia were exporting hundreds of horns annually, mainly from rhinos hunted in Sudan, while the small Ethiopian town of Lugh was said by the British legation in Addis Ababa to have exported 1,000 or more rhino horns in 1905–6, and 1,200 were exported from Borana in southern Ethiopia.[154]

During the nineteenth century, despite the declining power of the Ottoman Empire, its rulers in Egypt sought wealth through trade in ivory, rhino horn and slaves obtained from southern Sudan by Arab merchants from Khartoum and the Nile Valley of northern Sudan. They established networks that reached as far south as Gondokoro (just north of Juba, now the capital of South Sudan), nearly 1,000 miles south of Khartoum, and mounted expeditions to the south, establishing fortified trading posts, known as *zaribas*, from which to gather slaves, cattle and wildlife products. The British soldier, hunter and explorer Samuel Baker journeyed up the Nile in 1861 and at first saw no wildlife around Gondokoro, but further south found evidence of rhinos, elephants and buffalos; he believed the rhinos were black rather than the northern white, claiming that they were 'extremely vicious'.[155] Travelling in the Kassala region of eastern Sudan, Baker hunted with the Hamran Arabs, who pursued rhinos on horseback and killed them with swords.[156] Baker took part in some of these hunts, though he was armed with a rifle.[157] The Baggara and Rizeigat Arabs of Darfur and Kordofan, who would become notorious as the Janjaweed militias during the early twenty-first-century Darfur conflict, hunted rhinos on horseback, hamstringing them with spears or swords and then killing them when they were immobilised.[158] Samuel Baker said that some hunters in Sudan used spiked wheels buried in the soil or sand on forest tracks as a means of wounding and slowing down rhinos so that hunters could catch and spear them – a method also used in parts of Ethiopia.[159] In 1864, the naturalist P.L. Sclater told the ZSL that black rhinos were common throughout Sudan, and white rhinos further south and west; his party later shot several white rhinos in the Karagwe region where Tanzania borders Rwanda.[160] In the 1890s, the northern white rhino was still common in the region encompassing Bahr al-Ghazal in southern Sudan and the Congo district now containing Garamba NP, and eastern Chad.[161]

In his report to the British government in 1902, Lord Cromer, the Consul-General for Egypt who had responsibility for the administration of Sudan, reported that four rhinos (species not given) had been legally shot on licences that year.[162] In the Sudan administration's game report for 1904–5, it was noted that two male rhinos (species not given) had been killed on licence.[163] In 1907, the game returns recorded that between 1901 and 1908, nine rhinos had been shot on licence.[164] During his mammoth East African safari, Theodore Roosevelt hunted in the Lado Enclave in what is now South Sudan. He reported the presence there of the northern white rhino. Even though he and his party were to shoot white rhinos, he rather hypocritically stated that 'although there has not hitherto been much slaughter of the mighty beast, it would certainly be well if all killing of it were prohibited until careful inquiry has been made as to its numbers and exact distribution'.[165] Clearly he didn't view any proposed prohibition as applying to him, and a couple of pages later said that 'The morning after making camp we started on a rhinoceros hunt'; they killed nine rhinos, including cows and a calf, claiming they were needed by the museum that they supplied with skeletons and skins.[166]

A zoologist who corresponded with the Society for the Preservation of the Fauna of the Empire, Cuthbert Christy, wrote in the *Journal* in 1923:

> In the British Sudan very few [white rhino] individuals
> remain. Those along the west bank of the Nile can … not
> exceed half-a-dozen pairs … in the Bahr-el-Ghazal, they
> are more numerous, especially in that section of the divide
> between Meridi and Yambio. In 1916 on the Congo side of
> the Divide, especially in the district opposite the Meridi-
> Yambio section, I found the species individually was much

> more common than anywhere on the British side ... early
> in that year, the natives had speared two rhinos within sight
> of the station. The animals were both young males ... In a
> Greek store at Aba, on the same occasion, I was shown a
> pile of at least a hundred rhino horns, worth from £1 to £3
> apiece ... but which he could not sell owing to the restric-
> tions put upon their sale in, or transit through, the Sudan.[167]

Christy called on the British authorities in Sudan and the Belgians in the Congo to protect the rhino and requested the British to remove the white rhino from the list of game that could be shot on licence in Sudan. However, in 1924 an official in the Anglo-Egyptian administration in Sudan noted that the northern white rhino was hunted for meat by some communities and was an important source of their protein intake.[168] It is interesting to note that while hunting of white rhinos was prohibited, live animals could be exported to zoos on payment of an export tax of £24.[169] In 1946, one report in the *Journal* said that white rhinos were still plentiful in western Sudan between the Nile and the border with the French territories of Central Africa and Chad.[170] However, a survey of the southern Sudan and Uganda border area found only 190 white rhinos there.[171]

Mackenzie recorded in 1953 that in Bahr el-Ghazal he had come across 'a curious man-made wooden frame ... that consisted of an upright beam in the ground, about 12 feet from the bole of a growing tamarind tree, and a cross beam joining the top of the upright to the tree about 12 feet from the ground'. He was told by a Dinka guide that it was a white rhino trap. He added 'A man sits above the cross-piece on a small platform and spears the rhino as it walks underneath. A very big heavy elephant-spear is used and it is driven down into the backbone of the passing rhino.'[172] Cloudsley-Thompson also reported the use of this method by the Dinka to kill white rhinos on well-used tracks through woodland.[173] The meat was consumed and considered very good by the Dinka, while horns could be made into clubs and smaller pieces were made into rings or beads for necklaces.[174]

In the neighbouring French-controlled Central African colony, there was a small surviving population of white rhinos in the early twentieth century, near the borders with Sudan and the Belgian Congo.[175] In Chad, the northern white rhino survived, though not in large numbers, across the arid, open grassland or bush areas right up to Lake Chad in the west. Professor Trouessart of the Museum of Natural History in Paris wrote to the ZSL: 'According to current information on the geographical distribution of the northern form of species, it exists not only in the Lado enclave, but also in Bahr-el-Ghazal and Waddei, probably to Lake Chad.'[176] He added that historically the northern white rhino had been found across a much wider area.[177] There was frequent hunting of rhinos in French territory around Lake Chad in the 1927–31 period, according to the big game hunter Marcus Daly. He wrote in his book on big game hunting in the early twentieth century that he was shocked by the extent of rhino hunting there, with French hunters using modern rifles and arming Chadian hunters with cap-and-powder guns in order to kill as many rhinos as possible 'to make fortunes from selling the horn; some hunters returned from expeditions with up to 3 tons of horns, which represented 300 animals. As many as 10,000 rhinos were shot, Daly believed, in this period.[178] Adding to the impetus for hunting rhinos was the need to supply meat for French military garrisons in Chad. Blancou writes that professional hunters and French soldiers were engaged in extensive hunting of rhinos during French colonial rule there (1900–60).[179]

In the Belgian Congo, northern white rhinos were common in the north-east grassland and open woodland districts. Hunting reserves were established there between 1899 and 1910 both to protect wildlife and so that colonial officials and visiting hunters had game to shoot. The first NP established was Albert NP in 1925 (now Virunga NP), followed by Garamba in 1938 and Upemba in 1939.[180] Robert Foran reported seeing northern white rhinos on a hunting trip in north-east Congo, but 'realizing how extremely rare and how near to complete extinction is this variety … I refused to shoot a single specimen'.[181] Although some hunting took place in the region of Garamba NP, as a result of increased protection in the 22 years between the establishment of the park and independence, numbers rose from about 100 animals to 1,190 in 1960.[182]

West Africa

The extent of the historical distribution of the black rhino in West Africa has been subject to debate, but, according to Rookmaaker, 'In recent times, the occurrence of the rhinoceros has been substantiated in northern Nigeria and eastern Niger on the shores of Lake Chad. However, evidence from regions farther west is circumstantial.'[183] Whatever the western boundary of black rhino distribution was, they were definitely present in Nigeria in the early decades of the twentieth century. In 1932, Colonel Haywood produced a report on Nigerian wildlife for the Society for the Preservation of the Fauna of the Empire. He noted that there were many species that were worthy of conservation, and a great diversity of species across the vegetation zones. In what he described as orchard and more open zones, black rhinos were still to be found along with elephants, buffalos, giraffes, a variety of antelopes and lions. However, he was quite pessimistic about rhino numbers, saying they were only found in Borno, Adamawa and Bauchi districts, and probably numbered around 50 in total.[184] He recommended that because of their scarcity rhinos should not be killed under any circumstances.[185] In his survey of Nigerian mammals, Rosevear said that the western black rhino was present in the savanna regions of northern Nigeria but was 'very rare and local', and his distribution map showed a limited presence south-east and south-west of Maiduguri and on the border with Cameroon north-east of Yola.[186]

CHAPTER 4

Rhino populations decimated by the arrival of Europeans and firearms in southern Africa

Both white and black rhinos were distributed widely across southern Africa at the start of the nineteenth century, but had already been wiped out in most of the Western Cape by hunting and the expansion of European settlement. As late as 1828, Dr Andrew Smith, founder of the Grahamstown Museum in the Eastern Cape, reported seeing between 100 and 150 rhinos on a single day in the north-western Cape.[1] Their numbers were then massively reduced by hunting from the early decades of the nineteenth century onwards. In less than 100 years, black rhinos were substantially depleted in numbers, while the southern white rhino was almost rendered extinct. The conservationist credited with saving the southern white rhino, Ian Player, wrote:

> When first seen by white man, the southern square-lipped rhinoceros ranged throughout southern Africa from the Orange and Umfolozi rivers in the south, to the Zambesi and Cunene rivers in the north. Between 1812, when Burchell first described it from Kuruman in the northern Cape Province, and the end of the century, the white rhino was exterminated throughout its range except in the relatively minute area at the junction of the Black and the White Umfolozi rivers in Zululand.[2]

Settlement by the Dutch started the process of wildlife extermination through hunting for meat and to clear land for livestock farming, especially sheep. The chance to generate income from horn, hides, ivory and skins also played a role. The seizure of Cape Town from the Dutch by the British during the Napoleonic Wars and its incorporation into the British empire led to further European settlement and, later, to the migration of Dutch settlers inland to what became the Free State and Transvaal to escape British control. European hunting took its toll on Cape wildlife, from rhinos and elephants to quagga, blaubok, black wildebeest, blesbok, bontebok, lions, cheetahs and wild dogs. Black and southern white rhinos were hunted to near extinction in southern Africa. The progressive destruction of wildlife took hunters further north and east from the Cape, reducing wildlife numbers steadily in a way that thousands of years of indigenous hunting had not. African hunters had killed rhinos for their meat and hides, though not in large enough numbers to reduce populations. Some peoples, such as the Zulus, did not eat rhino meat, and rhinos were rarely hunted in Zululand until white hunters arrived and nearly wiped them out.[3]

To get an idea of the scale of killing, Selous reported that in 1847–8 several hunters in southern Africa had killed 89 rhinos between them, most of them white rhinos.[4] Botswana had lost all its white rhinos by 1870, and only 13 were left in Zululand by 1904. Harris, Gordon-Cumming, Selous and other hunters wrote exhaustive accounts of their hunting, stressing the danger and implied heroism of it all. Returning hunters, Selous in particular, gave public lectures boasting of their adventures and displaying horns, tusks, skins and skulls. Selous bemoaned the disappearance of the white rhino in southern Africa, and wrote in 1881 that 'Twenty years ago this animal seems to have been very plentiful in the western half of Southern Africa; now, unless it is still to be found between the Okavango and Cunene river, it must be almost extinct in that portion of the country; he also criticised Charles Andersson and other hunters who had written of shooting as many as eight rhinos in one night.[5] By the end of his hunting career Selous listed that he had shot 26 black rhinos, 23 white rhinos, 106 elephants and 175 buffalos, often hunting the rhinos on horseback and shooting them at shortrange.[6] Much of Selous hunting, and that of others who will be mentioned later, was to sell specimens to museums. He was commissioned by Albert Günther, the head of zoology at London's Natural History Museum, to get a specimen of a large white rhino. Günther specified:

> The skin with the horns well above two feet long in the
> central axis should fetch £200 [£28,740 in today's money].
> If the horn be considerably longer than 2½ feet, I may feel
> justified in recommending the purchase of the Trustees
> at a higher figure; but I do not think we should pay more
> than £250.[7]

When he first arrived in South Africa, in May 1899, the hunter Abel Chapman said that while many species remained between the Cape and the Limpopo, the 'elephant, it is true, had finally disappeared; so had the rhino, buffalo, giraffe and eland — all of these abundant but a generation before ... Though the Boers, being the most numerous, were the chief instruments of slaughter, yet other settlers were only less to blame.'[8] He added in a paper on the fate of the white rhino: 'Within the lifetime of many of us still living the white rhinoceros literally abounded throughout the whole southern half of the African continent-from the Vaal to the Zambesi. Today it is virtually extinct-extirpated within sixty or seventy years.' He added that white rhinos had been so abundant in most of southern Africa that visiting European hunters 'seem to have shot them wholesale, as we go out to shoot rabbits.'[9]

Angola

Detailed references to wildlife – species, numbers and distribution – are few and far between for the colonial period in Angola.[10] One of the first written accounts of Angola's wildlife came from Joachim John Monteiro, a Portuguese colonial official and naturalist. Working in Angola from 1858 to 1876, he was a keen collector of wildlife specimens. He wrote of many species in Angola (elands, buffalos, zebras, lions and hyenas, to name a few), but does not mention rhinos in the two volumes about his travels.[11] Much of northern, north-eastern and north-central Angola had terrain and vegetation unsuitable for rhinos, with dense woodland and rainforest. The central plateau had a mix of savanna and open woodland, but was the most suitable area for food production and the raising of cattle, leading to the progressive depletion of wildlife. The arid southern coast merged into the Namib Desert, which could support small numbers of wildlife, as is the case in northern coastal Namibia. The most suitable area for rhinos to thrive was the thorn bush, savanna and dry woodland of south-west Angola bordering Zambia and Namibia and close to wildlife-rich areas of northern Botswana. Brian Huntley, chief ecologist of Angola's NPs at the end of Portuguese rule, noted: 'The shrubs and grasses of arid savannas, occurring on rich soils, are highly nutritious and in past centuries supported large herds of herbivores such as wildebeest, buffalo, zebra, kudu, eland, impala, elephant and black rhino. These herds have been severely reduced, and the herbivore niche has now been filled by cattle and goats.'[12] In the 1880s, black rhinos were present in what became the Bicuar NP (it was made a hunting reserve in 1938 and an NP in 1964).[13]

The naturalist Shortridge said that black rhinos were present at the end of the nineteenth and in the early twentieth century in southern Angola, and were relatively plentiful in the south-west and north of the Caprivi Strip.[14] Newton da Silva, writing in 1952, recorded that the black rhino was 'still abundant in the district of Mocamedes [also known as Namibe], the region of Mupa, and certain areas in the south-east of Angola, [but] tends to decrease in number, and it is hoped that hunting these animals will be controlled more closely in future'.[15] Hunting was common, with hunting blocks taking up substantial areas of savanna and open woodland in southern Angola, bordering Namibia and Zambia. As one of the Big Five, rhinos were sought-after trophies for visiting hunters. Some travellers and naturalists reported the presence of white rhino in the extreme south-east of Angola in the nineteenth and early twentieth centuries, but Da Silva said that 'not the slightest concrete proof of its survival in those distant regions could be delivered since, so all fear is justified that this species is entirely extinct now'.[16]

Botswana

At the beginning of the nineteenth century, both black and white rhinos were found in northern and south-eastern Botswana, according to the accounts of European hunters, missionaries and traders. They had not been hunted much by the San or Tswana communities – the former were still hunter-gatherers and the latter were mainly pastoralists who supplemented this with hunting for meats and hides, but began to trade wildlife products to visiting Europeans. The Tswana became middlemen, buying horns, tusks, hides and skins from San hunters and selling them on to Europeans. Subsistence hunting was integral to the subsistence strategies of both groups, but hunting for commercial reasons increased as demand from Europeans and the availability of firearms increased.[17] Pre-nineteenth-century faunal remains from Tswana settlements show mainly animals killed for their meat or hides, but in the nineteenth century this changed, with more rhino and elephant remains being found.[18]

The British traveller and naturalist, William Burchell, in his account of journeys in Tswana areas of southern Africa in 1812, said that the Tswana, like the Batlhaping, were starting to carry out huge hunts involving as many as 500 men to create a ring around large groups of animals that were then speared as they tried to escape.[19] Rhino meat was consumed by hunting communities, the hide was used to make thongs and horns were made into war clubs. By the 1840s, increasing demand for wildlife products meant that Tswana hunters and traders were bartering 'high value animal parts, including ivory, rhinoceros horn, ostrich feathers, and skins of predators ... in exchange for manufactured goods such as tools ... guns and gunpowder'.[20] One trader, David Hume, established a post at Kuruman in the Northern Cape to trade with the Bakwena of Botswana and northern South Africa in ivory, rhino horn and ostrich feathers.[21] In addition to hunting with guns, the Tswana also used the San almost as slaves to hunt for ivory, rhino horn, hides and ostrich feathers.[22]

In 1838, Harris counted 60 black and 22 white rhinos in one day within half a mile of the Limpopo, and ten years later, Gordon Cumming saw 12 white rhinos in one day in north-eastern Botswana and said that black rhinos were common there.[23] Oswell shot two white rhinos on the banks of the Molokwe River in 1846.[24] During his trek to north-western Botswana in 1853–4, the Swedish naturalist Charles Andersson reported that the Bechuana (Tswana) people barter 'The only marketable articles … ostrich feathers, furs, and skins of various sorts, rhinoceros horns, and ivory (elephant and hippopotamus). The staple articles of exchange are beads, and more especially ammunition.'[25] They used guns and pitfall traps to kill rhino. Where Afrikaners settled in parts of Botswana, they hunted rhinos for horn, but also for hide to make thongs and whips.[26] In the 1860s, white rhinos were still being regularly seen and shot by hunters from the Marico and Limpopo districts, where the borders of South Africa and Botswana meet, to the Thamelakane River (running through Maun on the edge of the Delta) and to the Mkgadikgadi Pans. The hunter William Baldwin killed two white rhino cows in the territory of Chief Sechele in eastern Botswana, and later killed several more rhinos.[27] In western Botswana, near Lake Ngami, he reported seeing white rhinos on a regular basis and shot another white rhino near Thamelakane River.[28] Selous hunted extensively in areas of northern and eastern Botswana in the 1870s and 1880s – particularly the Chobe region, where in 1879 he bemoaned the fact that he did not see a single rhino spoor, adding: 'Twenty years ago this animal seems to have been very plentiful in the western half of Southern Africa; now, unless it is still to be found between the Okavango and Cunene river, it must be almost extinct in that portion of the country.'[29] He also recorded that 'One European trader on the Bechuanaland border employed four hundred native hunters and rhino were exterminated in a big area through his agency alone.' Selous said the man had a huge store of hundreds of rhino horns, which he sold to other traders who exported them to Britain, where they were used to make knife handles and combs.[30]

In 1885, Britain proclaimed a protectorate over Bechuanaland, to prevent Afrikaners from the Transvaal Republic from encroaching on what the British considered to be their sphere of influence. Under British colonial occupation, game laws and restrictions were introduced. A protectorate administration edict established that:

> game regulations are embodied in a Proclamation No. 22 of
> 1904, dated September 21, 1904. From October 1 to the end

of February no large game may be hunted. Hippopotamus,
rhinoceros, buffalo, zebra, quagga, and all animals of the
antelope species except the rhebuck, klipspringer, duiker,
and steinbuck, are defined as large game. These may be
hunted in the open season upon payment of £20 for the full
season.[31]

However, as Morton and Hitchcock noted, firearms and the expanded
European and commercial hunting in the last decades of the nineteenth
century had involved 'transformations in technology and concomitant
hunting strategies [and] led to serious game depletion', and by the early
twentieth century had resulted in less trading in wildlife products.[32]

In the late 1950s and early 1960s, leading up to independence in 1966,
with support from Tswana traditional leaders, conservationists proposed the
establishment of protected areas where hunting would be prohibited, and
clearly delineated hunting blocs.[33] There was a fear among local conservationists that expansion of the hunting industry would deplete wildlife, as
professional hunters moved to southern Africa following the independence
of Kenya and Tanzania.[34] They worked with Tswana communities to agree
a way forward to conserve wildlife and assist local communities to benefit
from wildlife through income from either regulated hunting or tourism. The
Batawana regent, Elizabeth Moremi, persuaded her community to vote in
favour of the formation of the Moremi Wildlife Reserve in 1963, and to limit
and regulate hunting in areas surrounding the reserve.[35] White rhinos were
not found anywhere in Botswana by then, but black rhinos were present in
northern Botswana, near the border with Namibia and Chobe reserve and
the Okavango Delta.[36]

Malawi

The European demand for wildlife products also affected the lakeside
regions that became Malawi. The latter had a substantial wildlife population,
especially elephants but also black rhinos. In the nineteenth century,
demand for wildlife commodities and the increasing availability of firearms
enabled local hunters to supply the demands of Indian Ocean traders for
commodities like ivory and rhino horn. The descendants of the Arab
traders from Kilwa, who had established trading posts at the northern end
of the lake, were able to monopolise ivory and other wildlife commodities,
trading with the Ngonde and Tumbuka communities along the western
lakeshore as far as Karonga. The British increased their presence in the
region in the 1870s, moving round the western shore in search of trading

opportunities. Initially, they traded with the Arab merchants. The African Lakes Company established a post at Karonga in 1879. It sent traders to buy and hunt for ivory in the elephant-rich area around the Songwe River, where it flows into Lake Malawi north of Karonga. British criticism of the Arab traders mounted as the legacy of David Livingstone's public campaigns against slavery took effect, coinciding with the British desire to take control of regional trade. The British fought a series of battles with the Arab traders in 1886–7 and 1888, ending a long period of Arab commercial dominance there.

In 1889, the British established the Central African Protectorate, later called the Nyasaland Protectorate, encompassing what became independent Malawi. Under British rule, it was subject to the imposition of game laws and the establishment of fees for hunting licences, with restrictions on the traditional hunting by its African population. There were black rhinos in the Shire Valley and surrounding hills, as well as at Liwonde on the southern edge of the lake and in the Kasungu region, bordering wildlife-rich areas of Zambia. In 1904, the result of a game census by the Commissioner for the Protectorate, Sir Alfred Sharpe, indicated an estimated 125 rhinos, 605 elephants and 2,687 hippos among a wide diversity of wildlife.[37] A report on the mammals killed on licences during 1906–7 revealed that just one rhino was shot in the Lilongwe/Central area of Malawi, out of a total of 2,607 animals killed on licences.[38]

One hunter, corresponding with the Society for the Preservation of the Fauna of the Empire, noted that Elephant Marsh, on the Shire River in southern Malawi, had black rhinos present in 1897, but by 1912, although game was still abundant, the rhinos were down to a few scattered individuals.[39] Frederick Vaughan Kirby hunted along the Shire in southern Malawi and in areas of neighbouring Portuguese East Africa. He wrote of a great diversity of species and of numerous black rhinos on both sides of the border, which he hunted avidly.[40] On one occasion, near Chiromo in Malawi, Vaughan Kirby stalked and shot a bull rhino, taking six bullets to kill it, and then took a wild shot at a fleeing cow rhino and calf, injuring but not killing the cow.[41] Vaughan Kirby judged that it was of the black rhino subspecies then called *keitloa* – the current classification of rhinos would place it as the south-central black rhino, *Diceros bicornis minor*. In 1930, in his report on the preservation of fauna in British colonies in East and Central Africa, Hingston proposed that an NP should be set up joining the Kasungu GR in Malawi with Zambia's Luangwa valley, covering a total area of 8,000 sq. m. and providing protection for 'a very fine and typical assemblage of the wild fauna of South-Central Africa. There would be in it elephant, rhinoceros, hippopotamus, buffalo, and a large selection of the antelopes.'[42] This scheme did not take off, though Luangwa Valley had a north and a south GR established, and Kasungu became an NP in 1970.

Namibia

In his comprehensive overview of the presence of the black rhino in Namibia, Joubert said that in the nineteenth century it had two main ranges:

> one arm reaching north-west from Outjo to the Kaokoveld
> and Kunene River, and the other arm stretching to the east
> past the present day Gobabis. Isolated localities existed in
> the Kungveld and Okavango. The rest of the distribution
> reached south past Windhoek along the 16th longitude to
> the Orange river. If they ever did occur north of the Etosha
> pan on the Ovamboland plains, it must have been the
> first area in South West Africa where they were wiped out
> by man.[43]

Joubert believed that the black rhinos in what became Etosha NP were relatively recent arrivals from the south and south-west, notably the Ugab Valley.[44] There were some areas where even by the mid-nineteenth century hunting had reduced their distribution, whereas they had been very widespread in earlier years. Their demise was due mainly to European hunting, but not specifically for their horns as tradable commodities. As Sullivan et al. concluded, 'rhino instead were attacked relentlessly for their meat and hide, as well as just for "the sport" of killing them' by the increasing number of Afrikaner settlers and visiting European hunters and explorers, some of whom sent skeletons and hides to Europe to be displayed in museums.[45] The British army captain, James Edward Alexander, travelled in central Namibia in 1836–37 and characterised rhinos as brutes, whose behaviour is 'wicked, fiendish and spiteful', which presumably justified their slaughter and the glorification of the hunters.[46] Prior to the European onslaught with firearms, rhinos had been hunted occasionally by local communities for their meat and hides, but not in large numbers.[47]

The recording of sightings, hunts and reports from local people indicated that black rhinos were prevalent in southern and central Namibia in 1837.[48] The first rhinos encountered by Alexander's party, which were labelled by them as black rhino, were along the Chuntob River (Tsondab) at Bull's Mouth Pass. Alexander said that the pass was the 'domain' of the black rhino; in true, callous European hunter fashion, Alexander said they 'tickled them roughly' with gunshots but they escaped.[49] Later, they shot two rhinos, the meat of which was eagerly seized upon by their servants. The local Nama communities that they met ate rhino meat and were very keen to kill rhinos for food. Journeying through southern Namibia, Alexander noted that both black and white rhinos were to be found there, notably along the Fish River.[50] Local hunters killed rhinos with assegais or used pitfall traps. Alexander said that on one occasion

> We now saw miles of hedges, about three feet high, laid
> to direct the wild animals to pit-falls … the pit-falls for
> the rhinoceros were four feet deep and four broad, with
> branches and leaves over them, and were consequently
> not large enough to take in his whole bulk, but were only
> sufficient for his fore legs, which the people said was the
> best way of securing him, as his legs once in, they have no
> purchase with which to raise his body.[51]

As Sullivan and Muntifering note, 'It is again clear from Alexander's 1838 account that rhinos were not hunted for trade in their horn as a high-value commodity. Instead it was the meat and hides that tended to be in demand.'[52] As more European hunters, traders and mineral prospectors visited Namibia,

Rhino hunt in Bull's Mouth Pass, included in Alexander's narrative. Engraving by William Heath (1795–1840). (Reproduced with kind permission of Princeton University Library)

the nature of rhino hunting began to change, with more firearms available to the Nama and Damara peoples and the possibilities opening up of trade in rhino horn and hide. The British in the Cape did not want guns sold to the peoples in Namibia, but an illicit trade in guns and gunpowder developed, with bartering for rhino horn, hide, ivory, ostrich feathers and cattle stolen by armed Nama groups from Herero pastoralists.[53]

The 1851–2 expedition led by the scientist Francis Galton, which included the Swedish explorer Charles Andersson, was profligate in its shooting of rhinos – feeding its retinue with the meat, collecting hides, skeletons and horns, and often shooting several rhinos at a time for sport. At the start of the journey, they heard that a Nama chief and hunter, Amiral, had recently organised a hunt in which 40 rhinos were shot. Galton recorded: 'we arrived at the first great shooting place … Rhinoceros skulls were lying in every direction.'[54] As they journeyed through Namibia heading towards Lake Ngami in western Botswana,[55] they frequently encountered and all too frequently shot rhinos, often for no obvious reason other than that the rhinos were there and they had guns. They fed their servants with and themselves ate the meat of the rhinos, Galton commenting that 'I like rhinoceros flesh more than that of any other wild animal. A young calf, rolled up in a piece of spare hide, and baked in the earth is excellent.'[56] Andersson was not so taken with the rhino meat and said that it was frequently 'poor' though edible, but their servants ate it and they made leather from the hide.[57]

It would be boring for the reader if I was to list every single encounter and killing, so I will limit references to those that indicate the scale of the killing and distribution of rhinos across the route heading north and east from the coast; there will only be footnotes for events of particular note. At a place they called Tunobis, near the present-day border with Botswana, the party shot a white rhino, and Andersson complained that 'the Hottentots shot away a great many bullets at rhinoceroses and did, I dare say, a great deal of mischief', only killing a few but wounding many which may have died later. Andersson wrote: 'in the course of the few days we remained at Tunobis, our party shot, amongst other animals, upwards of thirty rhinoceroses. One night, indeed, when quite alone, I killed, in the space of five hours (independently of other game) no less than eight of those beasts, amongst which were three distinct species.'[58] One presumes that Andersson shot a white rhino and then what he considered to be two separate species of black rhino, following the incorrect identification of black rhino as belonging to a number of different species. He later wrote: 'Another day I had the good fortune to shoot a rhinoceros. He was probably a straggler; for these animals have long since disappeared from the part of the country where we were then encamped.' The scarcity of rhinos there did not stay his hand. His excessive killing was criticised by other hunters, including Selous, who thought it wasteful and unnecessary. Andersson claimed that it was not wasteful, and that meat from the animals shot was eaten by his party or by local people.[59]

Andersson recorded that much of Damaraland and the neighbouring arid areas was covered with what the locals called the Damara milkbush, a type of euphorbia (*Euphorbia damarana*), the sticky white sap of which is toxic to many animals and can cause major skin irritations. He noted that although it appeared to be poisonous to white rhino, it was consumed without harm by the black rhino.[60] The Herero tipped their arrows with this vegetable poison, and some Damara hunters put the sap into waterholes to poison game – surprisingly, the meat of the animals was still edible.

Prospecting for valuable minerals led to another influx of Europeans in the 1850s, who fed themselves largely by hunting, and 20 years later there was an influx of Afrikaner farmers known as the Thirstland Trekkers, who settled in areas where they could raise cattle, sheep and goats and hunt for meat, hide, horn and ivory to trade.[61] In 1884, Germany occupied parts of the territory and proclaimed it a German colony, to forestall British occupation, calling it German South West Africa. Between 1904 and 1907, the Germans fought a series of brutal wars of repression against the Herero and Namaqua peoples, who had resisted occupation. At this time, black rhinos were said by Joubert to be present in the Kaokoveld, in some of the river valleys such as the Ugab and Huab between Damaraland and the sea, at Kamanjab, and were occasionally seen in the Outjo area, at Etosha Pan, and as far north as the Kunene River, but were absent from southern Namibia.[62]

Under German colonial rule, economic activity centred on cattle and prospecting for minerals. Hunting continued for meat and sport. Wildlife, particularly larger mammals such as rhinos, elephants, lions, cheetahs and wild dogs, were killed in large numbers. As Botha identified, 'Game conservation in Namibia ... exhibits many of the features that characterised approaches to game during colonial times ... slaughter followed by preservation and conservation ... game conservation benefited when it was combined with commercial enterprise in the form of tourism, game farms and trophy hunting.'[63] The first wildlife conservation and hunting regulations were issued during German rule, which ended with the defeat of Germany in the First World War and the designation of South West Africa as a mandated territory under the League of Nations with South Africa, part of the British Empire, as the mandated power.

The South African administrators widened the scope of the game laws after 1920, and hunting was prohibited on what were deemed crown or state lands 'with exceptions for visiting dignitaries and officials on duty in rural areas', and some restrictions applied on settler farms, but there were no specific regulations on what were described as 'African Reserves' for the indigenous population.[64] During the first half of the twentieth century, hunting remained an important source of food and income for farmers, especially the sale of biltong made from game meat.[65] The German colonial administration had proclaimed Etosha Pan a GR in 1907, which it remained until the South African administration made it into a park in 1958 and a fully protected NP in 1967.[66] In 1928, the South African administration declared Damaraland (vital for the survival of deserted-adapted black rhinos and elephants) a GR.[67] Black rhinos were said by the resident commissioner for the territory, Charles Manning, to be 'plentiful' after the First World War in north-western areas.[68] In 1934, Shortridge, in his *Mammals of South West Africa*, estimated that there were between 40 and 80 black rhinos resident in the Kaokoveld between the Lower Ugab River and the Kunene, with sightings of them or their tracks more common north towards the Kunene river. He believed black rhino to be a rare visitor in western Ovamboland and almost non-existent in the north-east. It was found in the Namutoni GR (now the eastern section of Etosha NP). Along the Caprivi Strip between the Okavango and the Chobe River, black rhinos were scarce.[69]

South Africa

During his travels across South Africa at the end of the eighteenth and beginning of the nineteenth century, John Barrow, secretary to the governor of the British Cape Colony, noted the presence of rhinos (species not given) in the Karoo, which he said were the only animals with hides tough enough to resist the double thorns of the Karoo mimosa, though they were no defence against musket balls.[70] Twenty years later, the naturalist William Burchill, travelling in the same region, said that he had visited an area called Rhenoster, named by the Dutch because of the presence of many rhinos, recording that

> The animal from which it takes its name is becoming every
> day more scarce in this part of the country, and indeed, is at
> present rarely to be met with. It is fond of inhabiting open,
> dry country, such as this is, abundant in low bushes; but
> the advance of the colonists, and their destructive hunting,
> have alarmed it, and driven it more into the interior of the
> continent.[71]

Members of his party shot rhinos on the journey into the interior – on one occasion shooting two at once, at least one of which was a black rhino. He cautioned that you had to shoot them from close by, otherwise the lead shot would not penetrate the skin, which was so thick that it was valued locally for making shields that might be a good defence against muskets.[72] Burchill's party shot nine black rhinos during the trip, despite being told in the town of Graaf-Reinet that rhinos had almost disappeared from the Cape because of hunting with guns. North of Kuruman, in the northern Cape, he reported seeing both black and white rhinos at a spring, and shot one of them.[73] Foran believed that the last black rhino in Cape Province was killed near Port Elizabeth in 1858, and that they had been exterminated in the Free State in 1842, were still abundant along the Limpopo in South Africa and Bechuana-land (Botswana), but were being decimated there, with some hunters killing 90 white rhinos on a single expedition, which he said was 'sheer butchery'.[74]

To the east, in the Zulu kingdom, the king, Shaka, periodically held huge hunts in which animals were driven into fenced-off killing zones or pitfall traps, where Shaka and his warriors would kill them with spears and axes. Remains of traps have been found in areas where it is recorded that white and black rhinos were still plentiful in the early nineteenth century.[75] One anecdotal account said that Shaka hamstrung a large rhino that had been driven into a fenced area. The injured rhino charged him, but was killed by a large number of spears thrown by his hunters.[76] One hunter, William Baldwin, noted the presence of white and black rhinos in the Zululand.[77] He gave lurid accounts of his hunting, which showed him to be particularly brutal in his methods. Baldwin often shot as many as 24 bullets into a hunted animal such as a rhino or an elephant, and frequently wounded rhinos that he did not follow up.[78] The first rhino he shot, a black rhino cow, was not used for meat. He just took the horns, some strips of hide to make whips and the tongue to eat.

In 1871, Henry Drummond hunted along the Pongolo River in what is now KZN. He said the local people were very keen to get the fat from rhinos that were shot, rather than the meat.[79] He recounted killing several and wounding many others on his expedition. Drummond thought the black rhino was vicious and morose, whereas the white rhino had a 'gentle and inoffensive disposition' and was often to be seen in large groups at waterholes.[80] At one waterhole, Drummond shot three white rhinos in one evening. Perhaps because of this killing of multiple rhinos by European hunters, 20 years after Drummond's expedition only a few white rhinos were believed to be present in the region of the iMfolozi (also rendered Umfolozi) rivers, where they had survived because of the presence of tsetse flies, according to Ian Player.[81] Many people believed they were extinct. However, according to Rookmaaker, 'No sooner had the white rhino been declared extinct, than word spread in 1894 that six had been killed on the Umfolozi River in Zululand. This population had hitherto been overlooked, and European naturalists were so convinced of the rarity of the white rhino that they suggested that only a few animals could still survive there.[82]

In 1895, partly as a result of concerns over rhino numbers, Hluhluwe, iMfolozi and St Lucia were established as reserves. There were no surveys of wildlife, and rhino numbers were guesswork based on scattered sightings around Hluhluwe and iMfolozi. The estimates were of about 20 white rhino remaining, but Rookmaaker suggests that given the growth in numbers that occurred in the twentieth century the number was probably nearer 200,[83] concentrated in a small area of Zululand. When the Sabi Reserve (later to be amalgamated with Shingwedzi to the north and become KNP) was established, it was believed that there were no rhinos in the area. In 1904 the reserve warden, Colonel James Stevenson Hamilton, thought there might be one old rhino within the reserve boundaries, and that it had been shot at and wounded in the past.[84] In 1905, the *Journal* started publishing regular reports on the Sabi GR by Stevenson-Hamilton detailing conservation efforts and wildlife numbers. In his first report he wrote: 'at present the Transvaal Game Reserves run in a continuous line along the Portuguese border for some 300 miles … The animals found consist of all those indigenous to the country with the exception of the elephant, the rhinoceros, and the eland.'[85] In the Shingwedzi Reserve, he said that the last black rhino had been shot 'by the Boers during the war' – referring to the South African War, which ended in 1902.[86]

Stevenson-Hamilton was aware of the opposition of most Transvaal farmers to GRs, seeing them as sanctuaries for livestock killers and crop raiders. In answer to an enquiry, in 1909 he stated in a letter published in the *Journal* that at Sabi

> there is permanently in the Reserve a herd of some thirty
> elephants … a herd of some seventy buffalo … probably
> thirty black rhinoceros, of which nine or ten are in the
> original Reserve … In Portuguese territory there are …
> black rhino and giraffe still to be found in Gazaland. In time
> we shall probably get most of these over to our side.[87]

With protection from poaching, the numbers of large mammals, including black rhinos, increased over the next couple of decades, and there was growing interest among white South Africans in seeing game animals. By 1923, white South African and a few European tourists had started visiting the Sabi reserve.[88] Many drove into the park; others used South African Railways' popular Round in Nine tours, utilising the Selati railway line, which ran between Komatipoort on the Mozambican border and Tzaneen. Growing public interest in seeing the wildlife strengthened the case that Stevenson-Hamilton was making for the Sabi and Shingwedzi reserves to be joined together in a single large NP. This added to the feeling among some leading Afrikaner politicians that they should follow up on President Paul Kruger's establishment of the first reserve there (under Afrikaner rule prior to the South African War) by nationalising and upgrading the reserve to an NP carrying Kruger's name. The park was formally established in 1926 as the KNP. Within a year, regular visits by tourists in their cars had started, and the park authorities put in 617 km of tarred roads, opening up more areas of the park to visitors.

In the 1920s, iMfolozi in Zululand was the sole refuge for the southern white rhino, but even there it was not safe, despite the creation of the protected iMfolozi Reserve in 1895. The game conservator, Vaughan Kirby, regretted that during an anti-tsetse campaign in 1920 huge numbers of game were culled to restrict the food sources of the tsetse, including 'butchery' of half of the 50 white rhinos still believed to be in the iMfolozi region.[89] The surviving 20 or so rhinos continued to breed and often left the reserve's safety to give birth, putting themselves at risk of being poached. There were at least four white rhinos in the unprotected areas north of the Black iMfolozi River.[90] In 1924, one estimate put their numbers in the 90,000-acre reserve between the Black and White iMfolozi rivers at around 30 but possibly as high as 50.[91] Hobley was pessimistic about the survival of the southern white rhino, and in 1926, following a trip to Zululand, estimated the surviving number to be around 20. He noted that there was local opposition to the existence of the reserve, 'and periodically we hear of surreptitious incursions into the area to slaughter one or two of these rare creatures. The carcasses are said to be left where they fall, the object being to gradually annihilate the species, for when it is extinct they feel that the main obstacle to obtaining the land occupied by the reserve will have disappeared.'[92] Hobley believed there were 20 white and two or three black rhinos in iMfolozi, 150 black rhinos in Hluhluwe and 30–50 black rhinos in Mkuze.[93] With protection, and perhaps because of previous underestimates, by 1933 a Natal province game report said that the animals had bred well in the last year and there were now 200 white rhinos at iMfolozi and no reports of illegal killing there. However, two rhinos left the reserve and moved to neighbouring farms during a drought, one of them being killed by Zulu farmers there.[94] Some white rhinos had dispersed to neighbouring Hluhluwe Reserve, where there were an estimated 85 black and seven white rhinos.

The game report by the Conservator of Zululand GRs for 1935 noted that there had been a serious drought, and while no rhinos died, 'one white rhino strayed outside the Hluhluwe Reserve during August presumably in search of new grass and was shot by a European on his farm', while a black rhino had strayed into the neighbouring 'native reserve' and been killed after reportedly attacking people.[95] The report said that black rhinos appeared to be increasing in number in Hluhluwe, and several young calves had been seen. The number of black rhinos was put at a little over 100, while 17 white rhinos, including three calves, had been seen.[96] Four adult white rhinos were killed during 1937 when they wandered long distances from the iMfolozi Reserve, but 75 white rhinos were counted in the Corridor area adjacent to the reserve, where grazing was more suitable than inside the reserve and which was thought by the conservator to be an important part of the natural habitat of the white rhino. Numbers were said to be increasing satisfactorily, with 19 calves seen during December 1935.[97]

During the late 1930s, the Zululand Game Reserves and Parks Board campaigned to get land surrounding Hluhluwe Reserve added to the protected area, though this was held up by the continuing tsetse fly eradication programmes on surrounding land. The extension of the park was necessary as the growth in rhino numbers meant that increasing numbers were moving outside both the Hluhluwe and iMfolozi reserves to find food.[98] A report from the board noted the numbers and activities of rhinos based on rangers' reports:

> 18 adult white Rhino with three calves living in the
> Hluhluwe Reserve, and a further 10 or 11 come and go from
> the corridor. In three days during February, 75 White Rhino
> were actually seen in the corridor by reliable game guards.
> In March, however, in four days only 37 were counted in
> the same area … During February the patrol reported the
> presence of 34 White Rhino outside the Umfolozi Reserve
> [Imfolozi] in Mahlabatini area (Native Reserve).[99]

In 1943, the board reported that efforts were still underway to finalise the expansion of the reserve to increase its fenced area to 57,000 acres. The board believed that 20 white rhinos were resident in the area to be added.[100] The reserve report in 1944 listed Hluhluwe as having 12 white and 150 black rhinos. Some of the white rhinos present in Hluhluwe had dispersed from iMfolozi, where numbers had reached 200.[101] In 1951, a Zululand Game Reserves and Parks report put the numbers at 500 white rhino in iMfolozi and 20 in Hluhluwe.[102] These figures need to be treated with some caution as this rise within a few years seems huge, but it may be accounted for by better conditions for surveys in some years.

In the 1950s, farming communities were living right up to the boundaries of the reserves; people were squatting on reserve land and grazing their cattle there. Poaching was rife, using dogs, snares, spring traps and guns to kill for meat.[103] In 1953, Ian Player was working as a warden in the iMfolozi Reserve and conducted an aerial survey, finding 437 white rhinos, which was thought to be too many within a fairly small area.[104] By September 1957, it was clear something would have to be done about rhino overstocking on the reserve. Player and Dr Tony Harthoorn started experimenting with darting and sedating rhinos to aid capture for possible relocation to areas without white rhinos or to zoos. The plan was to restock areas that once had white rhino populations and to relocate some to Zambia and Kenya, not part of the original range.[105] As the plans developed, the Natal Parks Board agreed to relocations, and initial plans were for 100 rhinos to be captured and relocated to KNP.

Zambia

Prior to colonial rule, African communities in what became Zambia hunted mammals, birds and reptiles for meat and skins, gathered fruits and wild honey, and cut down trees for firewood and construction of buildings.[106] If rhinos were killed it was for meat and hides. It was only with the development of trading routes from the Indian Ocean coast into the interior of south-central Africa that hunting for horn, ivory and skins developed. External demand created trade and brought the possibility of bartering wildlife products for guns, gunpowder and manufactured goods. Demand for commodities such as rhino horn and ivory generated commerce between hunters in Zambia and the Portuguese, Arabic and Swahili traders on the coast. Where Zambia borders north-eastern Angola, the Chokwe gained a leading role as both hunters and suppliers of wildlife commodities bought from neighbouring peoples – they traded with merchants at Luanda on the Atlantic coast and east to ports on the Indian Ocean coast. The Tonga of southern Zambia and Zimbabwe and the Yao of Malawi and Mozambique became middlemen in the trade between inland hunters and foreign traders on the coast at ports such as Kilwa and Zanzibar.[107] After the imposition of British colonial rule, as in other colonies, game and hunting regulations were imposed, but they had only a minor effect on reducing indigenous utilisation of wild-caught meat, which continued to be important to rural communities.[108]

Prior to colonial occupation, the wildlife-rich Luangwa Valley region of eastern Zambia had been settled by the Kunda people. They lived by subsistence hunting alongside small-scale vegetable growing, and engaged in barter trade with other communities, trading meat, hides and other wildlife products for grain and other commodities. Some of the meat and other products of hunting were given to local chiefs as a form of tribute by hunters, establishing a tradition of chiefs benefiting from the utilisation of wildlife. Part of the rationale for hunting was the prevention of destruction of crops, but black rhinos, were not known as crop raiders.[109] The belief that the region held great mineral wealth led to exploration by Europeans to assess the potential for mining. In 1888, Cecil Rhodes's British South Africa Company (BSAC) negotiated exploration and exploitation rights from the Lozi chief in Barotseland. The BSAC presence led to Barotseland and then the rest of what was to become Zambia being incorporated into British colonial possessions. In 1911, the different areas under company control were merged into the Northern Rhodesia protectorate. In 1923, the British Colonial Office took over administration from the BSAC.[110]

Northern Rhodesia had a set of hunting laws and licences put in place by the BSAC, aimed at regulating the killing of wildlife, protecting certain species and generating income through the sale of licences. Regulations restricted African ownership of firearms. Black rhinos were in a class that required the issuing of a special licence.[111] Game regulations also banned many traditional hunting methods, and indigenous people were prosecuted and felt persecuted for following a traditional way of life, but this did not stop hunting to supplement crop production. The Luangwa Valley's wealth of wildlife meant that as early as 1902 the region was designated the Luangwa GR as part of the game management and licensing system, with the need for licences to hunt certain species, including rhinos. In a report by the colonial administrator R.T. Coryndon concerning game in North-Western Rhodesia, it was noted that under the licence system 'Schedule No. 3 consists of large and rare game, and includes elephant, rhinoceros, giraffe, eland, koodoo, mountain zebra ... the charge for an Administrator's licence, under which game mentioned under Schedule 3 can be shot, is £50 to visitors and residents alike.'[112] No figures were given for the number of rhinos shot on licence. The colonial authorities sought to derive income from trophy and commercial hunting by Europeans, introducing duties on the export of rhino horn, hippo tooth and ivory. Every pound of rhino horn exported was subject to a duty of two pence.[113] Those found trying to evade payment of the duty would be prosecuted and liable to six months' imprisonment or a fine of up to £100 for each offence.

By 1931, Northern Rhodesia had 33,237 firearms in a population of 1,330,000 Africans, 13,846 Europeans and 176 Asians. Darling wrote that this led to the depletion of wildlife in many areas, pushing the colonial government to establish three more GRs – the Victoria Falls Reserve, Kafue Gorge Reserve and the David Livingstone Memorial Reserve (near Lake Bangweulu).[114] In 1938, the Luangwa area was clearly delineated as two reserves with division between the northern section and the southern one, the latter being named the South Luangwa Game Reserve. In 1941, further game ordinances restricted hunting by indigenous communities, banning the use of game pits, snares, muzzle-loading guns and other traditional methods.[115] Game reports at the time do not give numbers for the rhino population, but treated rhino hunting as something to be controlled by strict licensing and high fees. In a report on Northern Rhodesia's fauna published in 1936, Pitman estimated there were 1,500 black rhinos in Zambia, adding that 'The black rhinoceros particularly is in urgent need of additional protection and should be included in strictly limited quantity on a special licence and a fee paid.'[116] Pitman's estimate seems to be very low, given that in the early 1970s the population in Zambia was put at 12,000.[117]

In the mid-1940s, the colonial authorities began to change access to wildlife resources outside fully protected areas. They introduced game management areas (GMAs), which were situated in areas where land ownership was communally based. The GMAs allowed the utilisation of wildlife in various ways, including hunting, to benefit the colonial economy and to enable local communities to legally hunt. Sport hunting by foreign tourists and photographic safaris could also be part of the GMA mix.[118] Initially, 36 GMAs were established. In April 1950, Kafue NP was created. According to a contemporaneous report, 'The park contains representatives of most species of the fauna of Northern Rhodesia. Primates are represented by the Rhodesian Baboon, the Vervet Monkey, and the Greater and Lesser Night-Apes. There are Elephant and Black Rhinoceros, Buffalo, Lion, Leopard, Cheetah, and numbers of the smaller carnivora.'[119] Kafue NP covered 8,650 sq. m, while GRs covered 10,080 sq. m and controlled hunting areas covered 105,530 sq. m.[120]

Shortly before independence, the Northern Rhodesian colonial authorities passed legislation severely restricting trading in rhino horn and ivory.[121] This was in addition to earlier regulations restricting traditional hunting methods and the sale of bushmeat. During this period, opposition to colonial hunting and wildlife regulations by Zambian nationalist movements led by Kenneth Kaunda and Harry Mwaanga Nkumbula was part of the campaign for independence and against all aspects of the colonial system. Kaunda, who later became a strong advocate of wildlife conservation, railed against restrictions on traditional hunting and trading in wildlife products, especially meat, and openly encouraged Zambians to break the colonial laws and even to resist arrest if caught.[122]

Zimbabwe

The territory that was to become the BSAC's Southern Rhodesia colony – and later the self-governing British territory of Southern Rhodesia – was a wildlife-rich region and a source of meat, hides and sport for the people of the Ndebele/Matabele kingdom and the Shona, Tonga and other chieftaincies that had replaced the empires of earlier centuries. From early in the nineteenth century, European hunters journeyed to hunt in Matabeleland, further north on the plateau areas and in the Zambezi valley. Baldwin, mentioned earlier as a hunter of rhinos, undertook an expedition to Matabeleland in 1857; his party killed five black rhinos, and in 1860 he and three other hunters killed 11 white and 12 black rhinos.[123] Frederick Selous is the hunter most associated with sport, commercial and specimen hunting in the latter half of the nineteenth century. The following events recorded in his various books give a good idea of the extent of rhino hunting in Zimbabwe. I will not weary the reader with references, but list the books in an endnote.[124] In 1873, he referred to one season's hunting in which he shot 'a goodly number of elephants, rhinoceros, and buffaloes, I seldom fired at anything smaller'. He noted on that trip that both black and white rhinos were to be found south of the Zambezi in northern Zimbabwe.

In a rare example of restraint, Selous wrote of the area near the Gwai River (which feeds into the Zambezi south of Lake Kariba):

> My boys were very anxious that I should go and shoot one,
> as the white species at this season of the year are always very
> fat and excellent eating … I would not do so, as I consider
> it a grievous sin to shoot these lumbering, stupid animals,
> unless having real need of meat, or when tempted by a
> particularly fine horn.

However, a few weeks later, the need for meat to feed his retainers led him to shoot a black rhino. He lists in this account that he regularly killed rhinos for meat – between June and December he shot five black rhinos and four white, in addition to 24 elephants and 60 other large mammals. Between January 1877 and December 1880, Selous shot 548 mammals, including two white and ten black rhinos. In his account of later expeditions in Zimbabwe, he said that, between 1882 and 1887,

> The black rhinoceros is still very plentiful throughout a
> large tract of country along the southern bank of the central
> Zambesi … and it will be many years, perhaps centuries,
> before it is altogether exterminated; whilst its congener,

the large, white, grass-eating rhinoceros, whose range was
always much more limited, as it was entirely confined to
those parts of Southern, South-eastern and South-western
Africa where were to be found the open grassy tracts
necessary to its existence, is upon the verge of extinction
without there being a single specimen, or even head of one,
in our national museum.[125]

In spite of this lament, Selous set off on an expedition to the Mazoe River, in
north-eastern Zimbabwe, to hunt white rhinos for the British Museum, in the
very area that he said might see their imminent disappearance. He shot two
white rhinos at Emhlangen, north of Bulawayo, but criticised two Afrikaner
hunters, Karl Weyand and Jan Engelbrecht, who shot ten white rhinos in
the area. Despite the evidence of large-scale killing mainly by Europeans, he
blamed their demise on 'thousands upon thousands of rhinos' being killed by
'natives' armed with 'the white man's weapons'.[126] In his account of hunting in
the British empire, Mackenzie noted that rhino horn was in great demand in
the latter three decades of the nineteenth century, and that this and hunting
for horn depleted the numbers of rhinos in Central and Southern Africa –
the killing increased in 1880 when the price of horn went up as rhinos were
becoming rarer.[127]

In the 1880s, European colonisation of the Ndebele and Shona areas of
what became Southern Rhodesia started with Cecil Rhodes trying to control
areas believed to be rich in minerals. In a very underhand manner that
was used to justify occupation, Rhodes representatives, Rudd and Moffat,
obtained permission from the Ndebele ruler, Lobengula, to prospect for
minerals in his territory and in Shona areas that he considered tributary to
him. Rhodes established the BSAC in 1889 and obtained a royal charter to
operate in Ndebele and Shona territory and claim it for Britain. In 1890 he
sent a heavily armed expedition to occupy the area, established settlers there
and started prospecting for gold and diamonds. The Ndebele resisted but
Lobengula's forces were defeated in the First Matabele war.[128] A further rising
by the Ndebele and Shona against colonial occupation was put down brutally
by Rhodes forces, which included Selous and other hunters.

In the early years of settlement, access to game for meat and hides helped
to support the colonial enterprise and attract settlers. Although there was
no specific ban on hunting by Africans, and they were permitted to kill
crop raiders, relatively few had access to firearms, and traditional forms
of trapping, pitfalls, spears and hunting using fire were declared illegal.[129]
The British colonial administrator and former private secretary to Rhodes,
R.T. Coryndon, reported on a visit to Mashonaland in 1894 that 'it is more

than probable that before the close of this century the White Rhinoceros, the largest of all the mammals after the Elephant, will be extinct'.[130] Ignoring the blindingly obvious fact that the rapid demise of the white rhino was a consequence of sport and commercial hunting by increasing numbers of Europeans, Coryndon adhered to the racist and self-serving colonialist view that the rhinos were chiefly disappearing because of hunting by the local communities.[131] The naturalist Shortridge wrote that in 1895, 15 white rhinos were reported to have been shot in Matrabeleland and one female was shot as far north as Mazoe in north-eastern Mashonaland, the latter being the last recorded so far north.[132] In 1909, in a survey of mammals for the Rhodesia Museum and the ZSL, the museum curator E.C. Chubb recorded that black rhinos were present in several areas of Southern Rhodesia, but did not include the white rhino in the faunal list.[133]

Wildlife policy under BSAC rule included the mass slaughter of game for livestock protection, land clearance, rabies prevention and tsetse fly eradication.[134] There was a concerted campaign to eradicate wildlife on European-owned farms. Over a million game animals were killed on and around European farms to create buffer zones to control tsetse fly.[135] This programme was enthusiastically implemented, especially during a drive during the First World War to increase beef production. After the war and the granting of self-government to Southern Rhodesia in 1923, there was a shift towards conservation, with the government in Salisbury realising that if some species, notably the Big Five (including black rhino), were to survive there would have to be habitat preservation and regulation of hunting. The first reserve was established at Matopos in 1926, followed in 1928 by Wankie (Hwange) and Victoria Falls in 1933.[136] Between 1942 and 1965, the Southern Rhodesian administration expanded Hwange and then made it, along with Victoria Falls, Matopos, Gonarezhou, Mana Pools and Matusadona, into an NP.[137] There was a drive to develop wildlife tourism domestically, from both South Africa and Europe. The building of hotels/lodges and improvement of roads helped to increase income from safari tourism, giving economic value to conservation. Income also increased through the state monopoly over the issuing of permits to hunt outside reserves.[138] This led one correspondent to the Fauna Society's journal to lament the disappearance of wildlife outside the reserves, estimating that between 1948 and 1951, 102,025 game animals were shot.[139]

CHAPTER 5

The modern trade in rhino horn from 1960 to the present, legal and illegal

There have been records of trade in rhino horn since the fourth century BCE and the continuation of that trade for millennia, with a vast increase taking place with the arrival in Africa of firearms, a growing demand for horn in the Middle East and the Far East, and the arrival of European hunters. There was never great demand within African rhino range states or elsewhere on the continent for rhino horn, and rhinos were killed if at all for meat and hides. Killing rhinos for horn developed to meet demand external to Africa for horn for its medicinal, decorative or prestige value. Whether used for dagger handles, cups, as a medicine or simply a commodity whose ownership implied wealth and status, rhino horn was in demand for millennia, but it was only with European penetration and the influx of huge numbers of firearms that the hunting of rhinos went from being a minor factor in determining population numbers and ranges to a serious threat to their very existence.

The legality of the trade was not an issue until colonial occupation and the passing of laws by European colonial powers to ban or restrict hunting by indigenous communities while encouraging hunting by settlers, colonial officials and visiting European or American hunters. African hunters were effectively criminalised, as were their hunting methods, while sport and commercial hunting was enabled through game regulations that limited who could hunt. Income was generated through taxes on exports and fees for hunting licences. The hunting of rhino and trade in their horns continued. Poaching became the term for hunting, generally by African communities such as the Wakamba and Waliangulu in Kenya, with a long tradition but now outside the regulations. African communities across the colonial territories in Africa were the chief targets of hunting laws and of the game wardens charged with enforcing them. Prior to the Convention on International Trade in Endangered Species of Wild Fauna and Flora (CITES) bans on the trade in rhino horn between member states – with horn from Asian species banned in 1975, white rhino in 1976 and black rhino in 1977[1] – the legal trade from colonial territories in Africa and then independent states continued alongside poaching and smuggling. External demand for horn in Yemen, China and other East Asian countries was the driving force behind legal and illegal trade in horn. Poaching happened because there was a market for horn, and high prices meant that, for many rural dwellers and unemployed/underemployed young men in rural or urban areas close to rhino populations, the lure of cash to be earned from killing rhinos was just too tempting.

CITES was signed in Washington DC in March 1973 and entered into force on 1 July 1975; 21 states had signed the convention, with South Africa the only African rhino range state initially signed up. It had come about due to increasing concern among mainly European and North American countries and wildlife/conservation non-governmental organisations (NGOs), including the IUCN, that trade in wildlife could be threatening endangered species and there was a need for an international body with powers to regulate it.[2] The first Conference of the Parties to CITES (CoP) took place in Bern in November 1976, with 28 countries represented, South Africa and Zaire (now the DRC) being the African rhino range states present.[3] The British delegates proposed that all rhino species be placed in Appendix 1 to the Convention, which includes 'all species threatened with extinction which are or may be affected by trade. Trade in specimens of these species must be subject to particularly strict regulation in order not to endanger further their survival and must only be authorized in exceptional circumstances.'[4] The CoP partially adopted the British proposal, with the white rhino and the three Asian species put on Appendix 1 and the black rhino put on Appendix 2, which includes 'all species which although not necessarily now threatened with extinction may become so unless trade in specimens of these species is subject to strict regulation in order to avoid utilization incompat- ible with their survival'.[5] Trophy hunting of rhinos and the export of trophies were legal, as long as the necessary permits and quotas were in place. It was assumed in the CITES regulations that trophy hunting would benefit conser- vation by bringing in income to aid protection and habitat preservation, and that demand for rhino horn would diminish progressively following the ban.[6]

Data on the rhino horn trade show that after CITES members had voted for a total ban on international trade in horn and all other rhino products from all five species by 1977, there was a sharp increase in consumer prices for horn in Asian markets, anticipating a reduction in available horn with the end of legal trade. Official import figures from Japan, Taiwan and South Korea showed a significant price rise in subsequent years. The profits to be made from illegal trade in horn rocketed, with wholesale prices for rhino horn in Taiwan jumping from $17 per kilogram in 1977 to $477 in 1980. The CITES bans did not end all legal trade, but only the international trade between CITES member states – Taiwan was not a member of CITES. In North Yemen, the price of horn used to make dagger handles leapt from $764 per kilogram in 1980 to $1,159 in 1985.[7] In the following years, little was achieved by CITES in either reducing demand or halting illegal trade. Since the CITES ban, the illegal trade supplied with rhino horn from dead rhinos or horn stolen from legally held stockpiles has continued to grow. Wholesale and retail prices have fluctuated, reacting to demand, with wholesale for the best-quality raw horn in Vietnam and China reaching a high of $65,000 per kilogram in 2012 and 2013 before falling back down again to $31,000 per kilogram in 2015, while retail has decreased from $95,000 per kilogram to $53,000 per kilogram over the same period.[8] Vigne and Martin report that the top price for the best-quality horn in Vietnam and China was down to $29,000–35,100 per kilogram in 2015, $26,000 in 2016 and $19,000–28,000 depending on quality in Vietnam in late 2017. In Laos, rhino horn was selling for $20,000 in 2017 and $19,000 per kilogram in 2016, mainly to Chinese buyers.[9] Demand ensured the continuation of poaching to meet it. Whatever the price fluctuations, it is clear that, as 't Sas Rolfes concluded from his research into the rhino horn market, the 1977 CITES ban had 'no discernible positive effect on poaching', with the black rhino population being decimated, and then white rhinos being hit very hard in the 2010s.[10]

There were many reasons why the ban failed to stop trade, not least because North Yemen, the main importer in the 1970s and 1980s, was not then a CITES member and not bound by its decisions. Taiwan, denied a place at the United Nations (UN) by China, could not be a CITES member and so did not have to follow its strictures. CITES member countries that had exported rhino horn legally in the past were bound to follow the trade ban, but not all did or even could. CITES adopted resolutions at a series of meetings in the 1980s to urge members and non-members to abide by the CITES rhino horn trade ban, but this had little effect.[11] Many CITES members and states outside the organisation held large stocks of rhino horn, and many, including major range states in southern Africa, retained a legal domestic trade in horn. As noted in Chapter 7, South Africa sold rhinos to Taiwan after the 1977 ban, with seeming impunity.

What was clear as long ago as 1995 (by which time many range states were parties to the convention) was that the CITES ban had not reduced the illegal trade in rhino horn. Rather, with continuing corruption and mis-management as well as inadequate law enforcement, in many rhino range states, transit and consumer countries it has handed a monopoly to the illegal trade – a trade in the hands of increasingly sophisticated poaching networks and criminal syndicates. The end of the legal trade and the control by criminals of the whole chain from killing in the bush to consumption by buyers of horn, along with speculation by those buying smuggled horn as an investment, pushed up prices astronomically. This in turn massively increased the incentives for poaching and smuggling within rhino range states. Some success has been achieved in saving rhinos thanks to anti-poaching operations, translocations and other measures in the remaining rhino range states, with numbers slowly increasing overall. Customs seizures vary greatly from state to state and are subject to both incompetence and corruption, and less progress has been made in demand-reduction campaigns than had been hoped for. Back in 1995, 't Sas Rolfes called for a bold approach to prevent the threat of extinction of rhinos, including a legal, regulated trade that he argued, if acceptable to CITES parties, would bring in significant income for conservation, anti-poaching and local community economic development.[12] The question of legalisation and regulation of trade is an even hotter and more controversial topic now than when it was first mooted, and efforts to get CITES approval have failed at successive CoPs in recent years.

It should be noted that there was very strong demand for rhino horn from the three Asian species historically, and poaching and habitat loss over many years have reduced the numbers of two species today to near extinction (with well under 34–47 Sumatran and 76 Javan rhinos surviving), while the number of greater one-horned rhino of India and Nepal has increased gradually to 4,018, thanks to political will in both countries to combat poaching.[13] However, from the rise in poaching for horn from the 1970s onwards, 'African rhinos have provided the market with over 50 times as much weight of rhino horn as have the Asian animals'.[14] The bans, David Western wrote back in 1989, haven't 'slowed the loss of rhinos and the trade in horn [is] … the only factor exterminating rhinos', just as relevant now as it was when the words were written.[15] In the 1970s and 1980s, there was a major increase in the poaching of rhinos across their range in sub-Saharan Africa as demand grew for rhino horn in North Yemen, where it was used to make dagger handles, and in Hong Kong, Japan, Taiwan and South Korea as their economies grew.

There was a short respite in the mid- to late 1990s. The outbreak of civil war in Yemen in 1994 damaged the economy, reducing the buying power of Yemenis, and limiting demand for rhino horn. At the same time, East Asian states began to yield to US and other Western political pressure, and to campaigning by international conservation and animal rights groups, and banned the domestic use of rhino horn in manufactured medicines.[16] In 1993, the Chinese government issued a 'Circular on Banning the Trade of Rhino Horns and Tiger Bones'. This strictly prohibited the import and export, sale, purchase and transport of these wildlife products, and was accompanied by the removal of them as approved ingredients in the official Chinese Pharmacopoeia governing traditional Chinese medicine (TCM), though it did very little to change the culture of rhino horn use in traditional cures.

Between the mid-1990s and 2007, poaching continued but at a fairly low level, and rhino numbers in Africa slowly rose, with improved anti-poaching strategies in most of their last strongholds. There were no major spikes in horn prices on the market or in the number of rhinos killed. 't Sas Rolfes writes that there was limited information on what was happening in consumer markets at this time other than indications of increased demand in Vietnam from 1999; 2003 was the first year in which Vietnamese nationals were recorded as visiting South Africa for the purposes of white rhino trophy hunting (the only legal way to export horn under CITES).[17] The supply of horn from these so-called pseudo-hunts rose sharply, and they became an important source of rhino horn for traders and consumers in Vietnam. A spike in demand in China and especially Vietnam from 2006 led to massively increased poaching in South Africa in the 2010s. It followed and was influenced by the South African government's moratorium on domestic trade in rhino horn in 2009 and the measures that ended the pseudo-hunts by Vietnamese nationals. The end of the hunts, which had supplied horn to traders in Vietnam against CITES regulations on the end use of legitimate trophies, coincided with the growth of Vietnam's and China's economies. More rhino horn had to be obtained by dealers to meet growing demand. The rising business/middle classes in ever more affluent Vietnam were able to pay the increasingly high prices for horn.[18] It is also probable that as prices rose speculators in Asia entered the market, buying horn to stockpile, taking it off the market, and increasing prices and the value of their investment as demand rose.

Rising incomes in China and Vietnam and modernisation of their economies did not lead to cultural Westernisation (which some had hoped for as a way of reducing demand for rhino horn and other wildlife products) when it came to TCM and demand for rhino horn as a prestige commodity in Vietnam. Horn continued to be used in TCM and in Vietnam as an alleged hangover or even cancer cure, and as a commodity conferring status on the owner.[19] It also made demand-reduction programmes far less effective than would have been hoped for (see Chapter 9 for more detail on demand reduction). Strong demand and a substantial pool of potential poachers and middlemen in rhino range states meant that from the mid-2000s to the late 2010s poaching and smuggling horn was a growth industry, with prices peaking for horn at $65,000 per kilogram in 2012, and high levels of poaching continuing for most of the 2010s. As demand, high prices and poaching continued, there was growing dissatisfaction in South Africa, eSwatini,

Namibia and Zimbabwe that they could not legally export stocks of legal horn to raise funds for rhino conservation, anti-poaching and community development in wildlife areas. This frustration grew in the 2010s as legal stocks grew and the costs of conservation and anti-poaching mushroomed. Little financial aid came from the USA and West European countries, which had pushed for and maintained the international trade ban and had called on range states to ban domestic trade. An example was the UN Environment Programme (UNEP) meeting in Nairobi in 1993, which was specifically tasked with raising funds for rhino conservation – a target figure of \$60m was suggested over a three-year period, but only \$5m was pledged.[20].

Poaching, combined with habitat loss, meant that rhino numbers declined from hundreds of thousands at the start of the twentieth century to 75,000 in the early 1970s and just 13,000 in the early 2000s, including the three Asian species.[21] With the successes in breeding and translocating southern white rhinos and some recovery in the numbers of black rhinos and Indian one-horned rhinos, the total is now about 27,420–27,433.[22]

Why are African rhinos killed for horns, hide and other body parts?

Although in some parts of Africa rhinos were hunted in relatively small numbers over the centuries for meat, with hide and horn having practical uses, hide for making shields and horn for war clubs, the primary reason for the centuries of rhino hunting has been to obtain the horn – whether to make ornate dagger handles in Yemen or for East Asian traditional medicines. Horn is also considered a desirable commodity among those who want status and, for some, the spiritual protection that they believe flows from possession. It is presented as gifts for people whom possessors of horn want to impress or influence when setting up business deals. The medical uses vary, but it should be stressed that apart from a small area of Gujerat in India long ago, rhino horn has not been used as an aphrodisiac, contrary to Western beliefs.[23] This was until some in Vietnam encouraged this idea (not part of traditional Vietnamese medicinal uses). Where horn is used for traditional medicinal purposes, it is believed to cure fevers and headaches, cleanse the blood and also has many other uses – in TCM and traditional Vietnamese medicine especially. In the twenty-first century in Vietnam it was also marketed as a cure for cancer and to prevent hangovers. The belief in the medical properties of rhino horn originated in Chinese traditional medicine and spread to Chinese communities all over Asia. More sought after in the past were products from all three Asian rhino species, with the horn believed to have far greater efficacy than African rhino horn. All parts of the Asian rhino were in great demand in traditional medicine in Myanmar, Thailand, Cambodia, Vietnam, Korea and Japan, partly through the presence of Chinese business communities throughout East and South-East Asia.[24]

Vigne and Martin write that rhino horn 'is one of the eight immortal "power tools" in Chinese mythology, revered by Taoists, and as such believed to ward off evil and misfortune, and to give protection, blessings and good luck'; they explain how the *Book of Songs*, dating from 500 BCE and attributed to Confucius, describes wine in rhino horn cups used as libations for long life.[25] The first recorded use of horn for medical reasons dates back to the Emperor Hunadi in 2,600 BCE, and references to its use can be found in the *Divine Peasant's Herbal* (*Shen Nong Ben Cao Jin*), the earliest Chinese medical work, dating from the first century BCE.[26] Rhino horn is prized in Asia as a material for exquisite carvings, including bracelets, cups, bowls and figurines.[27] This is not the major source of demand for horn, although a recent report found that there has been a rise in demand for rhino horn carvings, some carved by craftsmen but increasingly for mass-produced ones, too.[28] The demand for horn was fed at first by killing China's own rhinos, then by importing rhinos from elsewhere in Asia. As rhino numbers there fell, the alternative source in Africa became the main focus of trade. Rademeyer believes that China has been a major importer of rhino horn from South and East Asia and then from Africa from the eighth century onwards.[29]

When Mao Zedong and the Chinese Communist Party came to power after decades of war and formed the People's Republic of China in 1949, they intensified use of TCM, including rhino horn, to break away from Western cultural influence and provide cheap alternatives to expensive drugs. Specialist research institutes were established to examine and refine TCM methods. Demand for rhino horn increased and then ballooned in the post-Mao period of economic modernisation, growth and the development of rich business and political elites, many of whom were desirous of reviving Chinese traditional culture.[30] Between 1949 and Mao's death in 1976, China imported 13% of East Africa's legally recorded exports, with perhaps 2–4 tons imported annually.[31] Five years after the CITES ban, Rademeyer estimated that 10 tons of horn were smuggled into China between 1982 and 1986.[32]

Poaching, animals dying of natural causes, including droughts, and the pseudo-hunts in South Africa in the 2000s have been the main sources of horn for the illegal trade, but quite large quantities of horn stolen or illegally sold from legal stockpiles have been diverted into illegal trade. Legal stockpiles, both government and private, are the product of seizures from poachers/smugglers, as well as dehorning operations that are carried out as a deterrent to poaching on both private and state reserves, and horns from natural mortality. Stoddard, quoting a report by the Wildlife Justice Commission, states that since 2016 in South Africa 'at least 974 kg of rhino horns seized in 11 incidents were confirmed as originating from the theft or illegal sale of horns from legal stocks, including both privately owned and government-owned stockpiles … 18% of all rhino horns seized during the period from 2016–2021'.[33] Some horn taken from legal stocks is obtained through robbery, but, as Stoddard and many others believe, 'Corruption, of course, is a key enabler, which is no surprise as South Africa is the main supply source.'[34]

The trade in rhino horn from Africa

Prior to the independence period in Africa, huge quantities of rhino horn, ivory and other wildlife products had been exported legally for centuries before colonisation, with mainly Swahili, Omani and Indian traders running the ivory and horn auctions after colonial rule began. From 1849 to 1895 it is estimated that about 11,000 kg of rhino horn a year were exported from Africa, amounting to 170,000 rhinos killed.[35] Parker and Martin reported that exports from East Africa in the 1930s were around 1,600 kg p.a., from about 555 rhinos, falling to 500 kg p.a. during the war and picking up to 2,500 kg soon after 1945 – with around 870 rhinos killed annually. The export volume dropped to 1,800 kg in the 1950s and 1,300 kg in the 1960s, but rose to 3,400 kg in the 1970s.[36]

Prices rose along with demand, especially from Yemen and Japan, in the 1970s. Between 1951 and 1980, three years after the CITES ban, Japan recorded the importation of 14,631 kg, about 487 kg annually, reaching a peak in 1973 with imports of 1,792 kg. In the 1970s, Japan imported about 800 kg a year, Taiwan about 580 kg and South Korea about 200 kg.[37] Trade figures indicate that 5,460 kg came from Kenya in this period and 2,649 kg from South Africa. Between 1966 and 1985, Taiwan imported 9,522 kg, about a quarter of the volume coming via Hong Kong. Yemen imported 43,000 kg between 1970 and 1990; it was not part of CITES until 1996.[38] South Africa and South African-occupied Namibia were among the major sources of rhino horn for buyers in East Asia, particularly Taiwan. South Africa joined CITES in 1977 but still exported rhino horn well after then, much of it poached by the South African armed forces or the National Union for the Total Independence of Angola (UNITA) in Angola or northern Namibia and shipped to East Asia.[39] In 1983, the administration in South African-occupied Namibia sold 99 kg of horn to a South African company for $460 a kilogram; that company in turn exported it to a trader in Taiwan.[40] It was only when Namibia became independent in 1990 (and joined CITES) that exports of rhino horn from there stopped.

When countries became members of CITES and were then bound by its rhino horn trade ban, the commodity disappeared from their trade statistics, but this should not be taken as meaning that imports stopped because, as will become clear, they did not. China and Hong Kong did not record all of their horn imports, but figures from exporting countries prior to 1977 show that Hong Kong imported 1,225 kg p.a. from Kenya, Tanzania and Uganda in the 1970s and China imported 1,025 kg. Most horn imported by Hong Kong was re-exported to China and Taiwan, and some to Japan. Yemeni traders exported rhino horn chips and leftover pieces to China. Britain imported 1,686 kg and the USA imported 2,642 kg, destined for the Chinese communities resident there.[41] By 1987, all of the major importing countries, North Korea excepted, had enacted import, re-export and export bans on joining CITES or under pressure from Western trading partners. However, Milliken et al. note that these measures 'have done very little to stem pressure on rhinos in range states'.[42] One reason for the lack of effect could have been that some rhino horn traders in Asia were stockpiling horn in the belief that rhinos would become progressively rarer and therefore the price would rise.[43]

The sixth CoP, held in Ottawa in 1987, recognised that the ban was failing to substantially reduce poaching and smuggling. The conference passed Resolution 6.10, which called on all countries to completely halt all trade, domestic or international, in horn or horn shavings, the only exemption being trophies from legal hunts. The resolution had little or no effect, and Milliken, Nowell and Thomsen, referencing work by Martin, Vigne and Song, note that market surveys showed horn was still being sold widely in South Korea, China, Taiwan and Thailand, while North Korean diplomats had been caught smuggling horn out of Tanzania.[44] Taiwan banned imports in 1984, but did not ban internal trade until 1989, and only started registering stocks a year later. They are believed only to have logged about 30% of stocks, leaving plenty of leeway for replenishment by smuggling in poached horn. Although international trade in rhino horn was illegal between CITES member countries, all individual countries of course were and are able to determine their own laws on the sale of rhino horn within their own countries. Since 1977, some countries have banned domestic rhino horn trade. However, 'even the decision to impose a local ban has not been straightforward or uncontroversial, particularly in rhino-range states that have sustainable use policies'.[45] In 2017, for example, South Africa's Constitutional Court dismissed an appeal by the Department of Environmental Affairs to keep the 2009 moratorium on the domestic trade in rhino horn in place.[46]

The post-ban illegal international trade in rhino horn

Prior to and immediately after the CITES ban in 1977 for both species in Africa, Cumming et al. estimated that at least 8 tonnes of illegally obtained rhino horn from the continent entered the market illegally, falling to 3.5 tonnes p.a. between 1980 and 1984 as the rhino numbers in eastern, central and western Africa declined rapidly, while those in southern Africa were still comparatively well protected. Burundi (which joined CITES in 1988) was a major hub for rhino horn and ivory that had been poached in Uganda, Zambia, Zimbabwe, Mozambique, Tanzania and Kenya. Up until late 1987, most of the horn from rhinos killed illegally in Zambia moved into Burundi, from where it was exported via the United Arab Emirates to North Yemen and East Asia. Martin found that horn from Luangwa Valley was taken to the port of Mpulungu on Lake Tanganyika, from where it was taken on by boat to Bujumbura in Burundi: 'We know of one shipment in September 1985, consisting of 40 pieces of Tanzanian horn weighing 66 kg, which was exported from Bujumbura to Dubai. Some Zambian and Zimbabwean horn was smuggled out in diplomatic pouches from the capitals of these countries.'[47] The Burundi government, under international pressure, banned the trade in rhino horn and ivory in November 1987, and there was strong evidence that the horn from rhinos poached in the Zambezi Valley was then moved into South Africa for illegal export.[48]

In October 1988, a lorry carrying 94 rhino horns (loaded in Zambia) was stopped by Botswana Customs and Excise officials at the Kazungula border post, on its way to South Africa. Rhino horn poached in northern Namibia and southern Angola by the South African army and Angolan UNITA rebels was exported via Rundu in South African-occupied Namibia and through South Africa with the involvement of the South African government (see Chapter 7).[49] The smuggling of rhino horn from South Africa was highlighted in a press release on 3 November 1988 from the US Department of Justice after the arrest of three Americans for conspiring to import rhino horn into the USA from South Africa. Three South African nationals were charged for their roles in the conspiracy. One of the South Africans in August 1988 smuggled a rhino horn into the USA, where he was participating in an event as a member of the South African Defence Force (SADF) Parachute Team.[50] He was paid $1,800 for the horn. The defendants in this case had agreed to sell five to seven rhino horns, which they had obtained in Angola, to a US Fish and Wildlife Service undercover agent; 14 rhino horns had been acquired in Angola and were transported to Namibia in South African military vehicles for shipment to the USA.[51]

Having been one of the major East Asian hubs for importing rhino horn before the 1977 ban, Hong Kong remained a waypoint for illegal horn destined for China, Taiwan and Japan, despite being one of the first places in Asia to ban imports of horn from all five Asian and African species.[52] In 1985, most of the 46.8 kg of horn seized by the Hong Kong authorities was from South Africa. South African government officials had also approached the Hong Kong government to try to obtain permission to sell rhino products there, despite both being signatories to CITES.[53] However, commerce in horn and products, including traditional medical compounds containing rhino horn, was not banned in Hong Kong until 1988. Yet the illegal trade continued and the British government, which still controlled the territory, and the Hong Kong authorities claimed that they didn't have sufficient personnel to police a ban on traditional medicines containing rhino horn.[54] The CITES ban and the territory's restrictions on trade in rhino horn did not stop smuggling into Hong Kong, though more stringent controls did reduce the volume. By 1989 the retail price for rhino horn there had risen from US$14,282 to US$ 20,751 per kilogram, as demand remained high but restrictions slowed supply.[55]

The Portuguese enclave of Macau on the Chinese coast was until 1999 another hub for trade in rhino horn owing to a CITES error in its regulations that overlooked the territory. In December 1985, the Portuguese agreed to abide by the CITES ban. In March 1986, they seized 89 kg of rhino horn sent from South Africa and discovered that the same South African source had previously sent 500 kg of horn and rhino hide to Macau. The supplier was a South Africa-based Chinese businessman, who had been a major supplier of rhino horn and hide to buyers in East Asia in the 1970s, using Hong Kong and Macau as the ports from which to sell the horn in China for use in medicines.[56] There was a strong link between poachers and smugglers in Mozambique and the traders using Macau as a base from which to send horn to China or other East Asian destinations.

In the late 1980s and early 1990s, research suggested that horn consumption levels in the major consuming countries were lower than the amount of horn leaving Africa, giving rise to the idea that middlemen in the trade and horn vendors were stockpiling horn as a speculative investment, assuming that as poaching rendered rhinos rarer, the price would go up.[57] However, they noted that it was also the case that middlemen in the trade and local importers, who sold on to pharmacies and manufacturers of TCM, were the ones who were making the greatest short-term profits and might buy more horn than immediately necessary to meet anticipated demand in the future, thereby taking horn off the market, though not as a form of speculation. It was thought that media reports of demand and very high prices might have a negative effect on rhino conservation as poachers and the criminals who commission poaching might increase the level of poaching when they thought the price of horn was rising.[58]

The poachers, who risked their lives or freedom to provide horn, were paid far less than those who commissioned poaching and sold the horns to the middlemen in the trade, who in turn sold them to the criminal networks that operated on a global scale, supplying demand whether in Yemen or East Asia. Emslie and Brooks note that in Lusaka in 1992, horns were being purchased for $100–360 (paid in Zambian kwacha), while a Zimbabwean conservationist told them that poachers and guides were paid $135 in local currency for horns, payments varying according to horn size. The dealers who bought them could sell them on for $625. In 1996, horn in South Africa was being sold by local dealers for $1,215. The authors thought that trade bans could cause increases in price on the black market as they encouraged illegal trading, 'thereby inflating horn prices and possibly stimulating demand. Trade bans also reduce opportunities for parastatal government conservation departments and the private sector to capture much needed revenue, reducing their ability and incentives to successfully conserve rhinos', and that 'if trade were legalised, horn could be supplied from … legal stockpiles in some range states from horns recovered from natural deaths in the wild, and by routinely harvesting horn from live rhino'.[59]

The Yemeni connection

In the 1970s and early 1980s, North Yemen (before unification with the People's Republic of Yemen in 1990) was the largest importer of rhino horn. It was importing about 50% of all African rhino horn in the 1970s and early 1980s, much of it exported from eastern Africa via the airports of Khartoum and Addis Ababa, and from Djibouti by boat. The main use of rhino horn was to carve handles for traditional curved daggers called jambiyas worn by many North Yemeni men. The wearing of a jambiya signifies masculinity, and this dagger was used in the past by a man to defend himself and his family. Use of the jambiya in conflicts is rare, but it is waved in men's dances, especially on important occasions. Despite bans, and attempts to encourage the use of substitute materials such as synthetic horn, buffalo horn and agate, nothing so far has compared to rhino horn, which remains sought after today by those who can afford it.[60] Good-quality rhino horn handles are a statement of wealth and confer status on the owners. Unlike inexpensive materials such as water buffalo horn, rhino horn has an almost amber translucency, and after many years develops a patina (called *saifani* – a term applied to the best rhino-horn handles) while remaining durable. The patina means that jambiyas which have had long use fetch higher prices than newly carved ones.[61] This is the only use of rhino horn in Yemen.

In 1962, North Yemen's ruling imam was overthrown, and fighting between monarchists and republicans only ended in 1970. There soon followed a massive rise in the number of Yemenis working in the oil industry in Saudi Arabia and the Gulf states; the cash earned by oil workers boosted the Yemeni economy and increased the number of men able to afford rhino horn jambiyas. The annual per capita income in North Yemen increased from $80 in 1970 to $500 in 1978 as a result. In 1978, about a million Yemenis working abroad remitted $1,500m to North Yemen.[62] Those who became wealthy were keen to advertise their increase in status by having the best jambiyas they could afford, Demand and prices soared in the 1970s as a result of this change in the economic status of many North Yemeni men. The main Yemeni trader is known to have imported 36,700 kg of rhino horn between 1970 and 1986. This represents about 12,750 rhinos. Imports into North Yemen amounted to over 3,000 kg p.a. in the 1970s, and there were over 6,000 dagger handles produced annually at a rate of five handles from every 2 kg of horn.[63] Rhino horn leftovers and shavings were re-exported, destined for China's factories to be made into patented TCM. These exports were officially banned by the Yemeni government in 1987 but this had little effect, as the trade in offcuts was too lucrative. In late 1978, the greatest single shipment of horn into Hong Kong consisted of 1,000 kg of chippings and shavings from Yemen.[64]

A campaign by the US-based African Wildlife Foundation (AWF) pressured the North Yemen government into banning horn imports in 1982, though this did not stop trade. What had an effect on reducing imports was the plummeting number of rhinos in East Africa. From 1980 to 1984, about 1,500 kg of horn were imported annually, but this fell to 1,000 kg in 1985 and 500 kg in 1986.[65] The downturn in the Yemeni economy, especially after unification of North and South Yemen in 1990 and civil war in 1991, also reduced the horn trade to Yemen, with dealers in Taiwan, where the economy was booming, buying more horn. Taiwanese traders had links with South Africa, from where rhino horn from the last remaining populations in Zimbabwe was smuggled. In 1996, jambiyas were still an important part of Yemeni dress in the north. Those with water buffalo horn handles on average cost $9.20, plastic handles cost around $6, camel nail handles cost around $4, but new rhino horn handles on average cost $86. Those that were 10–20 years old cost $1,477 on average. Efforts to curb new rhino horn demand consisted of education awareness campaigns in Yemen. A fatwa was issued by the Grand Mufti of Yemen (the chief Islamic cleric), stating that the killing of rhinos was against the will of Allah.[66] By 2000, it was estimated that there were shrinking stockpiles of horn in Yemen, but prices remained stable at $1,200–1,400 per kilogram, and fewer horns were being smuggled in (only three in 1998).[67] Vigne and Martin counted 59 jambiya workshops and 100 jambiya artisans operating in Sana'a's souk in June 1999; numbers rose in the 2000s as more Yemeni young men wanted the knives, with the country's growing population – but very few were rhino horn, most being cheap water buffalo horn.[68] Traders in Yemen told Vigne and Martin that Yemenis and Sudanese resident in Kenya, Uganda, Sudan and Ethiopia were bringing in rhino horns by air. They were also getting to Yemen via Saudi Arabia and across the Red Sea from Djibouti, often hidden in sacks of goods such as sesame seeds and groundnuts, with traders in Sana'a paying wholesale about $1,200 per kilogram, compared with about $600–800 per kilogram in Djibouti.[69] In 2003, most rhino horn reaching Yemen came via Djibouti, in a trade involving Yemeni smugglers selling petrol in Djibouti and buying alcohol, firearms, cigarettes and rhino horn to smuggle back into Yemen.[70]

There was some reduction in smuggling across the Red Sea when navies from a number of countries started operating in the area to intercept pirates and suspected insurgents. Assistance was also being given to Yemen by the Italian and other European governments to improve its capability to stop smuggling by sea. Despite this, Vigne at al. estimated that 60–70 kg of rhino horn entered Yemen annually between 2004 and 2006, and the number of workshops rose to 74, with 124 craftsmen making jambiyas mostly from water buffalo horn and cheap synthetic materials, as new rhino horn was rare. About 200 rhino horn jambiyas were made in 2006, compared with 300,000 using water buffalo horn.[71] The wholesale price of unworked rhino horn in Sana'a had gone up from $1,200 per kilogram in 2003 to $1,700 per kilogram in 2007, and a large jambiya made from it then cost on average $1,670. In 2013, Vigne and Martin found that the wholesale price for rhino horn in Yemen was down slightly at $1,500 per kilogram, and leftover shavings were sold via illegal export for $940 per kilogram. There were 83 carvers in 58 workshops and the rhino horn trade seemed to be in decline, with carvers using water buffalo horn and forms of solidified gum imported from China to make jambiya hilts.[72]

Yemen's most recent civil war, which started in 2014, continues today (2024) and has made most Yemenis poorer but some richer. Lucy Vigne spoke to contacts in Yemen in October 2023, and they told her that high-quality new rhino horn jambiyas were being made in Sana'a and advertised online for the equivalent of several thousand US dollars each. Jambiya retail shops have spread around the city. It is not clear how much rhino horn is smuggled into the country currently, but these new rhino horn jambiyas are being bought by rich Yemenis who have the funds to afford them. Most craftsmen who used to work in the old souk in Sana'a, which was for centuries the centre of the jambiya industry, have moved out. Many now work in the more prosperous areas where richer Yemenis live.[73]

Taiwan, China, Vietnam and East Asia

The markets in Taiwan, China, South Korea and Japan were the main ones for rhino horn imports for traditional Chinese, South Korean and Japanese medicine until the initial clampdown in the 1980s and the early 1990s. During the second rhino poaching crisis that hit South Africa in the 2000s and 2010s, traders in Vietnam and other East Asian countries, especially those with large Chinese communities or with strong trade links to mainland China, smuggled in rhino horn. In the 1980s, prices soared, with Hong Kong and Macau (before they became part of China), Singapore, Malaysia, Bangkok and other trading centres in the region being hubs for the illegal wildlife trade for rhino products from all five species. Prices for rhino horn from Africa in the late 1980s were much higher in East Asia than in Yemen. Rhino horn shavings and powder fetched $18,772 per kilogram in TCM pharmacies and other outlets dealing in horn in the 1980s.[74] In Japan, when this was still legal, 44% of Tokyo's pharmacies sold horn or compounds containing horn, though this fell to 17% by 1986.

South Korea remained an importer, particularly for the manufacture of Chung Sim Hwan (also rendered as Chung Shim Won) medicine balls, used to treat a variety of ailments, which took about 90% of the rhino horn entering South Korea.[75] South Korea banned imports of rhino horn in 1984, but this did not stop illegal imports, especially as use in medicines was not banned and horn was imported for this purpose. Imports were banned totally in 1986. South Korea did not join CITES until 1993. Even after that, 'the extensive domestic practice of over-the-counter dispensation of rhino products goes unregulated and demand remains high', according to Song and Milliken, and horn was still being smuggled into the country.[76] They surveyed 111 traditional medicine shops and found that 71 were illegally selling horn or its derivatives, with at least 16 different compounds using rhino horn for ailments that included nosebleeds, rashes, eye diseases, stomach ulcers, some mental disorders and swollen feet.[77] Chung Sim Hwan balls were the most widely used rhino horn compounds, with high blood pressure, mental conditions such as hysteria, disorders of the autonomic nervous system and insomnia among other ailments for which they were taken (equivalent balls, called Jufang Niuhuang Quingxin, are taken for similar ailments in China).

Singapore had been a major hub for illegal exports from African and Asian rhino range states. It was heavily criticised by CITES for perpetuating the trade in illegal horn, and following US Congressional Hearings in September 1986 on the illegal wildlife trade, the US government banned all imports of wildlife products from Singapore.[78] Singapore banned imports and exports of all rhino horn on 24 October 1986, five months before it became a signatory to CITES and would be bound by its trade rules.[79] This had the effect of cutting the amount of poached rhino horn entering the market there, but it was still being traded in significant quantities in Hong Kong, and the price there was $15,000 per kilogram in the late 1980s.[80] By the early 1990s, Asian imports of African rhino horn, often through Hong Kong, Bangkok or Malaysia, exceeded those of Yemen despite the longer and more risky trade route. Mainland China and Chinese communities spread across East and South-East Asia were the major consumers.

Taiwan's insatiable appetite for rhino horn

In 1989, the leading rhino horn trade researcher Esmond Martin described Taiwan as 'a new peril for rhinos', having identified the increasing demand for rhino horn there, despite Taiwan banning imports and exports in 1985.[81] The ban appeared to have little effect and there was thriving illegal trade in the following years, with the quantities sold within the country rising and the island becoming the world's largest centre for trading rhino horn. Taiwanese businessmen obtained rhino horn from South Africa often via Malaysia, Hong Kong and Thailand. Customs statistics showed that Taiwan had imported 7,281 kg between 1972 and 1985, but the real figure is probably higher, given that not all traders declared imports to avoid paying tax.[82] Soon after the 1985 Taiwanese ban on imports and exports of rhino horn, Martin found hundreds of kilograms of horn on sale, with shops having whole horns as well as smaller pieces. He found 76% of pharmacies in Taipei, Kaohsiung and Tainan offering rhino horn for medicinal preparations. Much of the horn was white rhino horn from South Africa, which had strong trading and UN sanctions-busting links with Taiwan (UN sanctions were intended to stop military-related materiel and technology from reaching apartheid South Africa).[83] One pharmacy alone had 35 white rhino horns. In 1984, the US government imposed limited trade sanctions on Taiwan under the Pelly Amendment, passed by Congress in 1978 to authorise the president to embargo wildlife products and 'limit other imports from nations whose nationals are determined by the Secretary of the Interior or Commerce to be engaging in trade or take that undermines the effectiveness of any international treaty or convention for the protection of endangered or threatened species to which the United

States is a party'.[84] Taiwan was accused of failing to stop the illegal trade in rhino horn and tiger bones.

The retail price for African horn in Kaohsiung, the main port and second largest city, doubled between April and July 1988, from US$1,536 to US$3,347 per kilogram. Part of the driving force behind price rises was the increasing scarcity of rhinos owing to poaching, which led some to buy horn as an investment as the price was very likely to rise. The continued Taiwanese belief that rhino horn-based medicines were best for lowering fever helped to maintain demand. Much of Taiwan's horn came from South Africa and occupied Namibia. Despite South Africa formally saying it was stopping exports in 1985, the trade continued underground.[85]

Taiwan's role in sanctions busting and as a major trading partner meant there was a regular flow of Taiwanese businessmen and ships crewed by Taiwanese to and from South Africa, many of them smuggling horn and paying bribes to customs officers to avoid their baggage being searched. In Taiwan in 1992, licensed pharmacies had registered stocks of 3,735 kg of rhino horn, but total stocks could have been as high as 9,015 kg, and consumption for medical purposes was estimated at 486 kg annually. Consumers used only small quantities each year and stocks remained high.[86] Taiwan banned domestic trade in rhino horn in 1992, but there is no evidence that this had much effect on consumption; nor did the public burning of 19 horns along with 700 kg of ivory.

China: import routes, TCM and production of patented rhino horn medicines

By the mid- to late 1980s, China was importing large amounts of rhino horn through a variety of routes, including Singapore, Bangkok, Macau and Hong Kong. One route that developed was through Myanmar (formerly Burma), a pariah state riven by ethnic/regional conflicts that has continued to be a smuggling hub in the region, with drugs, weapons and illegal wildlife products traded across its borders. Shepherd, Gray and Nijman made five visits in the 2000s and 2010s to the town of Mong La on the border with China, which was a major route for horn and other contraband to enter China. Rhino horn was openly on sale there and in increasing volumes from 2006 to 2015.[87] In 2015, rhino horn was available whole and processed into 30 g discs and bangles or leftover powder for medicinal use. In 2017 and 2020, Vigne and colleagues documented a growing number of rhino horn processed items openly for sale in several of Mong La's Chinese shops selling retail for about US$116 per gram in 2017 and about US$92 per gram in 2020. Popular with Chinese buyers are big-beaded bracelets, bangles (especially for men) and large pendants.[88]

Rhino horn was also consumed in Thailand, being much sold for medical use in Bangkok's Chinatown. In 1972, the Thai government banned the use of horn from Sumatran rhinos, because of their highly endangered status, but this had little effect on trade, and dried Sumatran rhino penises were still openly for sale in 1990. The main buyers of African and Asian rhino horn in Bangkok were Chinese, Taiwanese and South Koreans. Some horn was re-exported to China from Bangkok in 1987.[89] In the early 1990s, traders sold rhino horn to Chinese customers from Laos.[90] All trade in rhino horn or ivory is banned in Laos, but in 2013 Vigne found the laws were rarely enforced.[91] Sumatran rhino horn in its whole form is used traditionally by Lao people, being placed on family altars, but there is only very limited consumption of rhino horn. However, Laos had become by 2015 and 2016 a conduit for African rhino horn destined for China. Chinese nationals visiting Laos were the main buyers for both rhino horn and ivory ornaments and jewellery that sold in mostly Chinese-owned shops. Vigne found that rhino horn was priced in Chinese renminbi or US dollars for Chinese customers, and the country remains a hub for the illegal wildlife trade and home to a number of major illegal wildlife traders, who provide rhino horn to buyers in both China and Vietnam.[92] Xaysavang Trading Export-Import Co. Ltd is the leading company dealing in the illegal wildlife trade, and its director, Vixay Keosavang, is said to have been a major player in the international rhino horn trade.[93]

Keosavang's network, and others such as the Vannasang syndicate, operate from Laos and Thailand, from where horn is smuggled overland or by air into China, Vietnam and other regional destinations. One South African rhino horn and lion bone trader, Johnny Olivier, worked with the Thai wildlife product dealer Chumlong Lemtongthai to smuggle rhino horn to Thailand, Laos and onwards to China on behalf of the Keosavang syndicate. Olivier became a whistleblower and gave information to South African private investigator Paul O'Sullivan and to the investigative journalist Julian Rademeyer, who revealed in detail the extent of smuggling networks dealing in rhino horn in his superb book, *Killing for Profit* (2012). Lemtongthai would commission pseudo-hunts, with Thai women registered as hunters, so that they could bring the horn back to Thailand with CITES certification as legal trophies, from where it was sent on to its final destinations for sale illegally, whether in China, Vietnam or Chinese communities across East Asia.[94]

Conservationists, legitimate hunting organisations and many wildlife law enforcement officers in South Africa were furious when, in September 2018, Chumlong Lemtongthai, who had been given a 40-year jail sentence in South Africa in November 2012 for dealing illegally in rhino horns and other wildlife products, was released ludicrously early – leaving prison in South Africa on 11 September 2018, after serving just under six years of his jail sentence.[95] In the original trial in 2012, the court heard that he had obtained 100 horns from pseudo-hunts he had arranged, and in addition to these hunts had been involved in the illegal killing of at least 26 rhinos.[96] The court was told that he was the director of the Xaysavang wildlife trafficking network operating from Laos. Vigne recorded in 2018 that Laos remained an important market for rhino horn and a route for smuggling horn in both raw and worked form and as shavings into China. On a previous visit to check on rhino horn sales there, Vigne had seen 163 horn items on display openly for sale in Vientiane, Luang Prabang and the Kings Roman casino complex in the Golden Triangle Special Economic Zone.[97]

It was estimated that in the late 1980s about 600–700 kg of rhino horn, in whatever form, was being used in China each year for TCM production, and China imported about 2,000 kg annually between 1982 and 1986.[98] The majority was used in factories producing TCM cures that were exported for much-needed hard currency. In 1998, the registration of stocks held by TCM manufacturers, pharmacies, shops, museums and in private hands amounted to nearly 10,000 kg, and it is likely that there were substantial unregistered stocks.[99] At that time prices were declining but large stocks were held, something that may relate to doses of horn used in TCM being a fraction of the overall ingredients. A survey of use in 1989 found 'a staggering 9.875 kilo' in medicine shops, with evidence that about 650 kg were being used by pharmaceutical factories every year.[100] Overseas Chinese visitors to the mainland bought TCM products and horn to take back to the countries in which they lived. Martin found that the Shenzhen Special Economic Zone was a major rhino horn trade centre in the 1990s, with people from Hong Kong shopping there in large numbers and buying rhino horn and medicines, including horn.[101] Taiwan also served as a waypoint for horn to be smuggled into Hong Kong and China, where it supplied the factories making medicines in mainland China; these medicines were exported to the Chinese diaspora in eastern Asia for hard currency.[102] China was the biggest manufacturer of TCMs containing rhino horn, and although it became a party to CITES in 1991, and had in 1985 declared the international trade in these drugs illegal, it continued to market them abroad.[103]

The ban on trade in rhino horn issued by the State Council (the chief administrative body of the People's Republic of China) in 1993 had little long-term effect on the demand for rhino horn. It included the removal of TCMs from China's official pharmacopoeia, and placed a ban on the manufacturing and commercial sale of rhino horn medicines. However, researchers still found rhino horn or powdered horn in 11.4% of TCM dispensaries and in 14.8% of medicine stalls in markets between 1994 and 1996. The ban temporarily cut the price of rhino horn, and consumption appears to have gone down for a short period. However, the sale of TCM products was a huge business, and China's output of rhino horn-based drugs was then exported to much of eastern Asia, including Taiwan. The export of traditional medicines was one of China's most important foreign exchange earners, bringing in $700m in 1987 alone.[104]

Now that most of the poached horn reaching China comes from South African rhinos, one of the important connections involves well-organised and longstanding Chinese criminal organisations known as triads. One network links Chinese triads with the large Chinese community in South Africa, which numbers 250,000–350,000, and with triads in Hong Kong and Macau.[105] Rhino horn also reaches China from Mozambique, which has strong trade and investment links with China, and where an estimated 12,000 Chinese workers and businessmen are involved in Chinese-funded construction and infrastructure projects that give opportunities to engage in illegal trade. Chinese working in Mozambique frequently smuggle horn back to China. One part of the rhino horn smuggling network on the Indian Ocean coast is 'a Mozambican narcotics syndicate that also smuggles guns, ivory and rhino horn. This network has links with other syndicates along the Indian Ocean coast dealing in ivory, rhino horn, pangolin scales and meat, drugs, guns and other illegally traded commodities.'[106] Corrupt police and customs officials are involved, enabling safe routes to operate so that smugglers can transport goods from South African or Mozambican wildlife areas to the coast and then out of the country by sea. There has also been evidence of Chinese, Vietnamese and North Korean diplomats smuggling rhino horn in to their home countries.[107]

Another aspect of trafficking that has become evident through law enforcement investigations of the rhino horn trade between China and South Africa since 2016 has been the growth in Chinese-run horn workshops in South Africa. Here, horn is cut into smaller pieces, beads and bracelets are manufactured, and shavings and horn powder are kept to smuggle into China.[108] On 29 October 2018, there was a potentially dangerous development when the Chinese State Council announced that the 1993 ban on all trade and use of rhino horn in mainland China could be repealed to allow trade in antique and collectable items of rhino horn, and to allow licensed use of rhino horn in medicines through approved TCM doctors in accredited hospitals – with the horn used restricted to that obtained from captive bred animals.[109] However, protests from Western countries and conservation NGOs as well as concern expressed through CITES led to the official Chinese news agency Xinhua reporting on 12 November 2018 that implementation of the lifting of the ban had been postponed. On 13 December, Xinhua reported that the restrictions would remain in place and that China would continue to 'strictly ban the sales, purchasing, transporting, carrying and mailing of rhinos, tigers and their byproducts; and strictly ban the use of rhino horns and tiger bones in medicine'.[110]

Interestingly, some rhino horn trade researchers considered that the Chinese bid to lift restrictions on certain domestic trade and certain uses of rhino horn could provide an opportunity to start a dialogue between African rhino range states, particularly those in southern Africa that favoured a legal international trade to a willing buying nation, namely China, something currently banned under CITES regulations.[111] Duan Biggs said a dialogue was necessary, but warned that 'There is a lot of concern around domestic legalisation because if it's not done properly with careful consideration it could lead to laundering of poached rhino horn, and increase rhino poaching', adding that 'The implementation of this renewed legalisation needs to be done with deep and extensive consultation with African and Asian countries with rhino … If the trade commences with thorough and adequate evaluation, enforcement and regulation, this could actually be a good thing.'[112] His view coincided with those of groups such as PROA and wildlife economists such as Michael 't Sas Rolfes, some members of the IUCN AfRSG, and the governments of Namibia, South Africa and Zimbabwe, who favour legal trade, which is fiercely opposed by animal rights groups involved in the debates on conservation such as Born Free, the International Fund for Animal Welfare (IFAW) and the AWF (see the section in Chapter 9 on the arguments around legalisation of the rhino horn trade).

Illegal trade in rhino horn since 2006: increasing demand in Vietnam

The catastrophic rise in rhino poaching in South Africa, which started in 2009 and peaked in 2015, but still continues at a dangerously high level, supplied substantial quantities of horn, much of it to meet booming demand in Vietnam for use in traditional medicine and as a prestige commodity for the newly wealthy middle and business classes as Vietnam's economy boomed. During the poaching epidemic, it was estimated that 4,757 rhino horns entered the illegal trade in 2016–17, with about 2,378 horns entering the trade each year. Customs and law enforcement agencies seized 1,093 horns in the same period, and there was a doubling of the number of horns seized from poachers and smugglers in Africa, but 3,664 horns avoided detection and were smuggled to eventual buyers.[113] Between 2009 and 2018, a total of 2,733 horns weighing 6,349 kg were recovered from poachers or smugglers. South Africa was the location of or was implicated in 33% of all seizures between 2014 and 2018, with Mozambique accounting for 10%. The horn seized in or coming from Mozambique was mainly South African in origin, as the weight of horn seized was significantly greater than could possibly have been taken from the small number of rhinos in Mozambique.[114] Mozambicans still play a substantial role in poaching within South Africa, in KNP in particular. The horns entering the illegal trade consist of 4,531 (95.2%) from poached rhinos, an estimated 85 stolen from animals that died naturally, 52 stolen from government or privately held stocks, 57 sold illegally from private stocks and 32 obtained legally from trophy hunts.[115] A cautious estimate of the weight of rhino horn being held in stockpiles in Africa is 52.16 tonnes.[116]

About 55% of horns seized during the 2010s were destined for or recovered in China, but during that decade Vietnam became a major destination and transit spot for smuggled horn and a key driver of demand, with an estimated increase of 24% in trade flows between Africa and Vietnam and a rise of 35% in the quantities of horn seized in the country. Chinese and Vietnamese nationals are heavily engaged in horn trafficking, playing major roles in obtaining and transporting rhino horn out of Africa to Asia. In 141 smuggling cases, 219 Asian nationals were arrested in either Africa or Asia in the process of smuggling horn poached in African range states – 101 of those arrested were Chinese and 53 were Vietnamese. From 2009 to 2018, Chinese accounted for 57% of Asians arrested while engaged in the trade and Vietnamese accounted for 40%.[117]

Vietnam became a major participant in the rhino horn trade from the mid-2000s onwards, as increasing economic growth and the rise of a prosperous business and middle class produced the wealth necessary to sustain trade in this increasingly high-priced commodity. Use of horn in Vietnam is nothing new – it has been used there for centuries, linked with the assimilation of TCM usages into the country's own traditional medicine.[118] Over 48 hospitals and 9,000 health centres in Vietnam are licensed to practise traditional medicine, with rhino horn as a component. As in China it is used to reduce fevers, purge the body and blood of toxins and treat headaches, convulsions, epilepsy and strokes, among other uses. Several of the major Vietnamese-language traditional medicine pharmacopoeias feature sections on rhino horn as medicine.[119] Between 2010 and 2012, 24 out of 43 Asians arrested in South Africa for dealing in rhino horn were Vietnamese, and diplomats at the Vietnamese Embassy in Pretoria were documented as participants in horn smuggling.[120]

Wealthy or ambitious Vietnamese, especially businessmen keen to win favours from potential business partners and officials or demonstrate their status, would make gifts of rhino horn in some form. At business dinners, people might be given powdered rhino horn in wine to reduce the effects of overconsumption of food and alcohol. Medical uses continued, with the narrative emerging that rhino horn could both cure hangovers and treat cancer. Growing demand among the newly prosperous but poorly educated led to much higher prices.[121] For centuries, demand for horn was met by obtaining the horns of the Asian rhino species, especially the Javan rhino, which had been present in Vietnam, with the last one believed to have been poached in 2010. However, the scarcity of Asian rhinos and rising demand for rhino horn in Vietnam led to the importing of poached horn from Africa as a medicine and as a luxury commodity denoting wealth and status. Both peer group competition and social media promoted rhino horn to show off wealth, like an expensive luxury car, with rhino horn in wine being thought of as 'the alcoholic drink of millionaires'.[122]

Initially, the pseudo-hunts in South Africa, which ironically provided funds to private owners enabling them to increase the number of rhinos that they could support, were the major source of horn and 'by 2007, South African law enforcement officers had identified at least five separate Vietnamese-run syndicates of close-knit networks of operatives that actively probed the country's sport hunting industry for opportunities to come into the possession of rhino horns'.[123] South African rhino owners and professional hunters were drawn into the pseudo-hunting business, arranging hunts where Vietnamese, usually women working in bars or the sex industry, were taken to game farms to take part in hunts (usually as spectators, even though they were theoretically the hunters). The horns from the rhinos killed were then issued with export permits in the name of the supposed hunter so that they could be legally exported to Vietnam. In theory they could not be sold on and were only to be used as trophies of the hunt. However, the reality was very different, and the horns once in Vietnam became part of the trade to supply medical and luxury commodity demand and for a growing business in making men's bangles and other ornaments to be smuggled across the border to China.[124] Occasionally, the professional hunters who fired the fatal shot, which legally should have been fired by the Vietnamese rhino permit holder, were caught and fined but rarely jailed, as in the case of Chris van Wyk.[125]

Van Wyk was found guilty of arranging a fake hunt. He told Rademeyer that there were three or four agents who fixed up the pseudo-hunts and as many as 18 registered hunters who worked with them to produce trophies for export to Vietnam – sometimes paying as much as R300,000 for a horn (about $34,883) and maybe R1.5m in total for a hunt (about $175,000).[126] One hunter estimated that in one year at the height of the pseudo-hunts, 150 rhinos were shot to meet demand by just one hunter, and several other hunters between them killed scores of rhinos, with 2,003 hunts carried out between 2005 and 2007 for trophies mostly to export.[127] It was legal for rhino horns also to be traded within South Africa, until the government placed a moratorium on domestic rhino horn trading in February 2009 – the ban was struck down by the Constitutional Court in April 2017. In 2012, the South African Department of Environmental Affairs ceased issuing rhino trophy hunting permits to Vietnamese nationals, and although there were attempts to circumvent this, with the horn trading syndicates using Czech and other East European nationals, the pseudo-hunts were brought to an end.

Ending the pseudo-hunts may have had the unintended consequence of increasing poaching as demand increased. The wildlife trade economist Michael 't Sas Rolfes told Rademeyer that he believed 'the pseudo-hunts were meeting some of the demand but also in a way creating demand, and that when the government put a stop to the hunts, poaching increased to meet demand in Asia. Suspension of the domestic trade in South Africa in legally obtained rhino horn and the measures to stop the Vietnamese hunts may actually have taken away a buffer that was preventing poaching, according to 't Sas Rolfes',[128] while also probably increasing trade connections. The Vietnamese government had already banned the import, domestic trade or commercial sale of rhino horn, with fines of up to $29,000 and prison sentences of up to seven years as the punishments. However, imports of rhino horn trophies with CITES export permits issued by the exporting country were still legal. Trade statistics for 2003 to 2010 show imports of 170 horns into Vietnam, even though South African statistics show 657 horns exported to Vietnam in the same period – demonstrating that 74% of horns imported were not declared or recorded in Vietnam, giving huge scope for illegal use – in addition to the fact that South African customs figures underestimate the number of horns from pseudo-hunts that were sent to Vietnam.[129]

In addition to the export of trophies, there was clearly horn from poaching entering Vietnam, and this increased hugely after the South African hunting restrictions, with illegally imported horn being the sole source for dealers in Vietnam. This was carried out through the illegal wildlife trade networks already detailed, often via Maputo, Hong Kong, Bangkok, Kuala Lumpur and Singapore, but also through Vietnamese diplomats in South Africa and Mozambique who were complicit in the trade, as were some senior government officials, according to Milliken and Shaw's detailed report on the trade to Vietnam.[130] Rademeyer gave examples in his account of the Vietnamese connections of diplomats who were caught by the police when they set traps to catch rhino horn smugglers.[131] Some horn was smuggled by air, and a lot was smuggled through land routes from Thailand, through Laos and across the border into Vietnam. Some seizures were made by Vietnamese customs, with 100 kg of horn recovered between 2004 and mid-2009.[132]

Another source of horns smuggled to Vietnam is believed to have been undeclared stocks of horns held by private individuals in South Africa, which were illegally sold to agents who smuggled them into Vietnam.[133] The Vietnamese syndicates, and other Asian syndicates supplying China, Taiwan and other markets, recruited couriers, mules (who carried the horn through customs) and middlemen who commissioned the poachers and got the horn to those who smuggled it out of the country. Often the transport of horn within South Africa and on flights out of Africa is made possible by the use of diplomatic vehicles and the cover provided by diplomatic status.[134] Very full accounts of the roles of specific Vietnamese diplomats in the trade can be found in Rademeyer's book, as well as Milliken and Shaw's 2012 TRAFFIC report on the Vietnamese link in the horn trade.

The global COVID-19 pandemic had an effect on the rhino horn market in Vietnam, changing the location of the market from shops to online dealing. The Director of Policy for Education for Nature Vietnam (ENV), Bui Thi Ha, said: 'The COVID-19 pandemic may have resulted in the closing of a large number of wildlife-related businesses such as wildlife farms, souvenir shops, and restaurants, but social media provided a new alternative retail market for wildlife,' with a growth in online markets. During 2021, at the height of COVID-19 restrictions in Vietnam, ENV recorded a 41.3% increase in online wildlife crimes compared with 2020.[135] Online markets offer a much lower risk of being caught for dealers in horn, while enabling them to reach a very wide audience and potential customers. Dung Nguyen of the ENV believes:

> In this 'anonymous' market, hundreds of opportunistic,
> amateur, and professional traders seized their opportunity
> to adapt to the new way of living, and this trend has
> continued … products derived from rhinos, bears, elephant
> ivory, tigers and pangolins are crowding online retail groups
> … we have been handling hundreds of new online cases
> every month, with 54.4% of all recorded cases in the first
> two quarters of 2023 alone being online.[136]

While ENV believe that law enforcement regarding illegal online trading has improved with seizures and arrests in the last two years, Dung Nguyen warns that 'The illegal wildlife trade will undoubtedly continue to evolve as traders find new ways to operate in the face of mounting obstacles. It's therefore critical that law-enforcement authorities can rapidly learn and understand wildlife traffickers' new tactics, so that they can promptly and continuously develop effective measures to fight wildlife crime.'[137]

Who are the poachers and smugglers?

This is not as simple a question as one might think, as the answer varies considerably. Although many of those who go into NPs, reserves and private ranches, community conservancies or game farms to shoot rhinos are recruited from 'communities living in close proximity to protected areas ... [others may be] former military personnel, police officials or game scouts, all of whom would have had specialized training to develop tracking or shooting skills'.[138] In South Africa there are also professional hunters, veterinarians, game capture specialists and others who work on game farms or have sometimes occupied senior positions in NPs who have become involved both in the killing of rhinos and in the horrific sedating of rhinos to remove the horn and base, leaving the animals to die slow and painful deaths. They may also have been involved in pseudo-hunts as well as dehorning for conservation. Douw Grobler, former head of game capture in KNP, crossed from conservation into rhino crime.[139] Many poachers who operate in South Africa, especially KNP, live and are recruited in Mozambique;[140] they include local people in need of more income and also former fighters from the Renamo rebel groups, former and current soldiers, government officials and police. They are recruited locally in Mozambique by the 'kingpins' of poaching gangs or syndicates, as the bosses are almost universally known, who provide them with hunting rifles and then pay them for the horn obtained. As Xolani Nicholus Funda, KNP's Chief Ranger, told me in August 2016 when I interviewed him at Skukuza, many poachers used Czech-made high-calibre CZ/Brno hunting rifles that often turned out to have been imported legally for use by government game wardens and scouts in Mozambique, and had been sold on to criminals for use in poaching.[141] The popular image of the poacher is a poor rural dweller armed with an AK-47 assault rifle – true in some cases, but this number is declining in South Africa and southern Africa. Increasingly over the last decade and a half, poachers have been commissioned and armed by middlemen/criminal syndicates, who send them to poach armed with high-calibre .357 or .458 hunting rifles, which 'ensure that the rhino is killed with one shot, two at most ... Poachers have also shown ingenuity in assembling homemade suppressors to try to reduce the sound of gunshots'.[142] Rifles are obtained through the criminal syndicates, who may steal them in farm robberies, buy them on the very active firearms black market in South Africa, often stolen or taken illegally by police officers from gun stores at police stations or from Mozambique. Glasson records that between 2014 and 2018, about a quarter of the guns seized from poachers were the CZ rifles mentioned earlier.[143]

In South Africa there has also been the use of immobilisation drugs fired either from a helicopter or from the ground; this is carried out by trained professionals who have access to veterinary drugs and specialised equipment. The use of helicopters means that poachers can get in and out of areas before anti-poaching units (APUs) have a chance to react. Since 2008, there have been increasing instances of game ranch owners, professional hunters, game capture specialists, pilots and wildlife veterinarians becoming active players in rhino poaching or illegal dehorning, something that appears to be 'unique to South Africa and is a significant factor not only behind the record levels of rhino losses since 2008, but also the insidious spread of rhino poaching across the country' in NPs, reserves and on private game ranches or farms.[144] These methods, and the involvement of what would be considered middle-class South African professionals in the illegal wildlife trade, indicate that there is a greater diversity and complexity to the poaching and smuggling of horn in South Africa than is often presumed.

Criminal gangs involved in poaching and smuggling are also frequently involved in other high-value, high-risk crimes such as drug and diamond smuggling, kidnappings, illegal mining, armed robberies and the bombing of ATMs.[145] Not all of the syndicates involve solely South Africans; at least five Vietnamese syndicates operate in South Africa along with Chinese triad gangs. Nelson has identified that rhino horn crime occurs alongside a variety of wildlife and other crimes in South Africa and beyond, noting that 'Crime convergence with wildlife trafficking sees the same transporters, dealers and brokers moving multiple illicit products – for example, the illicit gemstone trade and illegal wildlife trade from northern Mozambique to Malawi; abalone and drugs, and rhino horn and cash-in-transit heists in South Africa; and drugs and ivory in Tanzania.'[146] Nelson gives the examples of the Akasha organised crime family in Kenya, which was involved in drug trafficking for years and also became active in rhino horn and ivory smuggling, and linked up with ivory and horn smuggler Feisal Ali Mohammed and the Kromah cartel.[147] These networks involved a variety of criminals from Kenya, Uganda, Guinea and Liberia, and had connections through the smuggling networks to well-known horn, ivory and other illegal wildlife trade dealers such as Vixay Keosavang in Laos, the Shuidong criminal organisation in China and Teo Boon Ching of Malaysia.

The main motivation of the poachers and criminal syndicates is money. For the poachers, who are usually poor, unemployed or underemployed young men, the driving force is improvement of their living standards (including clothing and educating their children). In southern Africa, there is a growing number of peri-urban and urban youths who want to attain a sought-after lifestyle and status, including emulating those who have grown rich from crime. For the middlemen and kingpins of criminal syndicates, profit and greater wealth are the drivers, and the larger the expected profit, the greater the motivation to be involved in poaching at whatever level.[148] When demand for horn rises and prices are high, poachers and traders will poach more and sell more horn if they can. An exploding human population, increasing poverty and massive economic inequality in African countries with rhinos can also be a motivating factor to become involved in poaching and smuggling. There is a balance of income and risk, and clearly the lower poachers and smugglers feel the risk of death, arrest and imprisonment to be, the more likely they are to engage in these crimes; conversely, the more efficient anti-poaching operations and law enforcement are, the more likely potential poachers, traders and smugglers are to think twice about taking part.[149]

It is worth re-emphasising at this point that the simplistic poverty–poaching link needs to be picked apart to highlight the diversity of poachers' motives and the diversity of their backgrounds. Duffy has been particularly important in her work on the different and complex motivations for poaching and the different forms that poaching takes.[150] Poaching can be for high-value commodities such as rhino horn or for subsistence in the form of bushmeat for personal consumption or to sell or barter locally, and can be used to simply survive, to improve one's standard of living or to progress up the socio-economic scale in economies where more legitimate forms of social and economic upliftment are absent or very hard to access.[151] Poverty is most certainly one of the drivers of involvement in the lower levels of illegal wildlife trade, most notably poaching as the bottom rung. Those who poach to survive are clearly driven by need, and those recruited to poach high-value wildlife products may do so to feed their families; in such cases poverty alleviation programmes may reduce the need to hunt illegally.[152] However, simply increasing cash income does not offer a universal panacea for illegal hunting, so the provision of paid work as guides, rangers or employees of wildlife departments or safari tourism operations does not necessarily remove motivations to poach. It needs to be recognised that issues such as empowerment and agency when it comes to major life decisions or the circumstances in which people live, status, prestige and the social contexts in which people live their lives are also factors in the decision to engage in the illegal wildlife trade.[153]

In Mozambique, as Duffy notes, it is not poverty alone that may drive recruitment as a poacher, but also economic inequality, lack of employment and little or no government support for poor rural communities.[154] Status in one's community is also a factor, as poachers who bring income to poor communities have a higher status within those communities. There may also be resentment of the loss of land to NPs and the loss of communities' agency in the face of government decisions about land use and conservation programmes. There was a political, social and grievance-related aspect to the recruitment of poachers in areas of Mozambique near the Limpopo NP (LNP) and in South Africa around KNP and the linked private reserves, with people resentful of the past loss of land, current harassment by the authorities that deemed communities near the parks to be potential or actual poachers, the denial of access to grazing land, the evident wealth of tourists and investment in conservation/tourism rather than rural communities.[155]

One should also note that although poverty or a desire to better one's life are motivations for some to engage in poaching, the primary driving force of commercial poaching is the support of those higher up the supply chain who commission poaching and smuggle rhino horn to make money by meeting an existing and often growing demand. There was no substantial trade in rhino horn in Africa before external demand led to the hunting of rhinos for their horns as a commodity to sell to traders and for traders to pass on to overseas buyers. Duffy and St John put it very clearly: 'Individuals from poor communities would not engage in the poaching of commercially valuable species unless there was demand from wealthier communities.'[156] Although those recruited to hunt rhinos may get paid more in cash than they can earn from subsistence farming or pastoralism, and certainly more than unemployed or underemployed young men can get from the informal economies, it is those involved in the higher levels of the illegal trade networks and crime syndicates who garner the greatest value from the trade in rhino horn and other wildlife products.

The whole network of poaching, trading and smuggling out of Africa has a hierarchy, with those who go out to poach – whether as the shooter, tracker, porter or look-out – at the bottom and paid the least. Above them are the recruiters and those who commission the poaching teams; beyond them are the middlemen who connect the poaching gangs with those who move the horn on to international smuggling groups, who in turn get the horn out to buyers in trading hubs in Asia. In recent years, Naro et al. looked at how these hierarchies operated in Namibia's Kunene conservancies and in the adjacent ≠Khoadi-//Hôas Conservancy near Grootberg. Progressing up the hierarchy, they found higher levels of education, status and wealth. Those on the bottom rungs had little formal education and were usually herders or unemployed, neither having much disposable income – unemployment being particularly key.[157] The group above them, the recruiters, frequently ran local businesses, such as shebeens, and provided credit for local people or supplied cheap alcohol, knowing that the recipients had no money to pay and would be in their debt. This provided a way of persuading or coercing people on the bottom tier into becoming poachers.[158] Lack of work, growing alcohol dependence but no money to buy it directly fed this 'cycle of dependency … the money generated by Tier 1 individuals from poaching operations is often spent on alcohol, providing Tier 2 recruiters with the opportunity to perpetuate the cycle of debt, dependency, and coercion'.[159] This becomes a vicious, self-sustaining circle and the Namibian example is one repeated across areas of high poaching, I am sure. The need for money, getting into debt or striving for a better standard of life or higher social status can all be strong motivations to become a poacher when the opportunity or offer of money to poach is there.

CHAPTER 6

From independence to 2005

ost African countries achieved independence from their colonial
occupiers in the late 1950s and early to mid-1960s. The newly
independent states inherited conservation and hunting laws from
the colonial period, and surprisingly few changes were made after independence. They also inherited the legacy of seriously depleted wildlife, as a result
of legal hunting, poaching and habitat loss during the colonial era. There was
a legal trade in rhino horn and hide, and continuing legal hunting of rhinos –
whether sport hunting, commercial hunting or culling to remove problem
animals, or to clear land of wildlife for agricultural purposes. This, along with
illegal killing, severely depleted the black and northern white rhino, while
the southern white rhino was making a remarkable recovery in South Africa.

In 1960, there were estimated to be 100,000 black rhinos, perhaps
1,200–2,400 northern white rhinos (stretching from the West Nile District of
Uganda, through the parts of southern Sudan,[1] a few remaining in Chad and
the CAR, with the largest population in the Garamba NP in north-eastern
Belgian Congo) and about 1,000 white rhinos in South Africa.[2] Two decades
later, in 1980, Hillman estimated that there were 10–20,000 black rhinos left
in Africa, with ranges, densities and overall numbers in decline since the start
of the century, mainly as a result of legal hunting and poaching for horn. She
gave the example of Kenya, where in 1900 rhinos could be found across about
90% of the country, but this had dropped to 50% by 1963 and less than 25%
in 1980.[3] Between 1948 and 1968, southern white rhinos in South Africa, and
those translocated from there to Namibia, Zimbabwe and Kenya, increased
from 550 to 1,800.[4]

In the opening decades of African independence, poaching, droughts
and loss of habitat reduced black and northern white rhino numbers, the
latter to the point of extinction. Large-scale illegal killing of rhinos went in
phases, with huge depletion of black and northern white rhinos from the
mid-1970s through to the 1990s, and then again, this time including southern
white rhinos, from the mid-2000s onwards, driven by demand and rising
prices for horn. The first wave of post-independence poaching of rhinos

started in the 1970s, with the growing demand for horn in Yemen, Hong Kong, Macau, Japan, Taiwan and Chinese communities living across East and South-East Asia.[5] This led to over a decade of poaching and the steep decline in black rhino numbers in East Africa and parts of south-central Africa and the disastrous depletion of northern white rhinos. Black rhino numbers were down to 10,000–15,000 by 1981 and 3,800 by 1987; northern white rhino numbers plummeted from about 821 in 1980 in Congo (DRC), Sudan, CAR and Uganda to 22 in Garamba NP in DRC by 1988. Only the southern white rhino was increasing in numbers in South Africa, with around 3,020 in southern Africa and Kenya in 1980 (Table 6.1).

In many areas, civil wars, insurgencies and other violent conflicts introduced large numbers of military-grade weapons into regions with rhino and elephant populations, both of which could be killed for cash by armed groups or by peoples rendered destitute by conflicts. As Emslie and Brooks noted, 'Rhino populations in Angola, CAR, Chad, the DRC (formerly Zaire), Mozambique, Namibia, Rwanda, Somalia, Sudan, and Uganda have all suffered from the consequences of war and civil unrest since the 1960s.'[6] War, greedy and incompetent governments intent on personal enrichment of leaders and their cronies and unequal international terms of trade all created conditions in which poverty remained a fact of life for tens of millions. Unemployment was the curse of the younger generation, and informal economies thrived alongside the formal ones – fertile breeding grounds for illegal hunting and the smuggling of wildlife commodities for subsistence, economic and status uplift and greed. Many leaders and prominent public office holders became the gatekeepers of the illegal wildlife trade rather than protectors of their countries' wildlife heritage.[7] In many range states, 'Corruption was endemic and an underground economy took over in which poaching played a role … [and] wildlife departments have become increasingly short of funds and are unable to counter the opportunistic poaching by individuals and professional gangs … During the 1970s and 1980s many rhinos were poached by the very people employed to protect them.'[8]

The rhino trade researcher Esmond Martin believed that between 1970 and 1980, 50% of the world's rhino population disappeared. Black rhinos bore the brunt of poaching – 90% of black rhinos were killed in East Africa to meet the demand for rhino horn in the Yemen and East Asia.[9] The decline in black and northern white rhino owing to poaching was particularly heavy in the CAR, Zambia and Tanzania.[10] Of the black rhino subspecies, the south-central, *Diceros bicornis minor*, found from Kenya to South Africa and Namibia, declined between 1980 and 1984 from 6,895 to 5,840; the eastern, *Diceros bicornis michaeli*, in Kenya and Tanzania, declined from 3,480 to 1,975 (down 43%); the western, *Diceros bicornis longipes*, mainly in CAR, declined from 3,135 to 285 (a catastrophic 91%); however, the south-western black rhino, *Diceros bicornis bicornis*, in South Africa had risen from 300 to 400 and was still found in northern Namibia.[11] The survival of the northern white rhino was under extreme threat, with the rhinos of Garamba NP in Zaire (Congo) the only survivors.[12]

The black rhino had made 'rapid transition from abundant to endangered. The estimated 65,000 black rhino which roamed Africa's wildlands in 1970 have now been reduced by over 95%', Milliken, Nowell and Thomsen wrote in 1993 when surveying the dramatic decline.[13] However, the southern white rhino had grown from 'a single small relict population to 2,500 [in 1980] following effective conservation measures in South Africa'.[14] Two years later, Martin and Vigne reported that there had been a major reduction in poaching since 1993 and that black rhino numbers had started to increase, suggesting the success of anti-poaching and other conservation measures – including the militarised Operation Stronghold in Zimbabwe, Operation Uhai in Tanzania and the initial effects of Richard Leakey's shake-up of the Kenya Wildlife Service (KWS).[15]

In the period covered by Table 6.1, the CAR, Chad, Ethiopia, Malawi, Rwanda, Somalia, Sudan and Uganda all lost their black rhino populations, while Angola, Cameroon, Mozambique, Tanzania and Zambia were very close to losing theirs too. There was a decline in the entire wild black rhino population of 28% between 1991 and 1992.[16] The AfRSG of the IUCN in 1992 identified parts of Zimbabwe (the Zambezi Valley and Hwange), Namibia (Etosha and Damaraland), South Africa (KNP and Hluhluwe-iMfolozi GR) as priority areas for black rhino conservation. Cameroon's small rhino population, the last surviving western black rhino (*Diceros bicornis longipes*), was also identified as a conservation priority – with about ten individuals surviving in 1997, but these became extinct and with them the subspecies as a whole.[17] By the end of the 1990s, there were 83 identifiable black rhino populations remaining, 45 of which were in NPs or reserves, 23 were managed privately on a custodianship basis, one state reserve was privately managed for the state, there were seven privately owned black rhino populations in South Africa and in Namibia, and there was a free-ranging population in Damarland/Kaokoveld. In Zimbabwe there were more privately owned black rhinos, a total of 229, than the figure of 116 in NPs or reserves. A further 171 black rhinos were managed privately under state-sanctioned custodianships in Kenya, Namibia and Swaziland.

During the 1980s and 1990s, following the 1977 CITES ban on inter-national trade in rhino horn, successive CoPs adopted resolutions on rhino trade and conservation. In 1981, a resolution at the New Delhi CoP urged all member and non-member states to prevent the sale of and trade in

Table 6.1: Estimated numbers of black and white rhinos

Country	Black rhinos				White rhinos			
	1980	1984	1992	1997	1980	1984	1992	1997
Tanzania	3,795	3,130	127	46	80	200	249	167
Zimbabwe	1,400	1,680	430	339				
Zambia	2,750	1,650	40	0?	5	10	-	6
South Africa	630	640	819	1,043	2,500	3,330	5,297	7,913
Kenya	1,500	550	414	424	25	30	74	137
Namibia	300	400	489	707	150	70	91	141
CAR	3,000	170	0	- (Northern)	20	1	-	-
Mozambique	250	130	+/-50	13	30	20	-	0
Cameroon	110	110	35	10				
Sudan	300	100	0?	- (Northern)	400	10	0?	-
Somalia	300	90	0	-				
Angola	300	90	+/-50	0?				
Malawi	40	20	0?	3				
Rwanda	30	15	15?	4				
Botswana	30	10	5	0?	70	200	27	23
Ethiopia	20	10	0?	0?				

Chad	25	5	0?	0		1		-	-
Uganda	5	-	0	-		1	1	-	-
Zaire/DRC	0	0	0	-	(Northern)	400	15	30 (1991)	25 (1998)
Swaziland	-	-	6	10		60	60	46	50
Total	14,785	8,800	2,480	2,599	(Southern)	3,020	3,920	5,784	8,437
					(Northern)	821	28	30	25

? = situation unknown, Ivory Coast had five southern white rhinos relocated there in 1991 and the group had dropped to 4. The 1992 black rhino figure is a maximum but could be lower at 2,365

(Compiled from Western & Vigne, 1985, p. 216 and Milliken, Nowell & Thomsen, 1993, p. 5; Emslie & Brooks, 1999, pp. 5, 8–9).

rhino horn. Six years later, in Ottawa, a resolution called for the complete prohibition of sales and trade, domestic and international, of rhino body parts and derivatives, but excluding non-commercial exports of legitimate hunting trophies fully certificated under CITES regulations.[18] The exemption of hunting trophies recognised that in South Africa (and it was later to be the case for Namibia) there was evidence that regulated hunting 'serves to enhance the survival of the species concerned'.[19] The success of South African white rhino conservation was recognised in 1994 when the ninth CoP voted to transfer South African white rhino to CITES Appendix II, allowing trophy hunting, the export of trophies and trade in live animals between CITES members.[20]

Whatever the intentions of CITES resolutions, they had limited effect on the ground, where poaching was a serious problem. One exacerbating factor was the shortage of resources for anti-poaching operations across the major rhino range states, with the possible exception of South Africa. Tanzania, with its successful tourism industry based on the Serengeti, Ngorongoro Crater, Lake Manyara and Zanzibar, did not provide enough income to the wildlife department to meet the needs of conservation and anti-poaching. Bonner noted that the funding needed to conserve rhinos and elephants across huge protected areas was between $200 and $400 per square kilometre, amounting to $24–48m annually, but that in 1991 available funding was less than $5m. Serengeti alone needed $3m a year, but income from tourism that was passed on to the park was less than $400,000.[21] The actual money available was even less, due to losses through mismanagement of resources and corruption.[22] This was common across the rhino range, with tourism income, NGO funding and very low government budgeting for wildlife and conservation being well below what was needed – even when income from legal trophy hunting in countries such as Tanzania and Zambia was included. Emslie and Brooks wrote in a report in 1999 that 'in large open areas, where funding and staffing levels were very low, rhinos have already been largely poached out of existence'.[23] Against this level of inadequacy, one has to place the 'presence of professional, well-armed poaching gangs, which are prepared to cross international boundaries in search of rhinos', and a deep pool of potential poachers in range states.[24] In conflict areas, conservation was often rendered impossible by the presence of armed groups, refugees in dire need of food and other means of subsistence. Sudanese poachers, often members of or deserters from rebel groups such as the Sudan People's Liberation Army (SPLA), poached in the DRC (formerly known as Zaire) and the CAR for gain rather than to support their insurgency in southern Sudan. The lack of targeted legislation, focused law enforcement and efficient judicial systems exacerbated the fight against poaching; those who commissioned poachers and the middlemen who sold horn to overseas buyers were given ridiculously low fines or short prison sentences if caught, which were no deterrent.

After the whirlwind of poaching in the 1970s and 1980s, there was a period of stabilisation of black rhino numbers, an increase in southern white rhinos but no recovery of northern white rhinos in the mid-1990s. Black rhinos were thought to number 2,400–2,500 between 1992 and 1995, while there was a steady increase in southern white rhinos.[25] South Africa was seeing an increase in private ownership and breeding of rhinos, and the generation of income from rhino-based tourism, live sales and regulated trophy hunting.[26] By 1997, a fifth of South Africa's white rhinos were conserved on 163 private reserves – 1,785 out of the total white rhino population of 8,441 there.[27] Kenya, Namibia and Zimbabwe also had growing numbers of rhinos on private reserves, ranches or game farms. White rhinos had been translocated from South Africa to all three countries, even though Kenya was not historically part of the white rhino range. The sale of live rhinos, whether within South Africa or for transloca-tion, increased the income of the Natal Parks Board for rhino conservation. After the demise of apartheid and restructuring of South Africa's provinces, the Natal board was amalgamated with the KwaZulu Nature Department to form the KZN Nature Conservation Service. South Africa and Namibia both had well-managed conservation sectors and as a result escaped the ravages of the first rhino poaching crisis, but were to be hit harder in the 2000s and 2010s. In the 1990s, Namibia developed 'a strong alliance between non-governmental organizations (NGOs), local communities, private and public sector stakeholders to promote [rhino conservation and recovery of numbers of black rhino].[28]

A decline in poaching from the early 1990s onwards was sustained for nearly a decade, and black rhino numbers doubled to 4,880, while southern white rhino numbers and distribution continued to increase, although the northern white rhino did not survive the onslaught and became extinct when the last ones in Garamba NP were killed or died of natural causes. In the opening years of the new millennium, poaching began to increase again as demand grew, particularly in China, Taiwan and later Vietnam. Data presented to the fourteenth CITES CoP in 2007 revealed that the volume of horn being traded illegally had increased steadily after 2000. Between 2000 and 2005, 278 horns were seized from poachers or smugglers globally, but 386 were thought to have entered the illegal trade, and at least 252 rhinos were thought to have been poached in that period and horns were illegally taken or sold from private stocks. The DRC and Zimbabwe experienced rises in poaching, with the former losing 59% of their rhinos between 2003 and 2005 and the latter losing 12%. There was also an increase in poaching in

Kenya and South Africa.[29] The numbers of rhino poaching incidents reported between 2000 and 2005 were as follows: DRC, 14; Kenya, 64; Namibia, 20; South Africa, 86; Swaziland, 2; Tanzania, 3; Zimbabwe, 105. At the same time, levels of recovery of poached horn fell.[30]

East Africa

From the 1970s to the 1990s, East Africa was a main focus of rhino poaching, with Kenya hit hard in the 1970s and 1980s, and Tanzania from the 1970s through to the late 1990s.[31] In 1980, Tanzania had an estimated population of 3,795 black rhinos; by 1993 this was down to 275 and by 1997 to 46. In 1987, 95% of the illegally traded horn was from Tanzania.[32] Kenya had a temporary respite in the mid-1980s, with numbers of births thought to be 3.5 times greater than deaths, and a 5% population increase, though this failed to make up for the losses of the 1970s and early 1980s, when the population dropped from 1,500 in 1980 to 424 in 1997. East Africa's share of the overall African rhino population dropped from 44% in 1984 to 17% in 1993 through poaching.[33]

Kenya

During the colonial period, Kenyan African nationalist parties seeking independence were strongly opposed to colonial wildlife and hunting regulations. However, this did not lead to major changes after independence and the assumption of power by Jomo Kenyatta and the Kenyan African National Union (KANU) in December 1963. Amid the demands for restitution of land alienated from indigenous communities during the colonial period, and competition between different communities for access to land, there was no attempt to end the exclusion of communities from land designated as NPs or GRs, and considerable areas of farming land and land suitable for wildlife (hunting or tourism) remained in the hands of the white settler community. Nor were hunting restrictions on traditional forms of hunting lifted – communities that hunted as they had done for centuries were still deemed to be poachers.[34]

Before looking at the fortunes of Kenya's rhinos, it is worth looking at the conservation system there, the promotion of tourism as a hard currency earner and the role of political and business elites in poaching, smuggling and the corrupt use of income from the wildlife sector. Until the CITES trade ban in 1977 and Kenya's own ban on hunting in that year, it legally exported rhino horn from animals that had died naturally in protected areas, what was termed 'found horn' (handed in to the wildlife authorities by Kenyans and for which they were paid – a way of laundering poached horn), horn from culled problem animals, horns seized from poachers or trophies from legal hunts.

From 1960 to 1969, legal exports were around 955 kg p.a., and between 1970 and 1976 an average of 3,296 kg p.a.[35] The last rhino horn auction was held in 1976. Licensed hunters, until the bans in 1977 (both the CITES ban and Kenya's own ban on hunting), could hunt rhinos and legally sell their horn to dealers in Mombasa, through the government-controlled Mombasa Ivory Room. Indigenous people could not, and their hunting was still deemed to be poaching. This did not end illegal subsistence hunting for meat and poaching for horn, ivory and skins, with the latter often involving patronage of or protection from senior politicians, government, wildlife and police officials, who skimmed off income in the form of bribes or a share in poaching profits.

Rhino horn commerce became increasingly lucrative for those engaged in the legal trade and for those commissioning poaching or smuggling. Between 1960 and 1976, the recorded weight of rhino horn exported through the Mombasa Ivory Room was 5,818 kg, worth $425,006, from about 2,963 rhinos – the product of killing rhinos or the collecting of horns from natural deaths. Most horns from protected areas came from Tsavo, where there was a massive die-off during the 1961 drought. Between 1960 and 1976, the international value of rhino horn rose from $25.56 in 1960 ($261 in current values) to $230.58 in 1976 ($1,229 in current values). For most of the 1960s, the horn price was in the mid-$30s and started rising steeply after 1970, with a spike in sales after the drought of the early 1970s, when demand was increasing.[36] Between 1969 and 1976, the price of horn in Kenya rose by 446%, from KSh (Kenyan shillings) 8.24 per kilogram to KSh44.60, and the average amount legally exported annually represented the killing of 1,197 rhinos.[37] Most went to Hong Kong but increasing amounts went to North Yemen. The figures do not include poached horn that was illegally exported. Parker and Martin believed that illegal exports exceeded legal exports.[38]

In the first years of independence there was growing evidence of corruption in high places,[39] including sharing in the illicit proceeds from poaching or taking bribes from poaching/smuggling gangs. This included the involvement in poaching and smuggling of members of the president's family, senior KANU politicians, wildlife staff and the police. There was a two-track policy, with Kenya's political, public service and business elites publicly promoting conservation and wildlife-based tourism while covertly benefiting from poaching.[40] By the 1970s, tourism ranked in the top three hard hyphen currency earners along with tea and coffee. Under Jomo Kenyatta, power was concentrated in the president's hands, which gave him massive influence over all aspects of Kenyan life, including conservation and the tourism sector. Public bodies and government departments were controlled by Kenyatta and his elite circle of relatives, political allies and businessmen, who benefited by skimming off income from a wide range of economic concerns, including tourism, the Wildlife Conservation and Management Department (WCMD) and the legal and illegal wildlife trades.[41] President Kenyatta was involved through signing backdated permits to legalise what were clearly illegally obtained wildlife products such as ivory and rhino horn.[42] Gibson writes that there was evidence that WCMD staff, including game guards, were responsible for a third of all rhino poaching in protected areas.[43] In his analysis of the global ivory trade for the US Fish and Wildlife Service in 1979, Parker notes that the breakdown of social structures and hierarchies in African communities during the colonial period played a role in changing traditional attitudes to hunting. The relatively low-profile hunting of rhinos and elephants that Sheldrick and Woodley had stamped out in Tsavo was largely replaced by a more rapacious form of hunting driven by the rising global prices for rhino horn and ivory.[44] The contradictory approach of supporting tourism and conservation while using wildlife as a form of rent-seeking and patronage has been maintained from independence in December 1963 to the present, despite the existence in Kenya of a strong and vocal pro-conservation lobby.[45]

Much of the poaching and illegal trafficking in rhino horn and ivory was linked to the country's political, government, police, wildlife service and business elites. The former game warden and ivory trade expert Ian Parker compiled what was known as the EBUR report on the ivory trade and corruption in Kenya, highlighting evidence of the role of political patronage and high-level corruption in illicit wildlife commodity trading.[46] One of President Kenyatta's wives, Mama Ngina, was a major player in the organisation of poaching and the smuggling of ivory and other wildlife commodities, and his daughter Margaret, once Mayor of Nairobi and then Kenyan Ambassador to the UN and head of the United Africa Corporation, which traded in legal ivory, was being issued permits on the orders of her father to launder illegally obtained wildlife products, so enabling her to legally export what were the products of poaching.[47] In 1969, Kenya had at least 16,000 black rhinos, perhaps as many as 20,000, with 6,000–9,000 in the Tsavo ecosystem.[48] By the early 1970s, there was growing concern among conservationists, game wardens and wildlife researchers that there had been a decline in rhino and elephant numbers through poaching; it was later estimated that rhino numbers had fallen to 1,500 by 1982.[49] Mombasa was the main exit point for poached rhino horn, with corruption in the wildlife department, police and customs meaning that there were few serious obstacles in the way of the determined smuggler. Horns were trafficked to the port and dealers would covertly ship them out, often on dhows, either to ports on the Arabian Peninsula or to freighters in the Indian Ocean, which took them to Dubai or East Asia.[50]

An example of the protection of smugglers was given by Parker. In June 1978, an anti-smuggling operation involving the Kenyan and German police discovered large parcels were being sent to West Germany from Nairobi. Rodney Elliott, of the Kenya Police Criminal Investigation Department (CID), thought the packages were suspicious, not least because one address from which parcels were sent was rented by one of Nairobi's largest game dealers. A total of 636 parcels weighing 6,320 kg had been sent to Germany. The German police seized several parcels, which were found to contain rhino horn and ivory.[51] Parker identified that the horn and ivory had come from the Kenyan paramilitary General Service Unit (GSU) store of seized horn and ivory, and had been sold by GSU officials or through secret auctions held by the Ministry of Tourism and Wildlife, with the income shown as just 13% of the actual market value of the rhino horns. He believed the difference between the recorded price and actual income was $133,760 ($616,633 in 2022 values) and had been pocketed by the officials involved in the illegal trade.[52] When Elliott tried to stop it, senior CID senior officers tried to block him. He had to get an order from the Attorney General, Charles Njonjo, to overrule them. The Minister of Tourism and Wildlife, Mathews Ogutu, was involved in the operation. The minister, senior judges and CID officers had paid small amounts to buy horn and ivory from government stocks and it sold abroad for a large profit. The scam was stopped, but no one was tried.[53]

In 1976, the NP department had been amalgamated with the notoriously inefficient and corrupt Game Department to form the WCMD under the overall authority of the Ministry of Tourism and Wildlife. The parks were poorly managed, inadequately funded and lacked fuel for patrol vehicles.[54] The high level of corruption was confirmed in personal communications with Ian Parker and Richard Leakey. President Moi appointed Leakey head of the WCMD in 1989, and he was the first head of its successor, the KWS, which was created in November 1989 to be responsible for managing Kenya's wildlife inside and outside NPs and reserves. In an interview in 2013, Leakey went over the longstanding problems of corruption within the WCMD and then in KWS. He repeated what he had told staff when he first headed the WCMD in 1989, aware that there had been over a decade of corruption and incompetence, which enabled substantial poaching: 'I want to make it absolutely clear that things are going to change around here. Many of you are corrupt, all of you are lazy, and before long the worst of you will be gone.'[55]

The corruption was why Tsavo is said to have become an open killing field, with poachers able to operate with impunity, and evidence of direct involvement of wildlife department personnel in poaching and transporting rhino horn and ivory.[56] In his book *Wildlife Wars*, Leakey detailed the corruption and incompetence he discovered on becoming head of the

WCMD in 1989.[57] The worst case he found was that, in Meru NP, 'senior park officers were almost certainly involved in the killing of the five white rhinos in their *boma*'.[58] The poaching of the prized white rhino (originally from South Africa) was what had made President Moi sack WCMD head Perez Olindo and replace him with Leakey.[59] Meru NP had already, by the early 1980s, lost most of its black rhinos, with the population of 127 in 1976 dropping to 35 by 1981, and evidence seen by the park warden, Peter Jenkins, of large gangs of Somali poachers in the park using AK-47s to shoot the rhinos.[60] After the poaching onslaught in the 1970s and 1980s and Leakey's more robust approach when he was first in charge of KWS, before his opponents weakened his ability to take a tough line, there was some recovery in black rhino numbers, rising from 380 in 1987 to 424 in 1997, with much of the increase brought about by protection and better breeding rates in fenced reserves or sanctuaries.[61] Leakey was not ultimately successful in rooting out corruption and incompetence at all levels of the KWS. Criticism of him, and what was seen by politicians and senior civil servants with whom he clashed as his high-handed approach, led to the waning of presidential support and a consequent inability to fight the corruption in the wildlife sector. Leakey resigned, but Moi kept him in office while reducing his freedom of action, his control over tourism income and his direct authority over KWS armed APUs.[62]

On 25 January 1990, President Moi, with Leakey's support, carried out a public burning of rhino horn, setting ablaze 283 rhino horns and 13,950 game skins worth $324,000 ($745,770.16 in current values) in NNP.[63] This was a very public gesture, but one not backed up with any lasting success in the fight against corruption that enabled poaching of rhinos. Leakey continued to be unhappy with his lack of power to fight corruption and finally resigned in 1994, to be replaced by David Western. President Uhuru Kenyatta, Jomo's son, later appointed Leakey as chairman of the KWS board in 2015. Right until the end of his life in January 2022, he lamented that he had been unable to eradicate corruption in the KWS. In his 2013 interview, he said that the corruption he found in 1989 was still there in 2013, with rangers and senior staff involved in poaching or turning a blind eye to it. He believed that the driving force was the poor pay and conditions of rangers:

> an arduous life of long hours, isolation and low pay … If you've got school fees that rise … parents that are dying, or kids that want to go to university, and the government seems totally deaf to your situation, then a lucrative asset like ivory is very hard not to say 'If you shoot it, I won't arrest you if you give me 30%'. This is the case in Kenya.[64]

Amboseli and Tsavo: protest killings, elephants and poaching

In 1951, the warden of Amboseli National Reserve had seen 31 rhinos in just two hours in the Ol Tukai area.[65] The reserve was considered in the 1950s 'to support one of the highest known densities of black rhino anywhere'.[66] It was estimated that 120 rhinos were there, mainly in the woodlands and swamp margins. However, rhino numbers fell owing to human–wildlife conflict linked to the loss of Maasai pastoralists' grazing land, and to poaching.[67] Collett said that 'the Maasai were systematically slaughtering rhinoceros in Amboseli as a protest against land alienation for wildlife preservation'.[68] By 1967 the population was estimated to have fallen to 55, and 'thereafter declined at 12% annually to approximately 35 animals in 1971. Over 75% of the decline was attributed to spearing' by Maasai.[69] Commercial gain became part of the motivation for the killing of rhinos in Amboseli in the early 1970s, as rising demand and prices coincided with a serious drought from 1972 to 1977; prices rose from $24 per kilogram in the 1960s to $300 per kilogram in 1978. This attracted poaching gangs in addition to the Maasai, the latter being drawn into poaching by the effects of the drought on their livelihoods and by simmering resentment over the loss of grazing.

Poaching of rhinos rose substantially between 1974 and 1977.[70] About 80% of killings took place outside the main tourist areas and involved the use of spears rather than guns, suggesting Maasai responsibility.[71] Western warned: 'Once the numbers drop as low as 50, it is doubtful whether this does constitute an effective breeding pool since animals tend to disperse over a considerable area such that few males actively contribute to the breeding stock'; the population was down to eight animals in 1977, with only two breeding males and three mature females.[72] However, these remaining animals did breed, and numbers increased to 14 in 1981. Spearing of rhinos fluctuated over time, according to Lindsay, and 'may have been linked with Maasai warrior age grades and the desire of young men transitioning to the warrior caste to hunt, with spearing of rhinos and elephants rising with the appearance of new generations of warriors', as well as being a form of political protest.[73] The last recorded spearing of rhinos by the Maasai was in 1991, when three of the last five rhinos in the park were killed. They had disappeared from the park by 1994. Another key aspect of the conflict was that after Amboseli was declared a national reserve and later an NP (from 1974), increasing numbers of tourists paid to visit and film company crews made documentaries and films there, because of the spectacular backdrop of Mount Kilimanjaro. The Maasai thought that part of the tourism income belonged to them, 'because they had protected the wildlife for centuries … only a very small proportion of the tourism income had been allocated to the Maasai representative body, the Kajiado County Council'. The government and wildlife department convinced themselves that the Maasai would accept the pittance passed on to them from tourism earnings and considered 'the influx of visitors to Amboseli to be of benefit to them, supporting local development of clinics, schools and water sources'.[74] However, little tourist income trickled down to the Maasai, and they were often vilified for their pastoral practices and supposed overstocking of cattle on rangelands prone to drought.

Tsavo was one of the major locations for black rhinos in Kenya. Their numbers were hit hard by drought in the early 1960s and destruction by elephants of the browse and shade on which they depended. By the mid- to late 1960s, elephant numbers in the Tsavo–Galana–Mkomazi ecosystem were rising substantially, with visible destruction of woodland. Parker estimated elephant numbers at 40,000.[75] Culling was being suggested as the solution to environmental damage, but the warden, David Sheldrick, believed culling would not be necessary to stop the damage, as vegetation would recover and the grasslands created would encourage diversity with rising numbers of grazers, such as buffalos and zebras. However, this would be to the detriment of browsers such as rhinos, with woodland and bush transformed into grassland with less cover or browse. Dr Richard Laws, who had worked on reducing elephant numbers in Murchison Falls NP in Uganda, was appointed by the Kenyan government to study the Tsavo elephants, assisted by Ian Parker. They recommended a cull of 2,700 to enable recovery of bush and woodland.

Sheldrick vehemently opposed this, and there was criticism of the plan in the press in Kenya and Britain. The park trustees supported Sheldrick and no cull took place. When the drought lifted, elephant numbers recovered – but with the looming threat that in any future drought the elephant population would destroy vegetation and rhinos would starve. Drought hit again in 1970–2, with greater destruction than before. Commiphora trees, which were an important food for the rhinos, were destroyed in large numbers by the elephants.[76] Parker said in 1971 that Tsavo had thousands of rhino and elephant carcasses scattered across the whole park.[77] And the drought brought a new wave of poaching – probably starting as opportunistic recovery of horn and tusks from animals that died during the drought, but then moving into killing for horn and ivory. Local pastoralists had been rendered destitute by the drought as livestock died in huge numbers without food or water. They moved their remaining cattle into the parks, and engaged in horn/ tusk collection and poaching to survive. Tsavo became a killing field, with poachers able to operate with impunity, and there was growing evidence of direct involvement of wildlife department personnel in poaching and trans-porting both rhino horn and ivory.[78] The drought of 1970–1 killed about 1,000 rhinos and poaching killed many thousands more, with only 200 left in Tsavo in 1981.[79] Most had been killed by Wakamba or Somali poachers.

Another factor in resurgent poaching was the availability of guns and the presence of groups of Somali pastoralists, resulting partly from the Somali insurgency called the Shifta War, in northern Kenya in the mid- to late 1960s.[80] During that war, the Kenyan army and the paramilitary GSU were deployed in central and northern Kenya to combat the insurgents, but often spent their time shooting game, particularly rhinos, giraffes and oryxes, as was reported by the Kenyan Permanent Secretary for Tourism and Wildlife, Anthony Cullen, in October 1970.[81] There was the case of a Kenyan army major who shot five oryxes, two rhinos, seven rare Grevy's zebras, three giraffes and a lion while on duty.[82] The Somali poaching connection was one that had troubled pre-independence Kenya for many decades and continued to be a problem after independence, with smuggling, whether of wildlife products, drugs, weapons or commodities such as sugar and charcoal, a major source of income for Somali communities in northern Kenya. Sheldrick's APUs caught Somalis collecting rhino horn and ivory in 1973.[83] That year was a watershed in terms of rhino poaching. It had remained a constant problem but not at a high level. In 1973, it became a serious and growing problem as the price of horn rose, largely because of increasing demand in Yemen.

By the time Sheldrick was removed from his Tsavo post in 1976, Tsavo was regularly suffering major incursions by Somali poachers, who were driving out the Wakamba and Waliangulu traditional hunters and killing rhinos and elephants on a far greater scale using AK-47s.[84] The poorly managed, underfunded rangers, often lacking fuel for their vehicles and ammunition for their rifles, could not compete. There was also extensive poaching of rhinos and elephants in the NFD. The game ranger and wildlife biologist Ian Hughes headed an APU in the NFD from 1973, tasked with reducing poaching of rhinos and elephants.[85] On one patrol, they found a Somali poachers' camp, with several poachers there, one of whom fired on them and was shot dead. They recovered two rhino horns and a bundle of leopard skins. The dead poacher was carrying a Somali army identity card.[86] Just prior to the end of his period of duty in the late 1970s, Hughes discovered that two of his rangers had been caught trying to sell unregistered rhino horn and ivory to a wealthy Nairobi businessman who was dealing illegally in horn and ivory, and that rifle ammunition was being sold by one of his NCOs to poachers.[87]

Black rhinos survived in small numbers in NNP; some were descended from animals there when the park was established, and some were translocated there from elsewhere, because of better security in the park. There were 27–33 black rhinos in the park, 22 rhinos having been settled in the park by the Game Department Capture Unit since November 1966. The newcomers had only one serious conflict with a resident rhino and two collisions with motor vehicles. Hamilton and King believed that 'Despite the small size of the Park, the inadequacy of its boundaries and the legendary intolerance of the species, it would appear that the translocation operations have been a success.'[88]

Poaching, corruption and signs of recovery

With massive and well-publicised poaching of rhinos and elephants, the danger of damage to the lucrative tourism industry and increasingly public airing of the levels of corruption pushed President Kenyatta into banning commercial and sport hunting in 1977, the same year that CITES voted to ban the trade in rhino horn. However, he took no effective action against poaching and smuggling, especially the involvement of the political and business elite he had gathered around him.[89] In 1978, under pressure from the World Bank, which was helping to develop Kenya's wildlife tourism sector, the Kenyan government banned the sale of wildlife products, driving it all underground but again without any effective action to deal with the corruption that protected poachers and smugglers in return for pay-offs.[90] Kenyatta died in August 1978 and was succeeded as KANU head and president by Daniel arap Moi, who did nothing to dismantle the network of corruption around Kenyatta's family, and proceeded to build his own networks, so corruption continued unabated within the wildlife service, the police, government and the ruling KANU party.[91]

Starting in the 1960s, some southern white rhinos had been sent to Kenya from South Africa in a move that was more symbolic than of any great conservation value. A total of 51 white rhinos – six in 1965, 20 in the 1970s, five in 1992 and 20 in 1994 – were translocated to Kenya.[92] Most of those sent to Meru NP were poached in 1989. However, white rhino numbers in carefully guarded reserves rose slowly, as black rhino numbers were falling, before beginning to pick up again a little in the 1990s.

Table 6.2: Kenyan rhino populations[93]

Eastern black rhino	1980	1984	1987	1991	1992	1993–4	1995	1997
	1,500	550	381	398	414	417	420	424
Southern white rhino	1968	1984	1987	1991	1992	1993–4	1995	1997
	0	33	47	57	74	87	122	137

Gakaharu gave a breakdown in 1989 of the major rhino populations on private and state reserves/NPs. For private ranches these were as follows: Solio, 81; Lewa Downs, 12; Ol Jogi, 9; Laikipia, 45. For government-protected areas they were as follows: NNP, 51; Nakuru NP, 20; Ngulia in Tsavo West

NP, 8; Aberdare NP, 39.[94] He noted that in the early 1960s, the Kenya Game Capture Unit translocated black rhinos from places where poaching was rife to safer areas: 17 were translocated to NNP, where they successfully bred to reach 51; 20 were translocated to Solio Ranch, and grew to 81, with an annual increase of 9.3%. The Ngulia sanctuary is a completely fenced area of 88 km^2 in Tsavo West NP, which has been successful in protecting and breeding black rhinos. In the 1990s and 2000s it experienced a serious problem with elephants, attracted and remaining there because of reliable water sources. Detailed assessments of food available for rhinos indicated that between 1991 and 2005 there had been a 59% decline in browse at rhino height.[95] A combination of relatively high rhino density, loss of food and the pressure of competition from elephants reduced the annual growth in rhinos to below the Kenyan national target of 5%.[96] The population was about 65 in 2005 and, given declining browse, rhino reproduction and health were a cause for concern because of overstocking of large mammals and it was feared that the population would crash. Between October 2005 and October 2006, 255 elephants were removed and waterholes provided outside the sanctuary.[97]

Another ranch had been established in Laikipia, with 47 black rhinos – a 400 km^2 private ranch with a 40-man anti-poaching squad, funded by the World Wide Fund for Nature (WWF).[98] There were another 45 black rhinos on the nearby Ol ari Nyiro Ranch in Western Laikipia, owned by the settler and writer Kuki Gallmann.[99] The successful breeding and better security on private reserves enabled them to become custodians of Kenya's rhino population and the basis for later reintroductions to create new populations and avoid in-breeding – part of a long-term plan to restore black rhino numbers to 600 by the year 2000, and eventually to 2,000.[100] Rhinos in sanctuaries accounted for half of Kenya's population.[101] Writing in 1990, Brett gave his estimate of the situation:

> The total number of black rhinos remaining in Kenya is between 370 and 400 animals. The majority of these animals are located in 11 well protected areas which come under the general heading of rhino sanctuaries. None of these areas has more than 60 rhinos … the completed 93 km^2 Ol Pejeta Ranch Game Reserve has received only 4 males so far and the Tsavo Ngulia sanctuary, being extended this year to 73 km^2, has been stocked with six females and one male.[102]

The totally unfenced national reserves/parks of Maasai Mara (25 rhinos, down from an estimated 108 in 1977)[103] and Amboseli NP (9) had fragile populations that were not viable in the long term without relocations to bolster numbers, but they were the most vulnerable to poaching and to protest-killings by Maasai. Reproduction rates varied greatly between parks and reserves, with Solio having an 'exceptionally high annual birth rate of 15% from 1980-1986 … while NNP's rhino population has grown at an annual rate of only 3% since restocking ceased in 1968'.[104] The increasing use of small, fenced and well-protected sanctuaries in Kenya in the 1980s helped to reduce poaching. Only 20 rhinos were poached between 1987 and 1992,[105] which helped numbers to rise from 380 in 1987 to 458 in 2003 and 570 in 2007. Kenya was home to 84% of wild eastern black rhinos.[106] There are drawbacks to small, fenced sanctuaries – there is no chance for dispersal or recruitment through inward migration, so populations have to be carefully managed to avoid in-breeding and a shrinking gene pool, involving translocations of breeding animals between sanctuaries. Improvements in security in Maasai Mara enabled numbers to reach 40 by 1996, despite loss of habitat and browse as a result of increasing elephant numbers, as well as continuing but lowered levels of poaching.[107]

When the Sweetwaters GR was being established in 1988 as a rhino sanctuary, one of the problems was that there was a large elephant population in a relatively small area. Plans were made to remove some of the elephants, but this was not done, and by the early 2000s it was clear that the planned rhino population of 70 was not viable; it was kept at around 25 alongside 100 elephants. The only alternatives were to remove some elephants or expand the reserve.[108] The latter was chosen, and when Flora and Fauna International purchased Ol Pejeta in 2004, Sweetwaters was joined to it to create one conservancy, with both black and white rhinos.[109] It became famous for having the last northern white rhino, when four were relocated from a Czech zoo. Both males died, but two females were still alive in mid-2024 in a closely guarded area. By 2004, there were 46 black rhinos at Sweetwaters, and the area available had increased through the addition of Ol Pejeta habitat. At the time of writing it has 150 rhinos and 300 elephants on a larger area, as 20,000 acres that had belonged to President Moi were added, bringing the land area to 110,000 acres.[110]

Lake Nakuru NP, intended originally mainly to be a bird sanctuary, also became a rhino sanctuary. Both black and white rhinos, the latter from Solio GR, were relocated there. Solio was a source because its rhinos had bred well but poaching had increased and it was thought that the rhinos should be moved to better protected areas. Lake Nakuru took 15 white rhinos from Solio and two from the wildlife department, and there were two survivors from the original population, which had been decimated by poaching.[111] The Aberdare Mountains and the NP there had once been a major stronghold of black rhino in Kenya, with a population density thought to be one rhino per square kilometre.[112] A census in the early 1970s put the population at 450, but this was down to 50–60 by 1987 because of poaching. It was identified by the KWS as a priority area for rhino conservation, and an electric fence was put up along the park boundary that was adjacent to a farming area. Patton and Jones wrote in 2007 that surveys in recent years had suggested poaching and the effects of drought had reduced numbers to just over 30.[113]

Between 2003 and 2006, poachers killed at least 11 black rhinos, mainly in Tsavo and three more at Solio. Kenya's population of black rhino at the time was about 540.[114] The Kenyan white rhino population was at this time put at 235, but nine were poached at Solio between 2003 and 2006. Vigne, Martin and Okita-Ouma believed that poaching in Solio was linked to an individual in nearby Nyeri, who was paying poachers $285 for a small horn and $428 for a 3 kg horn; the middlemen sold horn on for $550–600 per kilogram and it was smuggled out through Mombasa.[115]

Tanzania

At independence on 6 December 1961, Tanzania had a large rhino population spread over a wide area, with populations in the SNP, NCA, Olduvai Gorge, Lake Manyara NP, Selous GR, Ruaha NP, Tarangire and Lake Manyara NPs, as well as some in hunting concessions and other areas outside state reserves and parks. Ngorongoro Crater, at a time when safari tourism was taking off, was a place where visitors could be more or less guaranteed to see black rhinos. The Crater floor, with a high density of herbivores and predators, had a rhino population of between 42 and 61,[116] with fluctuations due to movement in and out of the Crater. In the mid-1960s, there was little poaching for either horn or meat in the Crater, but rhinos were speared by the Maasai, who grazed their cattle legally in the Crater. Between 1961 and 1965, five rhinos were speared by Maasai morans (warriors). They claimed they acted in self-defence, spearing rhinos that charged them, but some were prosecuted for failing to report the spearing of rhinos, or for possession of horn.[117] A count of the Crater's wildlife carried out between January 1970 and October 1973 put the black rhino population at 34.[118] There were also black rhinos in Lake Manyara NP, which was estimated to have 23 black rhinos in the early 1970s.[119]

The overall rhino population of the SNP was estimated at 700 in the mid-1970s, primarily in woodland and thornbush areas.[120] At the start of the 1980s, Markus Borner carried out surveys for the Tanzanian Rhino Task Force and the Frankfurt Zoological Society of rhino populations across northern Tanzania, and found that extensive poaching had seriously depleted them. He wrote that between 1976 and 1981 poaching had become so widespread that several of the ranges had lost all their rhinos or had numbers drastically reduced.[121] In over 130 hours of low-level anti-poaching flights over the SNP in 1979–80 he saw only two rhinos near Ngare Nanyuki and several near Moru Kopjes. During aerial surveying in March 1980, covering 13.8% of the park, only one rhino was seen south of Ngare Nanyuki.[122] He concluded that the only viable population in the NP was around the Moru Kopjes, with 16 animals seen, leading to an estimate of 25–30 animals likely to be present and a possible 50–100 in the whole of the SNP, with perhaps a few in the Grumeti Controlled Area and 41 carcasses seen in the Maswa GR.[123] In the Ngorongoro Crater numbers had dropped to 30 in May 1980 as a result of poaching, although a wet-season survey in the Crater suggested that the population fluctuated there according to seasons, with 67 being seen.[124] Most of the poaching was now thought to be done by non-Maasai armed with rifles, rather than the past problem of Maasai spearing rhinos.[125] Olduvai had also lost a large number of rhinos since a 1966 estimate of 77,[126] with none being seen in 1979 and 1980.

Arusha NP, to the west of Serengeti, was at independence a stronghold of black rhino, and one sighting of rhinos in 1961 was of 17 together near the Ngurdoto Crater in the park.[127] No survey was done of the park's total rhino population, but as the waves of poaching hit Tanzania very hard in the 1970s and 1980s the Arusha population disappeared, with rangers finding large numbers of carcasses with the horns removed. By the mid-1980s there were none to be seen there. From that time there had not been a documented sighting of a rhino there, and fingers were being pointed at forest guards and game scouts as responsible for some of the poaching.[128] Neumann writes: 'several villagers made clear to me that they knew who the poachers were and how they operated. Since villagers find that the park is of no benefit to the village … they have little interesting in cooperating with state agencies in wildlife conservation efforts.'[129] In Tarangire NP, estimates from surveys in the late 1970s were of 20 surviving rhinos, down from the 1974 estimate of about 250 and the 1977 IUCN census estimate of 55. Heavy poaching between 1975 and 1977 was the cause of the decline.[130] Similarly, poaching had reduced Lake Manyara's rhinos from 35–40 in 1975 to fewer than ten in 1980. Rubondo NP in Lake Victoria still had a small population, 16 animals having been relocated there in the mid-1960s, and this increased to 30 by 1981. There was little evidence of poaching because the park is an island with thick forest cover. What Borner called a 'guesstimate' put the population in the Burigi GR at between 50 and 100.[131]

Rhino ranges in northern Tanzania that had once had viable rhino populations had lost most or all of their rhinos by 1981. In the two decades after independence, poaching or killing to make a point about loss of land by

Maasai had increased in northern Tanzania, and 'Somali middlemen have encouraged them by paying for trophies in Kenyan shillings which have a higher black-market value than the Tanzanian shilling ... Maasai have shown a growing interest in buying consumer goods and are increasingly raising the necessary cash by poaching', according to Martin.[132] However, there was also increasing poaching by well-armed gangs with automatic weapons, reacting to the growing demand for horn from traders on the coast supplying markets in Yemen and Asia. Between 1966 and 1981, poaching reduced the NCA rhino population by 70%, and Tanzania lost half its total rhino population in the 1970s, with northern Tanzania losing 80%. Corruption within the administration of the NCA enabled employees of the conservation area to get away with poaching rhinos in the Crater in 1980 and 1981.[133]

Ruaha NP in central Tanzania also experienced poaching in the 1970s. It had around 447 rhinos in 1973, but this was down to 94 by 1977. Poaching was occurring in Selous GR, but the vast area of mixed woodland and grassland still had an estimated 3,000–4,000 rhinos in the early 1980s. Cumming, Du Toit and Stuart estimated that Selous had the capacity to hold more than 18,000 black rhinos, but numbers in the reserve were falling as a result of poaching and were down to 300 in 1987. Inadequate anti-poaching capability enabled poaching to continue from the mid-1970s to the late 1980s.[134] In 1980, Kes Hillman had estimated that there were 4,000–9,000 black rhinos in Tanzania, with a maximum of 4,000 of them in Selous,[135] but by the mid- to late 1980s the overall population had been more than halved, with no end in sight to the poaching. A staggering example is that in Ngorongoro Crater in the 1986 dry season only two rhinos were seen.[136] A year later, Moehlman, Amato and Runyoro wrote that there was evidence of 13 rhinos being resident, but with perhaps as few as five able to breed, suggesting that it was not a viable population in the long term.[137] In 1988, a wet-season survey encountered 17 rhinos, but in the following years numbers fluctuated around ten or fewer, with 12 in the 1992 wet season but only three in the same year's dry season, indicating influxes of rhinos from outside during the wet season and departure of many of them in the dry season.[138]

The poaching epidemic across rhino and elephant ranges in Tanzania led to the launching in 1989 of Operesheni Uhai (a Swahili phrase meaning 'Operation Save Life'), by the Tanzanian government. This involved an undercover operation to gain intelligence about poaching gangs and then the use of well-armed rangers, the police, militia personnel and the army to catch, or at times kill, poachers.[139] During the operation, many illegal immigrants from Somalia who were involved in the illegal killing of rhinos and elephants were repatriated and illegal arms confiscated.[140] The remaining populations of rhinos in Ngorongoro Crater, SNP and Selous GR were given special monitoring systems and law enforcement was strengthened with donor support. This had the short-term effect of reducing poaching,[141] with 2,607 poachers arrested but no major traders or their political and business elite patrons.

When poaching was reduced, the government and wildlife authorities planned to improve security and management of the scattered populations to increase numbers. This was no easy matter, as demonstrated by the very slow recovery in the Ngorongoro Crater during the 1990s. Having been reduced to single figures in the dry seasons in the 1980s and early 1990s, and perhaps doubling in the wet season, by the early 2000s numbers had not increased. This was not due to poaching but, as Mills et al. believe, to 'ecological factors including: neonatal predation by hyaena *Crocuta crocuta*, loss of calving refuges because of a reduction in *Acacia xanthophloea*, competition for browse with elephant *Loxodonta africana* and buffalo *Syncerus caffer*, tick-borne disease, and disturbance from tourism';[142] the latter is a very serious issue as the Crater needs tourism to bring in revenue for conservation and anti-poaching, but tourists can harass and alarm species. I witnessed this in relation to black rhinos during visits to the Crater in 1986, 1988 and 2000, each time seeing safari drivers driving too close to rhinos with calves and deliberately trying to provoke them into charging in an attempt to elicit a thrilling experience to generate tips from tourists. Mills et al. put neonatal mortality at 25–45%, which is high for rhinos, largely because of predation by the large clans of spotted hyenas present in the Crater; the likely figure for hyena predation of neonatal rhinos was about 40%.[143]

In 2003, the Crater population was made up of ten adults (three breeding males and five breeding females), three subadults and three calves. In that year a rhino conservation workshop highlighted the problems listed above and recommended working to overcome the issues that were preventing population growth, including possible replacement of the main breeding bull in the Crater to avoid in-breeding, and introduction of more black rhino cows, 'to achieve maximum growth and through supplementation a minimum population of 20 animals'.[144] The increase in the Ngorongoro Crater rhino population was seen as vital to the recovery in Tanzania of the eastern black rhino (*Diceros bicornis michaeli*), which in 2003 only numbered about 60 in the whole country, with the Crater, Moru Kopjes and a section of the SNP near the Maasai Mara GR the surviving ranges.[145] By the late 1990s there were only three isolated populations of black rhino in Tanzania – in the Crater, SNP and Selous GR.[146] In 2006, the Singita-Grumeti Reserve acquired 32 black rhinos from a private game ranch, Thaba Tholo, in South Africa, and a year later two black rhinos from Port

Lympne Wild Animal Park, UK, were introduced into a sanctuary adjacent to Ikorongo-Grumeti GR. In 2007, it was estimated that Tanzania's black rhino population numbered about 101.[147]

Uganda

The number of northern white rhinos in western Uganda fluctuated over the years. In 1961, Sidney estimated that they exceeded 300 in the West Nile and West Madi districts, citing two surveys carried out prior to 1961 that suggested there were between 190 and 335 in Uganda as a whole.[148] By 1963, white rhinos were in steep decline. Cave wrote that numbers had fallen from 350 in 1955 to about 70–75 in 1963,[149] 50 of which were found in Madi county, with the remainder in Aringa and West Madi. A team capturing rhinos for relocation from Madi in 1961 found the remains of 40 rhinos which they believed had been killed within the previous 12 months, with poaching carried out by local people on behalf of Asian rhino horn traders, according to Cave, who said there was no history of hunting rhinos for meat among local communities.[150] In 1962, the Murchison Falls NP warden, John Savidge, said that 50 white rhinos were seen alive in West Nile, but 60 skeletons had been found.[151] Sidney was pessimistic about the survival of white rhinos in Uganda:

> The future of the northern race of White rhinoceroses in Uganda cannot be described as secure, since these animals are not protected in a National Park. Thus, as human cultivation and settlement expand, so will the rhinoceros habitat become more confined. The two sanctuaries set aside for the White rhinoceroses are in forested areas, and since these rhinoceroses inhabit open savannah country, these 'sanctuaries' are obviously quite inadequate.[152]

In 1961, the Ugandan game department was working to introduce white rhinos into Murchison Falls NP in an attempt to stem their rapid decline in numbers and because of poaching in West Nile. The northern white rhino range was shrinking at the time, limited to the West Nile districts of Uganda, Garamba NP in Congo and south-western Sudan. By mid-1961, ten white rhinos had been introduced to the park from West Madi, though four died soon after release as a result of injuries or stress during capture and relocation.[153] Six more were later successfully translocated. These 12 were the nucleus of a population considered to be the only hope for the species' survival in Uganda, a forlorn hope as it turned out.[154] Over the next ten years they had increased to about 30.[155] By 1975, the National Wildlife Committee of Uganda said that a few white rhinos survived in the Ajai GR, with perhaps four left in February 1979, but Edroma saw none in February 1980 during an aerial survey there.[156] The rhinos established in Murchison Falls were all wiped out during the slaughter of wildlife when Idi Amin was in power from 1971 to 1979.[157] In 1979 after the invasion by the Tanzanian army supporting Idi Amin's Ugandan opponents, Amin's troops retreated northwards through Murchison Falls, killing most of the white rhinos along with substantial numbers of other wildlife.[158]

The Ugandan conservationist Eric Edroma believes the northern white rhino had become extinct in Uganda by 1982.[159] As the white rhinos were being poached out, black rhino numbers were also falling, with about ten left in Kidepo NP in 1980, and perhaps a few in Murchison Falls, though most had been poached by 1983.[160] By the following year, black rhinos were deemed extinct in Uganda.[161] Poaching and the inability to police the parks during Amin's violent period in power and the years of war and insurgency that followed sealed the fate of the rhinos. However, by 2007, four rhinos had been reintroduced from Solio in Kenya, two went to the Uganda Wildlife Foundation Centre in 2001 and two more to the Ziwa Rhino Sanctuary in July 2005.[162]

Horn of Africa, Sudan and Central Africa

In 1985, the northern region of Africa, which covered 39% of the former black rhino range, had only 500 black rhinos, 5% of Africa's total, distributed in Cameroon, CAR, Chad, Sudan, Rwanda, Uganda, Ethiopia and Somalia, and less than 1% of Africa's white rhinos.[163] It was feared that the western black rhino and the northern white rhino would soon become extinct, which they did in just over 20 years. The main factor in their decline was poaching, worsened by long-running insurgencies in southern Sudan and the DRC, which displaced people and forced them into poaching for meat and wildlife products to trade, and introduced large numbers of military-grade weapons into the region, along with rebel and militia groups that engaged in poaching. The northern white rhino numbers had been 2,250 in 1960, 1,000 of them in southern Sudan and the rest in DRC, Uganda, Chad and CAR, but had fallen to 650 by 1970.[164] By 1984, they were extinct in Uganda, Chad and CAR.

The western black rhino historically had a large range across central and western Africa, with populations in Cameroon, Chad, the CAR, Sudan and South Sudan. As the black rhino was decimated by poaching across its ranges, the western subspecies was worst affected. It had been reduced in numbers by sports hunting during European colonial occupation, and further threatened by the expansion of human populations and agriculture. However, widespread poaching, exacerbated by conflicts in Sudan, Chad and later the

CAR, introduced military weapons and made conservation and effective anti-poaching impossible. By 1980, Cameroon had about 110 western black rhinos and Chad had just 25.[165] Ten years later, Chad's black rhinos were gone and northern Cameroon had just 35 in 1992, falling to ten in 1997. This major decline was confirmed when WWF surveys found only five rhinos in 2001. A later survey in 2004 suggested a recovery to 34, but it was found that the trackers had faked footprints to save their jobs.[166] Further surveys just after this failed to find black rhinos, and they were declared Probably Extinct.[167]

Hillman estimated in 1980 that there were about 1,000 western black rhinos in the CAR, perhaps more, with less than 50% in protected areas. It was possible, she wrote, that there were still a few northern white rhinos there.[168] In 1984, the WWF said that surveys had failed to find any northern white rhinos in western CAR, the only region of the country where they had been found.[169] Black rhino numbers were falling dramatically there as a result of poaching, often by heavily armed Sudanese gangs.[170] From about 3,000 in 1980, western black rhino was believed to be extinct in CAR by 1992.[171] The western black rhino and northern white rhino had also been present in Chad; however, in 1980 Hillman said no population or distribution figures were available but there were likely to be very few surviving.[172] Martin thought there might be 20 left at the start of the 1980s, but Emslie and Brooks believed black rhinos had disappeared by 1991 and northern white rhinos by 1984.[173] Black rhinos were present in Ethiopia in the 1960s, but in very small numbers. A pair had been sighted in the Omo Valley in 1967, but records are few and far between.[174] In 1980, it was estimated that the country was home to 20 eastern black rhinos; this was down to ten in 1984 and zero by 1997, when aerial surveys failed to find any.[175] Somalia was thought to have a population of about 300 black rhinos in 1980; this had dropped to 90 four years later, and they had disappeared completely by 1991.[176]

Southern Sudan had provided ideal habitat for the northern white rhino and black rhino. White rhinos were found in Shambe, Nimule NP and black rhinos in the Southern NP in the 1960s.[177] In 1980, the region, along with Garamba NP in DRC, was the main stronghold of the northern white rhino, with perhaps 500–600 in the far south bordering DRC and Uganda. There were also an estimated 500 black rhinos – but human encroachment and poaching were a growing threat. By 1986, Hillman-Smith, Oyisenzoo and Smith believed northern white rhinos were nearing extinction in Sudan – they had decreased from about 1,000 in 1979 to fewer than 700 in 1981, and in 1983 a survey found fewer than 50.[178] Their main ranges had been in Shambe and the Southern NP, but poaching had increased and taken a huge toll. Aerial surveys in Shambe in 1981 saw 57 rhino skeletons and another 714 large skeletons that could have been rhinos, but no live animals.[179] Civil war in the 1960s and 1970s that started again in 1983 negated conservation efforts and introduced large numbers of military weapons, and both rebel groups and the Sudanese army are believed to have engaged in poaching, with the Nimule population wiped out along with the few found near Juba and Yei district.[180] Poaching gangs were well organised, with donkeys, horses and camels to carry horn, ivory and meat from poached rhinos, elephants and buffalos. In 1984, there were believed to be 15–30 regularly used poaching camps with anywhere from 10 to 300 men in each, the well-armed poachers able to outgun the NP rangers – poaching gangs were often protected by the Sudanese army or rebel groups.[181] By 1984, the northern white rhino had been wiped out, followed ten years later by the black rhino.[182]

Black rhinos were once present in suitable habitat in lowland areas of Rwanda and, as noted in previous chapters, the northern white rhino might have been present in small numbers, but they had disappeared by the late 1950s. In 1958, the Belgian colonial regime reintroduced black rhinos to Akagera NP, in eastern Rwanda. Six were transferred from Karagwe in neighbouring Tanzania. One died in captivity before it could be released, but another was caught and released in its place.[183] By 1980, the numbers had risen through successful breeding to 38, but by 1987 human encroachment and poaching had reduced the numbers to 15. The Rwandan civil war and genocide saw a massive influx of refugees into the Akagera region to escape the fighting and later the genocidal attacks of the Hutu militias. From ten in 1993, the black rhino numbers dropped to four in 1995. After that it was feared that they had been wiped out completely in Rwanda, but sightings in 1996 and 1998 suggest that four, or possibly five, animals survive.[184]

Democratic Republic of Congo/DRC (Zaire)

Garamba NP became the final northern white rhino range to survive the decades of poaching in Central Africa. When the park first became a protected area in 1938, the white rhino population was estimated at 100. By the 1960s it had risen to between 1,000 and 3,000. However, Congolese Simba rebels occupied the park in 1963 during the Congolese civil war. Between then and 1972, about 1,000 rhinos had been killed, and an aerial survey in 1976 suggested that only 490 remained.[185] Despite the IUCN rating conservation of the northern white rhino a priority, warfare in and around Garamba and in neighbouring Sudan made conservation and anti-poaching almost impossible. Rebels killed wildlife for meat, and also for horn and ivory, which could be traded for food or weapons. In 1986, Hillman, Oyisenzoo and Smith thought that fewer than 50 rhinos were left in Garamba.[186] Those in neighbouring states had all been killed.[187] Despite aid being provided by the UN's Food and Agriculture Organization (FAO) to improve management of the park, funding from the wildlife department (the Institut Zairois pour la Conservation de la Nature) for staff salaries, park maintenance and running of vehicles was inadequate, and salaries were often not paid for months; the result, as Hillman-Smith recorded, was that 'Park staff were not only unable to control poaching but many were involved in it to support themselves.'[188] Rhino numbers continued to drop, and in 1999, Emslie and Brooks provided estimates of the numbers between 1960 and 1998 (Table 6.3).[189]

Table 6.3: Northern white rhino in DRC/Zaire[190]

1960	1971	1976	1981	1983	1984	1991	1995	1998
1,150	250	490	<50	13–20	15	30	31	25

Poaching remained the major threat to the surviving rhinos. Although it was brought under control in the south of the park by 1987, perhaps because there were very few rhinos there to poach, it remained a problem in the north as poaching by Congolese and Sudanese continued.[191] However, efforts were made to patrol the park and prevent poaching, with 24 patrol posts around the park and regular patrolling by game guards, despite the difficulties posed by dense and high grass growth in many areas.[192] There were suggestions raised at the time that the rhinos should be relocated elsewhere because of the conflict and poaching threats in Garamba, but Charles Mackie of the Garamba Rehabilitation Project did not think this was viable as it was against Zairean government policy and because there were some signs that security had improved and the rhinos were breeding.[193] In 1988, Hillman-Smith carried out aerial surveys and identified 19 rhinos, but reported that the population was thought to be 21, with five adult males, five adult females, two subadult males, two small subadult males and one subadult female, one male juvenile and one female juvenile that were born in 1985, two juveniles born after that and two young calves.[194] There was some optimism when surveys in the early 1990s suggested a population of 60, but then by 1993, after incursions from Sudan and poaching by Congolese, the estimate was down to 28, but no carcasses were seen.[195]

War erupted in Zaire in 1997 as a coalition of rebel forces backed by Rwanda and Uganda overthrew President Mobutu. The success of the rebellion led to the renaming of Zaire as the DRC, but this was followed by five more years of warfare and then decades of insurgency involving the army, local militias and rebel groups, and the involvement of Uganda and Rwanda on opposing sides. Fighting was to gain territory, and also control of valuable mineral deposits. There were 80,000 Sudanese refugees near Garamba, who had fled the fighting in southern Sudan. The fighting, incursions by rebels and the local militias all rendered conservation of the remaining rhinos dangerous. However, ground patrols continued in some areas, and aerial searches in mid-1998 put numbers at 25.[196] Conservation was hampered by a lack of vehicles, radio communication equipment and aerial support. Some military units of the new government were based in the park for a while to deter incursions and poaching, but were not hugely effective until they were replaced with better trained soldiers.[197] This enabled the dismantling of 49 poaching camps and the capture of ten groups of poachers. However, poaching was becoming intensive, particularly that of elephants and buffalos. The main motivation for poaching was for meat for militias and displaced people, and there was evidence of resumed poaching by park personnel for ivory and rhino horn to sell.[198] About 70–80% of the poachers were Sudanese, often deserters from the rebel SPLA.[199]

By 2003, heavy poaching had developed, with 15 poaching camps operating and five or six gangs in the park. In August 2003, 34 fresh elephant and two rhino carcasses were found and only 22 live rhinos were seen.[200] In 2004, nine rhino carcasses were found. Between 2003 and 2004 there were aerial sightings of only four live rhinos.[201] Poaching was being carried out by former SPLA guerrillas, Sudanese Baggara and Rizeigat horsemen and members of the Ugandan Lord's Resistance Army (LRA) rebel movement. Estimates of the number of surviving rhinos were now reduced to fewer than ten.[202] A 2008 survey found no white rhinos at all in Garamba, leading to the conclusion by the IUCN that the northern white rhino was extinct in the wild.[203] The AfRSG report in 2009 concluded that 'it is increasingly likely that the future of this subspecies is going to primarily rest with getting the remaining captive rhinos from Dvur Kralove Zoo in the Czech Republic to breed'.[204]

CHAPTER 7

Southern Africa, 1960–2005: armed conflict, crime and conservation

I t is not a simple matter to pick apart the southern African regional poaching and trade networks and treat as them discrete national issues, so there will inevitably be some overlap in the account given here for individual countries. A prime example is the involvement of the SADF, South African military intelligence, the UNITA rebel movement in Angola and the Mozambique National Resistance movement (Resistência Nacional Moçambicana, also known as Renamo) in rhino and elephant poaching, and horn and ivory smuggling. The details are included in the narrative on rhino poaching in each of the countries, with the material on the South African role drawn from the official South African Kumleben Inquiry report in a section before the country-by-country narrative.[1]

Rhino numbers, distribution and poaching threats in the region

During the decades covered here, southern white rhino numbers expanded impressively, this being based on the increase in numbers in the Hluhluwe-iMfolozi reserve in KZN. White rhino numbers had risen from as low as 20 at the end of the nineteenth century to between 437 and 520 by the mid-1950s, thanks to protection and careful management. By 1961, the Natal Parks Board was concerned that overgrazing by expanding rhino numbers and accelerated bush encroachment required relocation of some rhinos.[2] Many of the white rhinos bred there were relocated to other NPs and reserves, sold to private landowners in South Africa, or translocated to seven former range states and even to Kenya. However, black rhinos were very badly affected by poaching in the period under review.

Angola, Botswana, Malawi and Mozambique, which cover 13% of the black rhino's former range, in 1985 had only around 250 or 2% of Africa's black rhinos. Black rhino decline was about 60% between 1980 and 1985.[3] Angola, where rhino numbers had not been properly surveyed by the Portuguese, were, from 1975 to 2002, in a war zone in which the South African army, the UNITA rebel movements, landmines and poachers would account for hundreds, perhaps thousands, of black rhinos. Mozambique had about 250 black rhinos in 1980 but only 13 in 1997, and no white rhinos had been seen for years.[4] The southern region, South Africa, Zimbabwe and Namibia, covered 25% of the black rhino's former range, and had about 2,700 animals, or 30% of Africa's total. It showed a 10% increase between 1980 and 1985, with black rhinos growing in numbers and white rhinos increasing steadily.[5] Namibia's rhino numbers were massively reduced by illegal killing by South Africa administration officials, army officers, visiting National Party politicians from apartheid South Africa and local poachers. Between the 1960s and 1990s, Zambia's rhino population was almost wiped out. The black rhino population in the Luangwa Valley was reduced from about 8,000 in the early 1970s to less than 100 by the mid-1980s, and these animals were exterminated in the next two decades.[6]

In 1991, South Africa had 40% of Africa's black rhino population, with a similar percentage in Zimbabwe.[7] South Africa was leading the way in protecting black rhino, with relatively low levels of poaching and conservation in NPs, provincial reserves, private safari areas and game farms. In 1980, its numbers were put at 630, rising to 1,043 in 1997,[8] while white rhino numbers reached 7,913 in 1997. Estimates in 1990 showed that there were fewer than ten black rhinos and 100–150 white rhinos in Botswana, 25 black rhinos in Malawi, very few rhinos left in Mozambique with no estimate given, 440–58 black rhino and 63 white rhino in Namibia, 60–100 white rhino in Swaziland (eSwatini), 106 black rhino and 6 white rhino left in Zambia after the poaching onslaught, and 1,754 black rhinos and 208 white rhinos in Zimbabwe.[9] In the first decade of the 2000s, about 2,000 of the remaining 4,000 black rhinos in Africa were in South Africa, with some in Zimbabwe, Mozambique, Botswana and Namibia, and very few if any left in Zambia.[10]

One reason for the growth in numbers in South Africa and Zimbabwe – and later in Namibia – was the development of private safari areas, game ranches and conservancies, with rights of ownership and utilisation of wildlife granted to private owners and communities; this included hunting, tourism and legal trade in live animals. Lindsay, Romanach and Davies-Mostert write that this transformed wildlife from a problem into an asset, 'and there was a rapid shift from livestock to game ranching across large areas of southern

Africa.[11] By the early 2000s, there were 91,000 km² and 205,000 km² of game ranches in Namibia and South Africa, respectively, and 27,000 km² of game ranches in Zimbabwe, before the land seizures of the late 1990s and early 2000s.[12] This increased numbers and the diversity of wildlife on private land. The recovery in white rhino numbers in Hluhluwe-iMfolozi meant that by 1987, 1,291 rhinos from there had been relocated to 149 ranches or reserves on private land in South Africa, Zimbabwe and Namibia by the Natal Parks Board, out of 2,000 in total relocated within southern Africa.[13] Emslie and Brooks compiled a table of white rhino numbers in southern Africa covering the period 1895–1997 (Table 7.1).[14]

In an effort to keep track of the rhino populations, relocations and the health of the increasing population of white rhinos, and the rise in the poaching of black rhinos, representatives from Namibia and South Africa formed the Southern African Development Community (SADC) Rhino Management Group (RMG) in 1989, with Swaziland (eSwatini) and Zimbabwe joining seven years later. The focus was primarily the conservation of black rhino, owing to their Critically Endangered status.[15] RMG members in 1999 had 2,100 (81%) of Africa's remaining black rhinos (100% of Africa's southwestern black rhinos, 97% of Africa's known south-central black rhinos and 7% of Africa's eastern black rhinos).[16] Its role has rather dwindled over time, with countries developing their own conservation plans for rhinos.

South African use of rhino poaching in the wars of destabilisation

The utilisation of rhino horn, ivory and other natural or wildlife products was an integral part of the wars fought in Angola, Namibia, Mozambique and Zimbabwe, involving liberation or rebel movements within those countries, special forces and army units of the white minority Rhodesian regime and the apartheid era South African army and military intelligence. As the investigative journalist Stephen Ellis wrote:

> South Africa's policy of destabilisation of neighbouring
> countries was closely associated with the rise of South
> Africa as a leading middleman in the international ivory
> trade. South African-based traders, acting in partnership or
> with protection from officers of the South African Military
> Intelligence Directorate, imported raw ivory from Angola,
> Mozambique, re-exported it to markets in the Far East …
> The same trade routes were also used for trade in other
> goods, including rhino horn.[17]

Table 7.1: Numbers of southern white rhinos, by country, 1895–1997

	1895	1929	1948	1968	1984	1987	1991	1992	1993/4	1995	1997
Angola	0	0	0	0	?	0	0	0	0	0	0
Botswana	0	0	0	0	190	125	56	27	18	20	23
Kenya	0	0	0	0	33	47	57	74	87	122	137
Mozambique	0	0	0	0	1	–	–	–	–	–	–
Namibia	0	0	0	0	70	63	80	91	98	107	141
South Africa	20	150	550	1,800	3,234	4,137	5,057	5,297	6,376	7,095	7,913
Swaziland	0	0	0	0	60	80	60	46	133	41	50
Zambia	0	0	0	0	10	6	0	–	6	5	6
Zimbabwe	0	0	0	0	200	208	250	249	134	138	167
TOTAL	20	150	550	1,800	3,800	4,665	5,565	5,790	6,760	7,530	8,440

The investigative work by Ellis and Julian Rademeyer, and the accounts given by those involved such as Colonel Jan Breytenbach of the SADF and John Hanks, the WWF Africa Programme Director, all of whom I have communicated with, have served to broaden the narrative of the connection in southern Africa between conflict, South African and Rhodesian destabilisation of neighbouring countries and the illegal rhino horn trade.

In 1975–6, the Popular Movement for the Liberation of Angola (MPLA), with the support of Cuban troops and Soviet weapons, defeated rival liberation movements to take control of Angola following the withdrawal of the Portuguese. The National Front for the Liberation of Angola (FNLA), backed by Zaire, the USA and China, was defeated in the north of Angola. The war with UNITA was to last effectively until 2002, despite peace deals and brief ceasefires, ending with the death of UNITA leader Jonas Savimbi in February 2002.[18] In March 1976, an invading SADF column was defeated and pushed back by the MPLA and Cuban troops. Part of the SADF force, under Jan Breytenbach, was augmented by former FNLA soldiers and San trackers to form 32 (or Buffalo) Battalion, which remained active in Angola until the late 1980s, carrying out attacks against the forces and infrastructure of the MPLA government to support UNITA's insurgency in south-eastern Angola, and disrupting attempts by guerrillas of the South West African People's Organisation (SWAPO) to enter Namibia.[19] Breytenbach recalled that during the retreat in the face of the Cuban advance, he rescued Savimbi, which he regretted later, as 'Savimbi was destined to become a big, and very bad apple in the future conservation barrel, exerting a baleful influence on the rhino horn and ivory smuggling in particular'.[20] Breytenbach confirmed to me in a telephone interview in November 1990 that he was disgusted by the extermination of rhinos, elephants and other wildlife in south-eastern Angola to finance UNITA and to line the pockets of Savimbi, his commanders and some senior SADF and South African military intelligence personnel.

Breytenbach was sent by the SADF to assist Savimbi at his Jamba headquarters in south-eastern Cuando Cubango province in Angola, from where UNITA conducted its guerrilla war. He noted that the large herds of buffalos, antelopes, elephants and the numerous rhinos that had once inhabited the area had all disappeared – systematically wiped out for rhino horn, ivory, hides and meat.[21] The rhinos and elephants had been shot on a highly organised basis, Breytenbach discovered, and the horns and tusks stockpiled before being trucked out to Namibia and then to South Africa. He found that a Portuguese businessman dealing in fruit, vegetables and other fresh produce used his lorries to pick up rhino horn, ivory and Angolan timber and transport it to Rundu in Namibia. Illegal traders in Katima Mulilo were also involved, and received rhino horn and ivory smuggled in from Zambia and Zimbabwe as well.[22] Breytenbach's attempts to investigate and publicise the illegal trade involving smugglers and the South African military led to him being refused the post of warden for the Western Caprivi GR and warned against investigating further, but he was able to testify at an official inquiry after the end of apartheid in 1994.

The Kumleben Report, released in January 1996, was the result of the official inquiry in South Africa headed by Justice Kumleben into the clandestine ivory/horn smuggling network run by South African military intelligence and personnel of the SADF during South Africa's destabilisation of Angola and Mozambique, and the occupation of Namibia.[23] Suspicions about this had surfaced long before the demise of apartheid. Those monitoring the illegal trade in tusks and rhino horn and investigative journalists had begun to see signs of a South African role in poaching and selling illicit horn and ivory at the end of the 1970s. The ubiquitous Ian Parker, in his 1979 report on the ivory trade, reported that the Angolan FNLA and UNITA were selling ivory and horn to buy weapons. He saw 700 rhino horns owned by the SADF at a military store in Rundu in northern Namibia, which had been issued veterinary permits as legal horn in Namibia and could be exported via South Africa.[24] He believed that the horns had all come from Angola. This was corroborated by an SADF officer, Des Burman, who found 60 crates of rhino horn, ivory and animal skins at Rundu that were to be flown to Pretoria. Burman was ordered by his superior officer to say nothing about it, though he told Breytenbach. Breytenbach also learned that a Namibian wildlife official, Jan Muller, had uncovered the horn and ivory smuggling operation, but had died in a suspicious road accident when he started to investigate this.[25]

To add to the growing body of evidence of a UNITA/SADF smuggling operation, investigative journalists, including the editor of *Africa Confidential*, Stephen Ellis, uncovered a poaching and smuggling operation.[26] Questions were asked by South African members of parliament about the SADF involvement in smuggling, and an internal SADF inquiry was set up with Brigadier Ben de Wet Roos as its president. It was a cover-up, and no details were published of its findings.[27] What later emerged from the work of Ellis and the Kumleben Report was a picture of a highly organised poaching and smuggling ring involving UNITA, the SADF and criminals in South Africa and Namibia, who took rhino horn and ivory poached in Angola and some from northern Botswana, and transported it to South Africa, where it was certificated and sold to buyers in Asia. The operation was sanctioned by the commander-in-chief of the SADF, General Magnus Malan.[28] Breytenbach told the author that UNITA had used income from poached rhino horn and ivory to fund arms purchases and enrich Savimbi.[29] In Namibia, he discovered that South African and south-west African government officials and SADF personnel were illegally hunting game in the Caprivi Strip. Breytenbach said that a brigadier based in Rundu in Namibia was particularly active, but despite repeated complaints the SADF tolerated his illegal hunting, the selling of tusks and horn, and even the use of military helicopters for poaching.[30] Proceeds from the Angolan horn and ivory would go to UNITA, but some was retained by military intelligence, and it is believed that senior intelligence officers also siphoned off funds for themselves from this highly covert and illegal operation.[31]

Other aspects of the operation emerged through the work of the Environmental Investigation Agency (EIA) – a British-based NGO researching illegal wildlife trade. They reported that the export end of the ivory and rhino horn smuggling by the SADF involved a Hong Kong trader, said to be South Africa's biggest ivory and horn dealer, Cheong Pong. In 1988, Pong was known to have been involved in sanctions-busting operations to evade international trade sanctions against apartheid South Africa. In October 1988, a consignment of goods on its way from Zaire to South Africa via Zambia and Botswana was intercepted by Botswanan customs and found to contain 382 raw ivory tusks, 34 carved tusks, 94 black rhino horns and a load of copper ingots. It was in a vehicle owned by a man called Tony Vieira, but registered as belonging to the Pong company. Pong said that the ingots were his but he denied knowledge of the rhino horn or ivory.[32] The Kumleben Inquiry discovered that in 1983 the South African Police (SAP) confiscated about 100 rhino horns that had been illegally obtained by Pong. He 'wriggled out of the case by producing ancient Natal permits and claiming that the horns were old stock'.[33]

Angola

Angola had suitable habitat for both black and white rhinos, particularly along the border with Namibia in the south-east where rivers that fed the Okavango Delta flowed south through Angola. The savannas and open woodland there provided good habitat for diverse herbivores, including rhinos.[34] In 1966, white rhinos brought in from South Africa were introduced to Quiçama NP, on the coast south of Luanda, north of the known range of the white rhino. The white rhino specialist Ian Player visited and arranged the introduction of the breeding group of rhinos, but they had disappeared completely by 1973, presumably having been poached.[35] There were about 30 black rhinos in Iona NP, on the border with Namibia, but they were killed along with hundreds of the estimated 500–1,000 rhinos in Cuando Cubango during the civil war that started in 1975.[36] The remaining 200, believed to be in the south-west, were mostly killed by the SADF and UNITA during their organised poaching operations.[37] Estimates of black rhino numbers fell from 300 in 1980 to 10 in 1993–4 and zero by 1997.[38]

Botswana

Botswana had once been home to south-western and south central black rhinos, and white rhinos – the latter extinct by 1870 because of hunting by Europeans. By the 1960s, the south-western black rhino that had been found in western Botswana was presumed extinct, and the south-central black rhino was reduced to small numbers in the Okavango and north-eastern Botswana. There are no reliable records of black rhino numbers prior to 1980, but it is believed that many were poached from the Chobe/Moremi area between the 1970s and the end of the 1980s. From the late 1960s to 1981, 94 white rhinos were reintroduced into the country, with 50 released in Chobe in the late 1970s;[39] the population was increased through the introduction of 156 southern white rhinos from South Africa in 1980. These were then mostly poached during the 1980s.[40]

By 1992, heavy poaching had reduced numbers of southern white rhino in Botswana to 17–27 in Moremi and Chobe NP. The black rhino population was estimated at 30 in 1980; by 1982 it was down to 20 and dropping.[41] When I made a radio documentary on conservation in Botswana for the BBC World Service in 1993, I was told by wildlife officials that there were fewer than 30 rhinos left in northern Botswana. A year earlier, the Khama Rhino Sanctuary had been established near Serowe to try to save the country's few remaining rhinos. I visited the new sanctuary and saw six white rhinos that had been taken there from Chobe NP to safeguard them from poachers; a few more individuals were shipped in from Pilanesburg NP in South Africa. The plan, I was told, was to breed from these animals and provide a safe sanctuary until they could be released in Moremi and Chobe. There was a Botswana Defence Force (BDF) camp situated nearby to provide protection.

Table 7.2: Rhinos in Botswana[42]

Black rhinos	1980	1984	1987	1991	1992	1993–4	1995	1997		
	30	10	<10	<10	5	4	0?	0?		
White rhinos	1895	1948	1968	1984	1987	1991	1992	1993–4	1995	1997
	0	0	0	190	125	56	27	18	20	23

In 1994, fearing that the remaining rhinos in Moremi and Chobe would be poached, the Department of Wildlife and National Parks (DWNP) captured most of the white rhinos in Moremi and moved them to sanctuaries; two more were relocated in 1996.[43] Three small and well-protected sanctuaries had been formed to hold the rhinos until they could be released back into the wild.

The other strategy used in Botswana was the deployment of the BDF to protect rhinos and combat poaching. In 1987, when poaching of rhinos and elephants in northern Botswana was reaching its peak, President Masire deployed military units to Chobe, Savuti and Okavango to catch or, if necessary, kill armed poachers operating there – many of whom were Zambian, Namibian or Zimbabwean, with only a few Batswana involved.[44] Many of the Zambians were armed and commissioned by criminal syndicates that were active in poaching in Zambia and Zimbabwe. The BDF worked with wildlife rangers and had a mandate to shoot to kill if poachers fired on them.[45] This worked quickly to deter poachers crossing into Botswana. There was considerable support among the Batswana for the approach, as they were allowed to hunt in certain areas if they had a licence – individuals had to apply for a licence and many communities had hunting quotas that they could use to hunt for meat or sell to safari hunting operators, which brought in income and meat. Communities benefiting from this system did not want foreign poachers stealing their natural resources.[46]

The initiative to involve the military in anti-poaching came from the son of the first president of Botswana and a senior BDF officer, Ian Khama.[47] Khama had a strong commitment to wildlife conservation and huge influence within the ruling Botswana Democratic Party; he later became president. With investments in eco-tourism, he did not want to see the lucrative tourism industry damaged by the loss of elephants, rhinos and lions. He rightly believed that poachers sent by organised gangs were not deterred by wildlife rangers or the small police force. Henk concurred: 'The military was the only security agency of the state with the firepower and capability to confront the poaching gangs, and the poachers' foreign origins and predatory behaviour significantly diminished any sympathy that citizens might otherwise have felt for them.'[48] This was the impression I gained on a trip for the BBC World Service in 1993, when I interviewed government ministers, senior wildlife officials, Batswana conservationists, safari operators and people living in wildlife-inhabited areas in the Okavango region and the Tuli block, and on further trips in 2006 and 2013. There was an informal social contract between the wildlife authorities, safari operators and local communities that the latter could live with potentially destructive and dangerous wildlife if they benefited through selling hunting quotas, getting meat and employment at hunting camps and tourist lodges, and in the industries serving them.

In 1987, Khama was involved in the deployment of the BDF Commando Squadron in northern Botswana's wildlife areas. The BDF recruited San trackers, who had worked for the SADF's counter-insurgency units in Namibia and southern Angola, and who were able to lead BDF units to poachers. Within three days of being deployed, the BDF engaged in their first firefight with a group of poachers, and within a few months 'dozens of poachers had been killed or captured, and the amount of poaching fell off dramatically', though it was too late to save most of Botswana's rhinos.[49] BDF units were operating from Shakawe in the Okavango Panhandle east to Kasane, and in the Tuli Block. They deterred poachers, undoubtedly saved elephants from being poached for ivory, and reassured tourists and tour operators. However, they angered the Zambian and later the Namibian governments whose nationals – some of whom were poachers, but some just fishermen – were being killed.[50] Despite initial strains with the wildlife rangers, the anti-poaching operations with the BDF became more coordinated, and in 2003, computer chips were inserted in the horns of rhinos released in Okavango to revive the depleted population. In November 2003, poachers killed a chipped rhino and removed its horn. In a combined operation involving the BDF, police and game rangers, the poachers were tracked by helicopter – the horn was recovered from a house in Maun, and the poachers were arrested.[51]

From 2001 to 2003 there were several releases of white rhino into the Okavango – 33 white rhinos and 5 black rhinos translocated from Zimbabwe and South Africa to the Mombo private reserve on Chief's Island in the Okavango run by Wilderness Safaris, in an operation involving the DWNP, Zimbabwe NPs, South African National Parks (SANParks) and North West Parks Board;[52] all were fitted with radio transmitters and had their ears notched to aid identification.[53] Some dispersed across the Delta from the release site, with two travelling 250 km to the Makgadikgadi and Nxai Pan NPs, south of the reserve. Most dispersing individuals were captured and relocated to Mombo, but two subadults that followed the zebra migration routes to the Makgadikgadi NP were poached in Nxai Pan NP, while another was poached 35 km west of Maun. A breeding bull relocated to Makgadikgadi Pan NP was poached not far from the scout camp.[54] After the releases about 80 rhinos overall were thought to be surviving in Botswana, distributed across a wide area with low densities.[55]

One thing that the Botswana government hoped would aid the conservation of wildlife and encourage local communities not to poach but to get actively involved in conservation was the launching in 1989 of a community-based natural resource management (CBNRM) project, backed by foreign donors and under the auspices of the natural resource management programme of the SADC. It aimed to empower local communities to engage in income-generating wildlife operations – both safari hunting and eco-tourism, with hunting bringing in direct cash income and meat supplies.[56] Communities formed trusts that were granted hunting quotas on their land and were permitted to enter into joint venture contracts with hunting or tourism safari operators.[57] By 2003, 96 trusts had been formed covering 100 villages, and in 2007 new measures were taken to strengthen management and ensure distribution of income raised from joint ventures to guarantee that funds were used in part for community projects. One trust, Sankuyo Tshwaragano for the village of Sankuyo, north of Maun, raised about $600,000 annually from selling hunting quotas, in addition to income from employment at hunting camps and meat supplied from animals hunted for trophies.[58] No rhinos were hunted, but habitat was conserved through the trusts.

eSwatini (previously Swaziland)

The black and white rhino populations in Swaziland were wiped out by hunting and human encroachment during the late nineteenth and early twentieth centuries. They started to be reintroduced from the South African population in 1984; 60 white rhinos and six black rhinos were reintroduced to the country's NPs. However, poaching developed and between 1988 and 1992, 50 of the reintroduced rhinos and their progeny had been killed.[59] Six more black rhinos were sent to the country as a gift from the President of Taiwan, which maintained good diplomatic and business ties with the kingdom. They were purchased from owners in South Africa and resettled in Mkhaya NP.

Table 7.3: Rhinos in eSwatini/Swaziland[60]

South-central black rhino	1980	1984	1987	1991	1992	1993–4	1995	1997		
	0	0	6	6	6	4	9	10		
Southern white rhino	1895	1948	1968	1984	1987	1991	1992	1993–4	1995	1997
	0	0	0	60	80	60	46	33	41	50

eSwatini's rhinos are managed by Big Game Parks, headed by Ted Reilly, who told me in August 2016 at the Mlilwane GR in eSwatini of the importance attached by the king and the government to building up the numbers of both rhino species as part of overall conservation planning. Anti-poaching operations were good, with rangers granted the right to search and arrest suspects without warrant and, in performing their duties, 'to shoot to kill in life threatening circumstances with immunity from prosecution'. Between 1992 and 1999 there were no reported poaching incidents, though several rhinos died during a drought in 1993.[61]

Malawi

Recorded rhino numbers never seem to have been high in Malawi in recent times, and although there was poaching, especially with Zambians crossing the border into Malawi in the 1970s and 1980s, there are few data on numbers or the level of poaching. By 1980, Hillman estimated that there were 20–50 rhinos in the country, the majority in Kasungu NP.[62] When I worked in Malawi in 1981–2 and was a regular visitor to Kasungu, black rhinos were scarce and declining in numbers. The researcher Hugo Jachmann, who was studying Kasungu's elephants at the time, told me that Zambian poachers were 'solely responsible' for the extermination of rhinos and elephants there.[63] Indeed, elephants were almost extirpated in Kasungu by the early 1990s – numbers had fallen from 40 in 1980 to just two in 1995.[64]

The last five known rhinos in Malawi, in Mwabvi Wildlife Reserve in the southern Shire Valley, had all been poached by 1991 and there were no verified sightings after that. In 1993, two black rhinos were introduced into a fenced sanctuary in Liwonde NP from South Africa and by 1997 had produced a calf.[65] Two more black rhinos were introduced into the park from South Africa in 1998 and another calf was born in 1999 – the rhino sanctuary was doubled in size to accommodate the expanded group.[66] There were no rhinos in Majete NP in the Shire Valley, management of which was taken over by African Parks (a non-profit conservation organisation established in 2000) in 2003, which had agreed plans to restock the park with a diversity of wildlife, including black rhinos, with the Malawian government. Two black rhino bulls were taken there from Liwonde by African Parks in 2003,[67] and five females and another bull were moved there in 2007.[68]

Mozambique

Once home to both the south-central black rhino and southern white rhino, poaching and the effects of decades of war served to deplete the population. The liberation war fought by the Frente de Libertação de Moçambique (Frelimo) disrupted what little effort the Portuguese had put into conservation. The best preserved areas were the hunting zones, known as coutada, with the greatest diversity of wildlife being found in coutadas 10, 11, 12 and 14 in central Mozambique, along with the Marromeu GR in the Zambezi Delta. There are no survey details for these areas to indicate whether any black rhinos remained in them in the last half of the twentieth century. All the NPs, reserves and coutadas suffered from poaching during the liberation war and then the war of destabilisation that followed independence, sponsored by white-ruled Rhodesia and then apartheid South Africa.

During the liberation war in Zimbabwe/Rhodesia, the Rhodesian Selous Scouts military unit became involved in the illegal wildlife trade. As part of their insurgency operations, these units established relationships with communities on both sides of the Zimbabwe–Mozambique border who poached elephants and rhinos, and they collected natural mortality ivory or poached ivory and some rhino horn.[69] From around 1978, the Rhodesians and the South African military assisted Renamo in destabilising Mozambique and disrupting Zimbabwean guerrilla activity from bases there, to prevent it from becoming a base from which the African National Congress (ANC) could infiltrate activists and fighters into South Africa. From then until 1989, the SADF role in Mozambique included smuggling poached ivory and rhino horn from Mozambique. The SADF or military intelligence arranged its export to Taiwan or Hong Kong and used the money for covert operations, training and arming Renamo forces.[70] Rhino numbers had not been well documented before the withdrawal of the Portuguese in 1975, and Hillman believed only a few hundred remained in 1980, mainly in the north of the country and perhaps some in Gorongosa.[71] Emslie and Brooks put the number at about 250,[72] but most of the survivors were poached during the civil war, many killed by Renamo.[73] By 1984, the population – in regions of Mozambique adjacent to the Liwonde NP in Malawi, further south in the Gorongosa NP and Zinave NP, in the Mozambican portion of Gonarezhou NP bordering Zimbabwe, the Limpopo reserve now part of the wider Greater Limpopo Transfrontier Park, Marromeou, in the Niassa GR bordering Tanzania and the coutada – had been reduced to about 130.[74]

In 1972, 83 white rhinos had been reintroduced into Mozambique (71 into Maputo GR and 12 into Gorongosa NP), but as a result of the civil war and poaching only one remained by 1984. By 1987 the species was believed to have died out in Mozambique. After the end of the civil war, in 1992, it was believed that there were 50 rhinos remaining, none of them white, but this estimate was cut to 45 in 1994 and a mere 7–13 in 1997, as poaching continued in what was a country impoverished by 15 years of civil war and periodic South African military raids.[75] The EIA undertook research into the ivory and rhino horn traded in southern Africa, and reported in 1989 that 98 rhino horns and 100 leopard skins had been sold by Renamo with the help of former Rhodesian Selous Scouts soldiers on Bazaruto Island off the Mozambican coast. Former Selous soldier Ant White, closely bound up with illegal trading of a range of wildlife and other natural resources such as rosewood, was assisting Renamo and later corrupt Mozambican government officials in exporting the rhino horn, ivory and skins that they poached.[76]

Namibia

The south-western black rhino was the only species present in Namibia in 1960, until southern white rhinos were translocated to Etosha NP and private reserves or game ranches from the mid-1960s. In his detailed study of the distribution of rhinos in Namibia, Joubert recorded that they were found in Damaraland/Kaokoveld, as far north as the Kunene river, and at times in western Ovamboland, in areas with an altitude of between 2,000 and 4,000 ft, with broken ground that provided cover and 'offers a certain degree of protection against man'.[77] Legal and illegal hunting were the main non-natural causes of death. The regions they occupied were sparsely populated by Himba pastoralists who moved their herds regularly and did not have permanent settlements around vital sources of browse and water. Because of the sparsity of water sources and edible vegetation, the rhinos ranged over large areas and the population density was low. Joubert noted that the threat level to the rhino population increased in the late 1960s as Himba, Herero and other pastoralists moved into rhino areas in search of forage and water for their herds, and conflict over resources developed.[78]

Rhinos had a greater degree of safety in Etosha NP, east of Damaraland, where they were more numerous and the population density was higher because of the availability of water at man-made waterholes, and on white-owned private reserves and game ranches where they also had reliable water sources. Parts of the western boundary of Etosha were not fenced and rhinos could move from the park to the farms in the Grootberg area, including Palmwag and other areas that became community conservancies or safari concession areas in the 1990s. Some rhinos were found in areas to the east of Etosha NP, but in isolated groups because of the scarcity of water.[79] The southerly limit of the black rhino was the Erongo Mountains, north-west of Windhoek. Some were present in small numbers in the Western Caprivi Strip (now called the Zambezi region).[80] In the early 1970s, the distribution was 72% in Damaraland/Kaokoveld and some areas of Ovamboland, 11% on private ranches, reserves or farms and 17% in Etosha NP.[81] In his survey of the distribution of black rhinos in southern Africa, Hall-Martin thought that in the late 1960s there were as few as 90 in north-western Namibia, some vagrant animals in Caprivi which moved between Namibia, Angola and Botswana – to protect them, 43 were moved to Etosha NP from areas where they were threatened by poachers or posed a threat to farmers.[82]

During the 1970s, poaching reduced the numbers in Kaokoveld to as low as 20, with perhaps 30 in Damaraland.[83] In 1980, Clive Walker of the South Africa Endangered Wildlife Trust (EWT) concluded, after conducting aerial surveys, that there were 15 rhinos left in Kaokoveld.[84] Hillman painted a slightly more optimistic picture, with an estimate of 190 rhinos in Namibia as a whole, 150 of them in Etosha.[85] In 1982, there were believed to be 60 black rhinos in Kaokoveld and Damaraland, which rose to 136 in 2002, 12 years after independence.[86] After independence in 1990, the rhino numbers gradually increased, and the proliferation of private reserves and then community conservancies helped to achieve an increase in rhino numbers. By 1993, dozens of white rhinos had been translocated from South Africa to Namibia and 55, about 60% of the white rhinos in Namibia, were on private reserves or game farms.[87] By 1997, black rhino numbers were up to 707 and white rhino numbers up to 141.

Table 7.4: Rhino numbers in Namibia[88]

South-western black rhino	1980	1984	1987	1990	1991	1992	1993–4	1995	1997
	300	400	449	421	479	489	583	598	707
Southern white rhino	1948	1968	1984	1987	1991	1992	1993–4	1995	1997
	0	0	70	63	80	91	98	107	141

Poaching and the role of the SADF

Poaching had occurred at a low level in the 1960s, but it was in the 1970s that a combination of the rise in international price of rhino horn and illegal trophy and meat hunting increased levels of killing, bringing the black rhino in Namibia 'to the brink of extinction', according to the leading Namibian conservationist, Garth Owen-Smith.[89] While working as a government agricultural officer in Kunene region in 1969, Owen-Smith noted the antipathy of the local Herero pastoralists to rhinos and elephants there – they were afraid of them, thought the rhinos were aggressive and wanted them removed.[90] He also became aware of illegal hunting by government officials and visiting South African ministers and SADF officers – mainly for sport and meat – including the shooting of rhinos in Kaokoveld. When he reported this to the police, he was told it was not his business, and that he would lose his job if he told anyone about it.[91] This illegal hunting increased after the Kaokoveld GR was de-gazetted in 1970. South African hunters, especially senior National Party politicians and high-ranking SADF officers, would fly in by helicopter, shoot wildlife, take their trophies and meat and fly out.[92] This increased as more South African troops were garrisoned along Namibia's border with Angola to combat guerrillas of the South West African People's Organization (SWAPO), fighting for independence from South African rule.

SADF units operated in northern Namibia and southern Angola from 1975 until 1990, and some were involved in wildlife poaching and illegally trading in rhino horn, ivory and timber. Owen-Smith feared that this killing, if sustained, would wipe out elephants and rhinos in north-west Namibia.[93] A wildlife conservator in the Uchab Valley, Rudi Loutit, told him in the mid-1970s that the SADF, government officials and local people who had guns had 'shot the shit out of everything in Kaokoland' and were starting on Damaraland. Piet de Villiers, the nature conservator for Kaokoveld, said that poaching by SA and SWA politicians and officials was continuing, and any attempt to investigate was blocked.[94] The Kumleben Report detailed this illegal hunting and also the use of the military base at Rundu in northern Namibia to issue veterinary permits for rhino horn and ivory poached in Angola and northern Namibia.[95] The issuing of permits, effectively laundering poached horn, enabled the importing of the horns into South Africa, creating a semblance of legality.

Colonel Breytenbach discovered that South African and south-west African government officials and security personnel were regularly hunting elephants, rhinos and other game in the Caprivi Strip.[96] In the 1970s, the western Caprivi was supposed to be a GR, but SADF and SAP units stationed there used it as a private hunting area, with no interference from the sole conservation official, stationed 130 km away in Katima Mulilo. Breytenbach said that black rhinos, elephants and other game were plentiful in the strip and in areas of Angola and Botswana adjacent to Namibia.[97] The town of Katima Mulilo, in the eastern Caprivi, became a major trading centre, with white businessmen running the illicit wildlife trade under the cover of legitimate businesses. They organised poaching, supplying transport and weapons, and it began to develop into one of the biggest ivory and rhino horn 'smuggling rackets that the African continent has ever seen'.[98] Amid this onslaught, Owen-Smith was part of a team carrying out a black rhino census in Etosha in 1980 to check on black rhinos that had been relocated from Damaraland and Kaokoveld, where they were in greater danger of being poached – 42 from Damaraland and 10 from Purros in Kaokoveld. He believed that there were probably around 350 black rhinos in Etosha and adjacent districts, a huge increase over the estimate of 55 in 1970.[99] He thought that there could be as many as 150 still present in Kaokoveld. South of the Hoanib river in Damaraland he saw 23 live rhinos but 19 carcasses. With clear evidence of poaching, Owen-Smith then looked into the role of the SADF, and said that the South African Ministry of Defence had admitted that its helicopters had been used for hunting, claiming this was in response to reports from local communities about problem animals.[100]

Poaching by soldiers and government officials with impunity prompted local people to hunt illegally, including hunting for rhinos. Black Namibians lacked land rights and therefore the right to hunt on the land that they occupied under South African occupation. In 1983, continuing his conservation work, Owen-Smith found that Namibian poachers could get R200 for an average size pair of horns and R300 for a large pair – which was more than a month's wages if you were a government employee and far more than that if you were a labourer on a white farm or a small-scale herder.[101] Horns were sold to businessmen who dealt in horn, skins or ivory. One such dealer, garage owner Ziggi Goetz, was sent to prison for possession of 68 rhino horns and 17 tusks bought from poachers in Kaokoveld.[102] At around the same time, a former member of the Koevoet paramilitary group, established to fight SWAPO, Rupuree Koviti, was caught after commissioning two unemployed men to poach rhinos and shooting three rhinos himself in Etosha NP. This then led to the arrest of a Windhoek car salesman and an Okahandja businessman who had bought rhino horns and ivory from the poachers.[103] Sullivan and Muntifering believed that there were approximately 300 black rhinos in Etosha but only 46 black rhinos outside the NP by 1984, though a six-month survey of Damaraland in late 1986 found 89 adult rhinos.[104]

Poaching by South Africans increased dramatically in 1989 as Namibia was preparing for independence and for the withdrawal of South African troops and administrators; most of it was concentrated in Etosha and Kaokoveld, and some was being done by local herders and farmers.[105] Poaching was made possible for black Namibians in Kaokoveld in the 1980s when the SADF issued .303 rifles to those who were considered loyal black Namibians, for protection against SWAPO insurgents who entered their areas.[106] The regular loss of rhinos in Damaraland, Kaokoveld and even Etosha led to Namibia's first experiment in dehorning in the late 1980s and translocations to what were considered safer areas.[107] There was also a growth in conservation NGOs working to protect the rhino, through whose work auxiliary game guards were trained and employed to deter poaching.

Independence, conservancies and custodianship

During German and then South African occupation there was extensive settlement of white farmers on land from which the original inhabitants had been evicted and moved to what were termed communal lands. White-owned land was freehold, whereas African land was communal, without individual ownership.[108] After independence there was some redistribution of land, with black farmers gaining freehold land, but there remained huge areas of land under white ownership that were being used as cattle farms, game farms or safari reserves, on which there were privately owned white rhinos. In 1967, the South African administration had proclaimed that farm owners had the right to 'dispose of game' on their land as they saw fit.[109] This opened the way for game ranching, with wildlife hunted for trophies by visiting hunters paying large sums to professional hunting operators, or shot to meet game biltong demand. Trophy hunting became an increasingly lucrative industry and one that improved game conservation on land once used for cattle farming.[110] After independence, what had been communal lands often became community conservancies under sustainable-use policies championed by the SWAPO government elected in 1990, and by conservationists such as Garth Owen-Smith. This gave previously marginalised Namibian communities a sense of ownership, together with the ability to manage their land and to generate income from hunting and tourism.[111] As !Uri‡khob described the changes made in 1995:

> the right to utilise and benefit from wildlife on communal land should be devolved to a rural community that forms a conservancy in terms of the Ministry's policy on conservancies; each conservancy should have the rights to utilise wildlife within the bounds of the conservancy to the benefit of the community. Once a quota for each available species has been set, the conservancy members may decide how these animals may be utilised.[112]

Conservancies could decide whether community members hunted for their own use, to sell game meat, or whether they sold quotas to trophy hunting operators. Trophy hunting and eco-tourism had previously been the preserve of white landowners and the state, but now communities could sell hunting quotas, lease land for hunting camps and lodges or enter tourism contracts with large safari tourism operators such as Wilderness Safaris, bringing in income and creating jobs and demand for local services such as food provision and transport.[113] In 2003, income from hunting and tourism on the Torra Conservancy in Damaraland paid for community projects and employment of local game scouts, and also paid N$630 to each household, which was estimated to cover basic food costs for three months and was the equivalent of 8% of annual cash income. The Nyae Nyae conservancy, near the border with Botswana, earned $114,000 from hunting, selling hunting quotas and other natural resource utilisation.[114] The free-ranging black rhino, elephants and desert-adapted lions were major attractions for photographic tourists on the Damaraland conservancies, but hunting was too, though it should be noted that the limited number of permits issued annually to hunt old, male black rhinos were for those on private ranches or game farms and not free-ranging black rhinos in the conservancies. In 2004, CITES approved black rhino hunting quotas for South Africa and Namibia – the latter permitted to hunt five black rhino males annually.[115]

These land use changes, combined with the withdrawal of the SADF and South African administrators by independence in 1990, reduced the level of rhino poaching. There was a steady increase in black rhino numbers from 421 in 1990 to 707 in 1997.[116] Recovery was also aided by the development of the custodianship scheme, launched in April 1993, under which rhinos were relocated to private farms, ranches or reserves where owners became responsible for the rhinos under a contract with the government. White rhinos, translocated to Namibia from South Africa, had increased in numbers, and in 1993 there were 99 on six private game parks and in Etosha NP; with protection and careful management, the number had increased by 1997 to 141 in ten locations, with the majority privately owned and used to generate income from tourism and hunting of surplus old males.[117] Soon after independence work began on a rhino conservation plan, which set the long-term goal of having 2,000 black rhinos and 500 white rhinos, and by 1997 they had reached 35% of the target for black rhino and 28% for white rhino, with an aimed-for growth rate of 7.7% for black rhino to reach the target by 2011.[118]

One of the ways in which community conservancies were brought into national plans for rhino conservation was through the partnerships developed by the Save the Rhino Trust (SRT; founded in 1982 to research the desert-dwelling rhinos in the Kunene region).[119] Community conservancies and the tourism operator Wilderness Safaris sought to develop rhino-based tourism, with land leased to Wilderness Lodges and SRT working with the conservancies and lodges to develop rhino tracking for tourists,[120] along with Integrated Rural Development and Nature Conservation, the University of Namibia, and other NGOs and government ministries responsible for wildlife and rural development. These partnerships started surveys in north-western Namibia to identify human–wildlife conflict issues and development needs and to develop CBNRM projects that could offer ways in which rural communities could benefit from the various forms of income generation relating to wildlife in general and rhinos in particular. The plan was that communities formally recognised as conservancies would have greater control over the use of natural resources, including wildlife, on their land, empowering them to sustainably use these resources to generate income and employment.[121] The first conservancy formed in the black rhino range area was Torra, registered in 1998.[122]

As these plans proceeded, 40 black rhinos from the Palmwag concession were reintroduced into areas of north-western Namibia up towards the Kunene river between 2000 and 2010.[123] This created tourism potential but also the employment of local people as game guards, which helped to persuade communities to accept the introduction of potentially dangerous animals.[124] The conservancy movement took off well, and as more conservancies were established in Damaraland and Kunene (the district set up after independence including the Kaokoveld ecological region), it was clear that rhinos ranged across several unfenced conservancies and that joint management and so joint benefit of income from rhino tourism would be ideal in community and conservation terms.[125] Surveys conducted in conservancies in rhino ranges supported plans for wildlife conservation on the grounds that they could access income from tourism and that jobs would be created in districts with few opportunities other than pastoralism and some crop production.[126]

South Africa

From the establishment of KNP in 1926 onwards, safari tourism developed under pre-apartheid white minority rule, through the apartheid period and into the post-1994 democratic period. Wildlife tourism was a major money earner, as were sport and commercial hunting. Coexisting, sometimes uneasily, they succeeded in starting the recovery of South Africa's wildlife after the mass destruction of the nineteenth and early twentieth century. In 1930, there had been about 110 south-central black rhino left in South Africa in two parks in KZN. From there, as well as later south-western black rhino relocated from Namibia and eastern black rhino from Kenya, animals were distributed in NPs and reserves across South Africa with some going to KNP, which also received black rhino from KZN and Zimbabwe.[127] White rhinos were relocated from KZN to KNP and other reserves, and some were sold to private owners. From the 1960s through to 1997, the programme of relocations and protection from poaching enabled KNP to increase its black rhino population to 976, out of a national total of 1,043 black rhinos across 25 populations, 'a ninefold increase in numbers over 60 years'.[128] South Africa became home to about 40% of Africa's black rhinos and was initially the sole repository of the wild southern white rhino.[129] South Africa had all three extant black rhino subspecies – eastern, south-central and south-western. Poaching occurred, but not at serious levels, and both black and white rhino populations grew over time.

Table 7.5: South Africa's rhino population[130]

Black rhino	1980	1984	1987	1991	1992	1993–4	1995	1997
	630	640	577	771	819	897	1,024	1,043
Southern white rhino	1895	1929	1948	1968	1984	1987	1991	1992
	20	150	550	1,800	3,234	4,137	5,057	5,297
	1993–4	1995	1997					
	6,376	7,095	7,913					

The white rhino success story

In 1953, an aerial count showed that there were 437 white rhinos in the whole Hluhluwe-iMfolozi ecosystem, but with 269 in iMfolozi and fewer in the buffer zones around it and in Hluhluwe.[131] The population trend in this last stronghold for the white rhino was positive, with numbers increasing steadily between 1930 and 1960. There was a temporary downward movement in 1961 when lack of browse and other factors led to the deaths of 46 rhinos in northern Hluhluwe.[132] However, numbers increased in iMfolozi, and were such that translocations of white rhino to other areas of South and southern Africa, and even out-of-range relocations to Kenya, began in order to prevent overcrowding. In 1960, the boundaries of the reserves were being changed and some fencing erected, so many of the white rhinos would be left outside the fenced boundaries of the newly defined Hluhluwe-iMfolozi reserve area.[133] It was decided by warden Ian Player and the Natal Parks Board that they should be captured and sent to new locations.[134]

The first relocations were to Mkuze GR in Zululand, with the first to be moved being a young female captured in iMfolozi; it injured its foot after release and died of blood poisoning. Ten more rhinos were moved, but only four remained in the reserve, the others dispersing outside where poachers killed most of them.[135] In the same year, ten bulls and eight cows were moved to Ndumu GR in Zululand, initially being kept in a boma and then a large paddock before being set free to roam the reserve; six died on the way to the reserve, one escaped and was poached, but the others survived. A calf was born and killed by a crocodile, but six more calves were born.[136] Between June 1962 and September 1964, 92 rhinos were sent to KNP, and in the mid-1960s, 74 rhinos were sent to Rhodesia (Zimbabwe).[137] Over 12 years, 351 white rhinos were settled in KNP, and by 1988 numbers had increased to 1,229 as a result of successful acclimatisation and breeding.[138] By the end of 1970, 800 white rhinos had been relocated from the Zululand reserves; 56 rhinos were sent to Portuguese Mozambique, 30 to Maputa GR and 26 to Gorongosa NP. By March 1972, 1,109 rhinos had been captured and sent to reserves in southern and east Africa and to foreign zoos, with KNP receiving 203. Angola, Botswana, Kenya, Mozambique, Rhodesia and Zambia all received white rhinos from iMfolozi-Hluhluwe.[139]

The sale of rhinos from KNP and Hluhluwe-iMfolozi generated income for rhino and habitat conservation.[140] Those that remained in Hluhluwe-iMfolozi increased in numbers, but the removal of those translocated prevented overstocking, and the population was 429 in 1993 before declining, partly through continuing translocations and natural mortality, to 325 in 2000.[141] The success in increasing southern white rhino numbers in the 1960s meant that in 1968, with the white rhino population up to around 1,800, the South African government allowed sport hunting of them on a regulated basis, to raise money for conservation and give private landowners the incentive to buy and maintain rhinos.[142] Milliken and Shaw write:

> Rather than hindering population growth, trophy hunting is regarded as having positively influenced White Rhino numbers and population performance. From a biological perspective, sport hunting results in the elimination of 'surplus' male animals that otherwise might engage in fighting and/or kill other rhinos ... selective removals have served to stimulate breeding performance so that South Africa's sustained population growth rate for White Rhinos has remained high.[143]

Sport hunting of white rhinos created incentives for land-use changes over large areas of privately owned farmland, which became game farms, ranches or private reserves that maintained rhino populations, helping to boost numbers and providing 20.5m ha of former farmland for wildlife. The sale price of a live rhino was fixed by the Natal Parks Board at a low level of about R2,000 (about $900), well below the price charged by hunting businesses, which might be R35,000 in the 1980s, creating a large profit margin for rhino owners who offered trophy hunting.[144]

Over the decades of recovery, poaching occurred at a very low level and with no indication of there being organised gangs involved, according to John Hanks.[145] He was working for the Natal Parks Board from 1975 and noted that there was little poaching or demand for rhino horn:

> rhino horns picked up in the field were of course of interest but nothing beyond that, and in those years some of the reserves' staff would even use rhino horns as doorstops ... one of the rangers had come across a beautiful, long, thin black rhino horn that he mounted on a piece of wood and placed in his bathroom as a perfect holder for toilet rolls.[146]

As numbers increased, the Natal Parks Board started rhino auctions in 1986, rather than sales at fixed prices. Demand was high as many private owners had made quick money by allowing overhunting of their rhinos,[147] and needed to restock to continue bringing in wealthy hunters. The first auction brought in an average price of just above R10,000, a fivefold increase. Over time, prices rose at the auctions, reaching an average for a live white rhino of R48,732 ($18,600) in 1989.[148] The auction price rose to R80,000 in 1990, with the trophy price for white rhinos that were hunted about 60% higher than auction prices.[149] Milliken and Shaw estimated that about 10.5% of rhinos on private land were hunted annually until the prices rose at auction, after which the figure dropped to about 3%; between 1968 and 1994, about 820 rhinos were legally hunted, the vast majority by foreign hunters.

Auctions continued through the 1990s and into the 2000s, with 581 white rhinos sold between 2005 and 2008, earning $14m. Under CITES regulations, trophies from legal hunts could be exported; 938 horns and 1,638 rhino trophies were legally exported from South Africa between 1980 and 2010.[150] In 2004, CITES approved quotas for hunting of five specified individual black rhinos annually by South Africa, recognising success in increasing numbers and the viability of staging limited, high-cost hunts to bring in income for wildlife authorities and private rhino owners. As a result, between 2005 and 2015, South Africa hunted 40 males out of a possible 55 under the quota, an average of 3.6 p.a., amounting to 0.2% of the black rhino population.[151]

In addition to selling live rhinos, engaging in translocations of white rhinos and developing a lucrative trophy hunting sector, South Africa also conducted legal sales of rhino horn from natural mortality, legal hunting and seizures of poached horns. In 1978, 149.5 kg of rhino horn were sold to buyers in Hong Kong and 14 trophy horns were legally exported by hunters to France, Germany, Spain and Taiwan, despite the 1977 CITES ban. However, statistics for sales to Hong Kong do not match Hong Kong's figures for imports from South Africa – which were 344.7 kg – leading to the conclusion that horn illegally obtained from Angola, Namibia, Zambia and even Tanzania was being sold on by South Africa.[152] An illegal trade in poached horn was occurring, with an average of 14 rhinos poached annually between 1990 and 2005.[153] Poaching of rhinos and elephants led to the formation in June 1989 of the Endangered Species Protection Unit as a specialist police unit charged with combating wildlife poaching and smuggling. It became fully operative in 1991, and between then and 1995 it investigated 792 reports of ivory or rhino horn poaching and smuggling and prosecuted 529 people, with 478 of them convicted and 403 rhino horns recovered. This included the arrest of a Taiwanese citizen in possession of 115 rhino horns and the arrest of two more Taiwanese with 55 rhino horns.[154]

During the 1990s, the private ownership of white rhinos increased, and there were some black rhinos in private hands – 62 in seven groups in 1997. A small eastern black rhino population was established in Addo Elephant NP, with rhinos introduced there from Kenya in the 1960s, which bred well enough for six to be translocated later to Tanzania to help it to rebuild its numbers.[155] Rhinos were also introduced to Pilanesburg NP in the North West Province of South Africa (24 black rhinos were introduced between 1981 and 1989), which increased through breeding, allowing nine to be sent to Madikwe GR. Continued breeding success increased Pilanesburg's population to 55 by 2001.[156] Some black rhinos were settled in Marakele NP, north of Pilanesburg, with a first group of black rhinos released there in 1993, 12 in 1996 and a total of 28 between 1993 and 2015.[157] By 1999, South Africa had 1,043 black rhinos, over 70% of wild black rhinos in Africa, and had succeeded in increasing white rhino numbers to 7,913 by 1997. By 1996, the Natal Parks Board (renamed the KZN Nature Conservation Service after the end of apartheid, and later changed to Ezemvelo KZN Wildlife) had moved about 4,350 white rhinos and 340 black rhinos to other locations in South and southern Africa and Kenya.[158] This success in increasing numbers and distribution of white rhinos led to the ninth CITES CoP downlisting South Africa's white rhino from Appendix I to Appendix II, but only for trade in live animals to approved destinations, and for the continued export of legal hunting trophies – it was passed with 66 in favour, two against, but the CoP rejected a South Africa proposal for a legal, limited trade in white rhino horn, with Western animal rights and environmentalists fiercely opposing resumption of legal trade.[159]

Despite this setback to South Africa's attempts to provide income-generating opportunities for state and private rhino owners, rhino numbers continued to rise in the 2000s, and black rhinos began to be settled on communal or private land under custodianship arrangements to expand the diversity of locations of black rhino populations.[160] By December 1994, there were 1,742 rhinos (mainly white) on privately owned land, with the numbers constantly increasing and prices high – 45 white rhinos being sold at the Hluhluwe Game Auction in 1998 for R5,234m ($870,000), vital income as government conservation funding was declining. Live rhino sales in KZN generated a turnover of approximately $1.57m; in Hluhluwe-iMfolozi, annual turnover from all game sales from the park represented 22% of the total cost of running it in 1998–9. Another income stream for rhino conservation was trophy hunting, which was continuing with 0.5–0.6% of the white rhino population hunted annually.[161] The Natal Parks Board exported 658 southern white rhinos to destinations outside Africa between 1962 and 1994, bringing in more income. However, poaching was beginning to increase as demand rose in markets in Yemen and East Asia, especially Taiwan and China. Between 1993 and 1999, 61 rhinos were poached, 59 of them white rhinos and 32 of those in KNP.[162]

Zambia: poaching, corruption and the loss of Luangwa's rhinos

Zambia had a substantial population of rhinos at independence in 1964, but over 30 years later the estimated 12,000 black rhinos were gone, with none thought to be left in 1995 – the population having been destroyed by illegal hunting mainly in the 1970s and 1980s.[163] There had been only 2,750 left in 1980; these were mainly wiped out by 1991 and then finished off in the next four years.[164] At independence, Zambia had a number of NPs and reserves with significant numbers of rhinos and large herds of elephants. The rhinos had their main stronghold in the Luangwa Valley and along the Zambezi Valley. There were rhinos in the huge Kafue NP, but as the WWF's John Hanks reported from his time in Zambia studying wildlife in the late 1960s, rhinos were encountered there only occasionally, while sightings were very regular in Luangwa, where he believed there were about 4,000 black rhinos.[165] Hunting was permitted under licence and for crop protection, but poaching was taking place for rhino horn, ivory, meat and hides in and around the reserves across the whole of the country.[166]

Kenneth Kaunda, who became the first president of independent Zambia, had been highly critical of colonial restrictions that stopped Africans hunting and of the loss of land by communities to create NPs and reserves. By the time he led the first independence government he had modified his approach and retained colonial wildlife laws, taking a pro-conservation line. However, as Gibson pointed out clearly, 'few Zambians besides Kaunda favoured conservation during this time ... the safari hunting and tourism industries had not yet begun to earn significant amounts of revenue. Game cropping schemes had not convinced many politicians or rural residents about the value of protecting wild animals to "rationalise" meat supplies.'[167] In 1968, Kaunda's United National Independence Party (UNIP) government passed the National Parks and Wildlife Bill, establishing amended regulations for the management of NPs and reserves, and for the regulation of local and tourist hunting, supposedly to replace colonial wildlife legislation. It basically updated colonial regulations, and there was little change in the rights of Zambian communities over the wildlife with which they lived. Licensed sport and commercial hunting was allowed to continue, but most Zambians in rural areas remained excluded from ownership or utilisation of wildlife. Despite a UN Development Programme and FAO project centred on the Luangwa Valley area to improve rural livelihoods while furthering conservation, most rural communities had insufficient support for their agriculture and felt that they suffered from the presence of wildlife. Its only value for them was the meat that they could obtain from hunting. Most communities hunted to supplement food that they grew and, as Dalal-Clayton and Child point out, they gained 'about 30% of their nutrition from wildlife foods.'[168]

Despite being in Kaunda's home province, the Luangwa Valley, with its wealth of wildlife, including a large number of rhinos, was not an area that had received much government assistance, development of roads, markets,

educational or health facilities either in the colonial period or since independence, beyond infrastructure to support the growth in safari tourism.
There was some employment provided by the tourist sector, but no great
benefit to the population from income generated by the NP. People were
excluded from residence, hunting and from gathering wild foods or wood
in protected areas. Communities lived in the area surrounding the South
Luangwa NP, including the Lupande GMA (LGMA), which comprised six
chiefdoms: Jumbe, Kakumbi, Malama, Mnkhanya, Msoro and Nsefu.[169]
However, even in the GMA, where regulated hunting was allowed, Zambian
families benefited little from the wildlife, illegal hunting being the exception;
studies showed that less than 15% of household income came from legal
utilisation of wildlife.[170] In 1971, the South Luangwa National Park (SLNP)
was formed from the old GR, together with nine adjacent GMAs. The SLNP
was placed under the National Parks and Wildlife Service (NPWS).[171]

The SLNP and the neighbouring LGMA cover an area of 15,000 km^2
in eastern Zambia, rich in wildlife but in an area with a substantial human
population (around 45,000 in the 1980s) engaged in subsistence or small-scale
commercial farming, with a history of hunting to supplement agriculture
and also a history of conflict with wildlife – especially elephants, hippos and
buffalos – that raided crops and damaged water and other infrastructure.
In 1971, Kaunda's government established new regulations covering trophy
hunting, hunting licences, protected species and legal methods of hunting,
and 32 GMAs were established to allow regulated wildlife conservation and
sustainable utilisation. Eighteen NPs were also established, from which local
communities were excluded. Licensed hunting continued in designated areas,
while traditional hunting, considered in law as poaching, also continued.
During the debates over wildlife policies, the Minister of Lands and Natural
Resources, Simon Kalula, told parliament that he would instruct his game
guards 'to shoot at people who may be there in the country poaching'.[172]

The Luangwa Valley was the most important habitat for the south-central
black rhino – there were about 12,000 black rhinos in Zambia in 1973, the
third largest black rhino population in Africa.[173] In the 1970s, Hillman cited
estimates of 3,500–4,500 black rhinos in South Luangwa.[174] Save the Rhino
suggested that there were between 500 and 2,000 in the North Luangwa
ecosystem, in addition to the South Luangwa population.[175] South Luangwa
NP, along with Victoria Falls, was at the centre of the development of wildlife
tourism and, having the greatest concentration and diversity of wildlife
with large populations of rhinos and elephants, also became the focus of
poaching for rhino horn and ivory. Poaching in South Luangwa reduced
rhino numbers there from around 4,000 in 1973 to 2,000 by 1979 and none
by 1995.[176] Poaching was carried out by gangs of men commissioned by

middlemen in the rhino horn and ivory trade, often including people from local communities. Local communities had lost land and hunting opportunities to NPs and reserves, and, as Dalal-Clayton and Child emphasised concerning South Luangwa:

> There was little incentive for local people to resist
> commercial poachers or inform on their activities. Revenues
> from wildlife, such as hunting license fees, park entrance
> fees, and safari earnings, went to central government or
> businessmen living outside the area. Communities living in
> the Valley gained little direct legal benefit from local wildlife
> resources, and wildlife was seen mainly as meat on the hoof,
> a serious agricultural pest and the plaything of rich, white
> Westerners.[177]

The Department of Game and Fisheries was responsible for conservation, anti-poaching and problem animals. However, they did not have sufficient staff at independence – just 265 game guards, scouts and drivers. By 1967 the number had increased to 509 staff, of whom 492 were game guards.[178] Senior staff dealing with wildlife were mainly European and they were gradually replaced over time, often with unqualified political appointees. The standard of recruitment and training was poor.[179] Control over the licensing of hunting under the 1971 regulations, dealing in wildlife trophies or commodities such as rhino horn, ivory and skins, was often handed to politicians or businessmen close to UNIP. UNIP politicians were well aware that by ensuring their control over the issuing of hunting licences, permits for the export of rhino horn, ivory, hippo teeth, hides and skins, and influencing who got permission to hunt and utilise wildlife, they had garnered a 'distributable benefit' that could be used for political patronage.[180]

There was little incentive for game guards, despite Kalula's dire warnings about shooting poachers, to search out and prosecute poachers, many of whom had political godfathers right up to cabinet level protecting them and taking a cut of the income from poaching.[181] Rampant poaching in the 1970s and 1980s led to the extermination of black rhino in many areas and severe reductions in numbers across the country. Illicit hunting for bushmeat, as well as more organised poaching for horn or ivory, was widespread, and little serious effort was made by the wildlife authorities or police to catch poachers. Most Zambians, especially in rural areas, where hunting was commonplace for meat and commodities to sell, did not see poaching as a serious offence, and there was no great public or political pressure to stop it.[182] Prior to 1977 and the ban on the international trade in rhino horn, legally obtained as well as poached rhino horn was exported through Portuguese-controlled Mozambique (prior to Mozambique's independence), Burundi or Tanzania. Corrupt officials laundered poached horn and ivory through the legal system.[183]

Against the background of Zambia's economic problems in the 1970s – caused by global trade terms, the disastrous effects on the economy of economic sanctions against white minority-ruled Southern Rhodesia,[184] and economic mismanagement and substantial corruption – local people had little choice but to turn to harvesting wildlife resources through poaching, whether commercial or for personal consumption, to feed their families amid massive price rises for basic foods. In 1975, the country experienced an ever-worsening economic crisis, resulting from drastically falling copper prices caused by a world recession following the massive oil price rises after the Arab-Israeli War of 1973 and oil embargoes imposed by Arab oil-producing states. Inflation was 17.3% a year from 1975, reaching 37% in 1985.[185] The rural economy and people's livelihoods declined drastically, giving even more incentives to poach not only for bushmeat, for which there was growing demand in rural and urban areas, but also for the lucrative commodities of rhino horn and ivory, with villagers helping poaching gangs.[186] Falling living standards made it easier for the poaching middlemen (who were placed between the actual poachers, whom they recruited, and the international smuggling syndicates) to find men to become poachers and send them into Zambian parks and reserves or across the Zambezi into Zimbabwe. The base level porters, boatmen and other menial members of the gangs might be paid as little as $12.50–25 per trip, while the hunter or gang leader would get considerably more, though only if they came back with horn.[187] A horn in the late 1980s might sell for $100–360 in Lusaka or other trading centres in central or southern Africa.[188]

Poaching was increasingly carried out by well-organised gangs armed with easily obtainable automatic rifles. The APUs of the NPWS were poorly motivated, poorly armed and lacked training, vehicles and fuel. They were no match for well-armed poaching gangs, according to Richard Bell, who became head of the Luangwa Integrated Resource Development Project (LIRDP). He told me that much of the poaching was organised by Senegalese and Malian traders in Mpika (between North and South Luangwa and the Bangwelu National Reserve), who persuaded local people to hunt for rhino horn, ivory, leopard skins, hides and meat.[189] South Luangwa, with its substantial rhino population, was the focus of an explosive increase in poaching. Rhinos were sought out and killed, and also died in snares set by bushmeat hunters.[190] Gibson noted evidence found by the Wildlife Conservation Society that the army and police were involved in poaching and smuggling or were simply taking horn, ivory and meat from poachers at roadblocks set up in areas where poachers were active; it was also believed that Zambian soldiers were lending or renting their military-grade weapons to poachers.[191] Zambian and other middlemen provided guns and ammunition for poachers and then smuggled the horn, ivory and skins out of the country. There were more incentives for people to help the poachers in return for cash than to inform on them, and they felt no ownership of the wildlife.

Senior politicians and government officials were also receiving money from the poaching gangs in return for protecting them. Richard Bell told me that at one stage the minister responsible for wildlife was himself benefiting from proceeds from rhino horn and ivory poaching.[192] There were also many instances of senior politicians and army and police officers hunting without licences for sport and profit, through selling meat and wildlife products. During a parliamentary debate in the Zambian National Assembly on 13 August 1982, one MP said that when senior UNIP members and government officials visited his constituency it 'turned into a hunting camp'.[193] The likelihood of the NPWS enforcing poaching laws and catching those responsible was small, not just because of the high rank of the illegal hunters but also because corruption was rife in the service and poorly paid rangers and other NPWS staff worked with the poachers, as was revealed by the Zambian media,[194] but without any response in terms of prosecuting staff involved in poaching or smuggling horn. Local chiefs and senior district officials were heavily involved in organising and taking a share of meat from bushmeat and other poaching as wildlife products became ever more sought after to fill the gap caused by shortages and massive food price rises.[195] Poachers supplied game meat locally and worked with ivory and rhino poaching gangs.[196] The police and army were also involved, either directly in poaching, in confiscating and selling the products of poaching when they caught poachers at roadblocks, or in renting out weapons, ammunition and military vehicles to poaching gangs.[197]

By 1987, officials from CITES were getting increasingly concerned about the rhino poaching escalation in Zambia. Chris Huxley and Jacques Bernay from the CITES Secretariat met John Hanks of WWF International in April 1987 to look for ways to stem the tide of poaching both in Zambia and by Zambian poachers in Zimbabwe. Huxley and Bernay asked Hanks whether the WWF could fund an in-depth investigation into the links between the poaching syndicates and corrupt officials in government and wildlife management.[198] Hanks was aware that investigating corrupt officials and politicians in countries in which it was working on conservation projects could lead the WWF into very dangerous territory – so this was no doubt a motivation for Hanks and, by association, the WWF, to become involved in the whole sorry Operation Lock saga.[199] Operation Lock was a bizarre anti-poaching project involving Hanks, Prince Bernhard of the Netherlands, former special forces personnel recruited in Britain and, by association, the WWF. It was an almost surreal covert operation aimed at identifying and in some cases killing leading figures in the southern African ivory and rhino horn trades. Lock involved former British SAS soldiers working for a security company called KAS Enterprises, funded by Prince Bernhard, who had decided he wanted to smash the smuggling rings.

In 1987, John Hanks, Director of Africa Programmes for WWF and later head of the SA Nature Foundation, flew to London to meet Sir David Stirling, founder of the British SAS and of KAS, whose personnel would pose as smugglers, infiltrate smuggling rings in southern Africa, and help to arrest or assassinate key members.[200] Hanks agreed the deal in a letter to KAS, in which he said that the operation should not be regarded as a WWF-funded one. Hanks told the Kumleben Commission that funds from Prince Bernhard had been passed to KAS, though he continued to deny it was an official WWF operation. KAS was supplied with rhino horn by the Natal Parks Board to use as bait to catch smugglers, and they started trying to work up entrapment operations.[201] However, the whole project was infiltrated by South African agents working for a leading South African security agent, Craig Williamson, who tried to use Lock personnel to spy in Zambia and Zimbabwe. The head of Operation Lock, Colonel Ian Crooke, also identified a series of conservationists – including Ian Parker, Richard Bell and Rowan Martin – as possible targets because they had in some way annoyed him or stood in his way.[202] It all began to fall apart, and questions were asked about the rhino horn given to the operation and where it ended up, alleged spying activities in neighbouring countries and links with South African intelligence bodies. A number of journalists dug into the story, and despite KAS threats to sue and retractions by the publication *Africa Confidential* of a story about Lock, under pressure from the newsletter's owners, the publication's editor, Stephen Ellis, published an exposé of the entire botched operation in the London *Independent* on 8 January 1991, revealing the involvement of the WWF and effectively bringing the whole sorry episode to an end. At no point did it seem to have any success in catching or eliminating rhino poachers or smugglers.[203] It was an inept and very poorly conceived operation that became a huge embarrassment to Hanks and the WWF, but did not advance the cause of rhino conservation in any way.

As part of the context for the setting up of such a far-fetched operation as Lock, Hanks wrote that he became aware, through discussions with the anti-poaching commander in Zimbabwe's Zambezi Valley, Glen Tatham, and others who had knowledge of the poaching set-up in Zambia, that Zambian poachers were killing large numbers of rhinos in Zambia and in Zimbabwe's Zambezi Valley, and selling the horns to the networks of criminals operating in Zambia, who then smuggled the horns to Yemen and the Far East. As Zambia's own rhinos were being rapidly depleted, the Zambian poachers were crossing the Zambezi more and more often to poach in Zimbabwe across a river frontage of over 230 km, which was impossible to police in its entirety.[204] At around this time, the Zambian government proposed setting up a rhino sanctuary within South Luangwa NP to provide increased protection from poachers. The WWF, urged by Hanks, declined to go along with this plan, given that in the preceding 15 years the South Luangwa rhino population had been reduced by poaching from 4,000 to under 100, and of six white rhinos relocated from South Africa to Livingstone Game Park, four had been poached.[205]

On the face of it, the Zambian government, with funding and advice from the WWF, other international conservation groups and Western governments, made attempts to strengthen anti-poaching capabilities, but poaching continued unabated and there was mounting evidence of corrupt army personnel, wildlife officials, police and government officials profiting from it. As a result the NPWS, the WWF, businesses involved in tourism and locally based conservationists came together to form the SRT, primarily to protect the rhino population in the Luangwa Valley's NPs.[206] The huge mining company Zambia Consolidated Copper Mines and photographic and hunting safari operators started funding the SRT's anti-poaching efforts, with SRT and other private contributions amounting to 14% of the NPWS budget by 1979.[207] SRT was backed by the WWF, which aided its establishment and organisation. However, with just 18 scouts their ability to reduce poaching was tiny compared with the threat posed by poaching gangs, and their geographical reach was small. Patrols by the SRT revealed horrific levels of slaughter of elephants and rhinos for tusks and horns. The black rhino population in the Luangwa Valley was reduced from about 4,000 to less than 100 by the mid-1980s, and has now completely disappeared. Over the same period, elephant numbers in the Valley were reduced from 90,000 to fewer than 15,000.[208] Despite the assistance from SRT and the funding from businesses and the safari sector, the money available to curb poaching was still far less than was needed. The NPWS was still in a poor state, and had to return its one helicopter to the Wildlife Conservation Society of Zambia because it could not afford to maintain it.[209]

Despite the work of the SRT and of those wildlife officials who were committed to conservation, it was clear that something new was needed to stem poaching and gain community support for conservation, particularly in the Luangwa Valley, with its abundance of game and potential for wildlife utilisation and tourism. The Zambian Government and WWF had launched a $3m anti-poaching programme in the Luangwa Valley in 1980, but it proved ineffective and far more resources were needed along with a purge of corruption in government, which simply was not going to happen. The 1982 Zambian government ban on elephant and rhino hunting and all exports of ivory and horn was similarly ineffective. The poaching of rhinos in Luangwa and elsewhere in Zambia reached a peak in the early 1980s. By the early to mid-1980s, the SLNP had become an increasingly popular destination for foreign safari tourists because of its concentrations of game – elephants, rhinos, lions, leopards, wild dogs, buffalos, huge numbers of hippos and a diversity of antelopes, as well as rich birdlife. The Zambian government was keen to capitalise on the foreign exchange that could be earned from safari tourism, but also wanted to find ways to improve the opportunities for income generation for the local communities that were living on poor-quality land outside the park. Local development, it was hoped, would cut local support for commercial and bushmeat poaching and gain the support of communities for conservation and anti-poaching.

It began to be appreciated that communities felt little benefit from the income from safari tourism, apart from those working as cooks, waiters, domestic staff, drivers and guides on relatively low wages. Something new was needed. The answer that emerged from the conservationists working in and around the Luangwa Valley and from the Zambian government was the LIRDP, a community-based management scheme funded by the Norwegian Agency for Development Cooperation (NORAD), which had the support of President Kaunda – no doubt partly because it was in Kaunda's political heartland of Eastern Province. The concept of a community-linked conservation and local development programme in the Luangwa Valley was discussed and endorsed in 1983 at the Lupande Development Workshop, funded by NORAD and hosted by the NPWS to consider a management strategy for Lupande GMA, and officially launched in 1986. Initially, success was achieved in improving law enforcement and starting the establishment of agricultural support, water supply, small-scale credit and women's programmes,[210] LIRDP, the SLNP and the LGMA. The project involved the utilisation of wildlife through tourism and hunting to improve the livelihoods of the people who lived in the LGMA area. It was similar to CAMPFIRE (Communal Areas Management Programme for Indigenous Resources), which was launched in Zimbabwe, aiming to combat the rural deprivation that encouraged poaching by tying conservation goals to poverty alleviation through the generation of revenue from tourism and hunting to directly benefit local communities and give them income, a sense of ownership of wildlife resources and tangible benefits such as better roads, water supplies, and health and education provision.

The project was opposed before and after its launch by the NPWS, which lost power over the Luangwa region, and its personnel lost opportunities for corrupt access to income from hunting and tourism.[211] Local chiefs bought into the scheme as a way of trying to garner local resources for their patronage networks, taking on the role of local or regional gatekeepers, and skimming off funds coming in and income generated. There was a constant battle between those running LIRDP and the chiefs over the allocation of income, with the chiefs trying to get their hands on income to the detriment of cash and services that were due to local people, which over time served to reduce local support for the scheme and encouraged some people to go back to poaching.[212] Richard Bell, an experienced ecologist who had managed Kasungu NP in Malawi, and Fidelis Lungu, a Zambian resource economist with the NPWS, were appointed as co-directors of LIRDP. The government decreed that wildlife-related revenues within the project area would accrue to the project and be distributed locally for development and project work. About 60% was to be used for project operations and 40% allocated for local development initiatives, as determined by the representatives of local leaders and community representatives. The income came from NP entry fees, hunting licences, culling operations (buffalos, hippos and impalas) and also a commercial safari hunting venture which carried out safari hunting operations in LGMA from 1989 to 1991 'on behalf of local communities'.[213] Some key plans, as far as rhino poaching and conservation were concerned, were improved wildlife management, including 'a unified operational command, with increased scout densities, an investigations unit, and involvement of local communities with an expanded village scout program' and more scientific monitoring of populations, along with research that would be applied in practice.[214]

The biggest conservation challenge for LIRDP was controlling the poaching epidemic that by 1986 was threatening to wipe out the black rhino population. Fighting poaching on this scale took money, manpower and the buy-in of local communities that had benefited from poaching or turned a blind eye to it as they felt no benefit from wildlife conservation and the associated safari tourism industry. Anti-poaching teams were funded from the project and 200 local people were employed as rangers or game scouts, providing direct employment and income for the communities. Poaching was reduced by 90% to a level of ten elephants a year between 1988 and 1995, but by then rhino numbers were so low that improvements in anti-poaching were for them at best extremely marginal, while the skimming of income by chiefs and the less than enthusiastic support from the wildlife authorities hamstrung the project and reduced its effectiveness. President Kaunda's government was voted out of office in 1991 and his successor, Fred

Chiluba, did not have Kaunda's personal commitment to conservation; he was also suspicious of projects launched with Kaunda's backing, seeing them as designed to bolster UNIP's power in such areas as the Eastern Province.[215] In 1992, responsibility for LIRDP was transferred to the NPWS and brought under the control of the Administrative Management Design for Game Management (ADMADE) programme. It was more centrally controlled than LIRDP; any sense of ownership that had been developed was largely lost and there was less local control of income and its uses.[216] Whatever it could still achieve in fighting poaching, it was far too late to save the black rhino in Zambia by the early 1990s, and by the late 1990s it had been wiped out in the Luangwa Valley.[217]

The historical range of the southern white rhino had not included areas north of the Zambezi, and attempts in the 1980s and 1990s to introduce white rhinos had failed because of poaching. Small numbers of white rhinos had been relocated from South Africa to sanctuaries in Zambia, but by 1991 all of them had been killed. Six were introduced in 1993, but did not thrive.[218] In 2008, four more white rhinos were taken to Zambia from South Africa and released in Mosi-oa-Tunya NP, on the north bank of the Zambezi, just outside the historical range of the white rhino. They were kept under constant guard in a 26 km² sanctuary, which prevented any of them from being poached,[219] though they could hardly be described as a viable wild population. The development of a National Rhino Conservation Plan in 2003 included reintroduction of the black rhino to try to re-establish a national population, with five black rhinos being translocated to North Luangwa NP in 2003, and further rhinos being brought in to bring the population in a fenced sanctuary in the park up to 25 in May 2010.[220]

Zimbabwe: increasing numbers despite poaching

By the early 1960s, Southern Rhodesia was heading towards its Unilateral Declaration of Independence from Britain, which led to a 15-year war with the movements that were seeking liberation from white minority rule. Conservation measures adopted in the decades leading up to the 1960s had enabled some recovery in wildlife populations in protected areas, following the mass slaughter of the anti-tsetse campaigns and the clearing of wildlife from huge areas of farmland on which whites had been settled. Most of the surviving black rhinos were in areas around the Zambezi Valley, and south of Lake Kariba. Some were in parks/reserves such as Mana Pools and Matusadona, the latter receiving 39 black rhinos translocated between 1960 and 1963 that had been marooned on islands during the flooding of Lake Kariba. Some were still present in Wankie (Hwange) NP in the west and

Gonarezhou in the south-east. Hwange received 13 black rhinos rescued from Kariba.[221]

In 1967, Roth of the Southern Rhodesia wildlife department said that a 'very rough estimate' of the territory's population was 1,000–1,400 rhinos, with the largest concentrations around Mana Pools NP and the Urungwe hunting area (270–300) and in the Chewore safari area (260–320). He believed that 'about 400–650 rhinoceros could optimistically be expected to live south of Lake Kariba, of which only about 200–240 may be inhabiting national land'. The large numbers in the north along the Zambezi prompted the wildlife department to start relocating some to Hwange NP.[222] In 1962, the transloca-tion of white rhinos from Natal to Zimbabwe had begun with eight released in the fenced Whovi Game Park in Matobo NP and in Kyle Dam GR, with plans to send a total of 100 rhinos to Rhodesia.[223] Eight more white rhinos were moved to Whovi in 1966–7, and to these were added in 1987 four white rhinos from eSwatini.[224] The population increased and became too large, with increased fighting between males, leading to the relocation of 20 to other Zimbabwean reserves. At the same time that conservation and translocation were proceeding, a limited number of black rhinos were allowed to be shot in designated hunting areas, 18 being shot between 1961 and 1964.[225] Licences cost £100 for residents and £150 for non-residents. In 1965, to conserve and increase rhino numbers, all hunting of rhinos was banned with the exception of shooting in self-defence or in cases of rhino damage to crops, fences or water systems. The sale of rhino horn that had been registered with the wildlife department as legally obtained was permitted. Roth recorded that dealers in Rhodesia had also imported 113 rhino horns from Angola and Zambia in 1963, 80% being exported to Zanzibar.[226]

In 1971, Kerr and Fothergill noted that no detailed survey of the black rhino population had been carried out to ascertain numbers; estimating the population was very hard due to the thick bush that they inhabited.[227] Between 1967 and 1969, while surveying the Zambezi Valley from Lake Kariba to the Mozambican border, they estimated the Dande Tribal Trust Lands to have about 40 rhinos, Chewore Wilderness about 125 (believing the previous estimate, 260–320, to be far too high), Sapi Controlled Hunting Area about 60, and at least 60 in Mana Pools NP. They thought heavy hunting and human pressure on land had reduced Urungwe's population to about 40. Matusadona had at least 65, possibly more because the difficult terrain made a precise estimate impossible. Overall, in the area south of Lake Kariba they estimated the population at 255, way down from Roth's estimate of 450–600, which they thought over-optimistic, and they almost halved his national estimate, putting it at 740, with 440 in protected areas.[228] Poaching began to increase in the late 1960s. One cause was a rise in snaring by Batonga hunters in and around the

Chizarira NP, using steel cables, with 60 rhinos snared there between 1965 and 1968.[229] Auctions of rhino horn began to be restricted at this time, and only three took place in the 1970s – two in 1978 and one in 1979. In the 1978 auctions, 57 horns were sold – 24 each from the Zambezi Valley and Hwange NP. The total weight sold was 78.65 kg at $195 per kilogram. In 1979, one South African dealer bought all the horn for sale, at $186 per kilogram.[230] No rhino horn auctions were held after independence in 1980.

In the last decade of white minority rule in Zimbabwe, land and wildlife laws were passed that changed ownership of wildlife. Wildlife on privately owned white land had become an increasingly valuable resource for meat, hides and other commodities obtained from hunting. In 1975, the government enacted the Parks and Wildlife Act, fully transferring control over wildlife on private lands from the state to landowners. The measure benefited white landowners, but not Africans, who had been forced off land that was given to whites and into the Tribal Trust Lands, where land was not individually owned.[231] Ironically, the war of liberation fought by the black nationalist movements (the Zimbabwe African People's Union, ZAPU, and the Zimbabwe African National Union, ZANU) from 1964 to 1990 led to a decline in poaching in some areas, because of the dangers of hunting in remote areas where nationalist guerrillas and the Rhodesian army were operating.[232] However, towards the end of the war, the conflict provided opportunities for rhino and elephant poaching by units of the Rhodesian army in the country and in neighbouring Mozambique.[233] The main area where the liberation war, and later the civil war in Mozambique, enabled poaching was Gonarezhou NP along the border with Mozambique. Gonarezhou once had a large black rhino population, but they had been poached and few remained. In 1971, 72 black rhinos were relocated there from other parts of the country. Over several years, the numbers had increased naturally to 150, but poaching by criminal gangs, Mozambican rebels and the Zimbabwean national army after 1980 had, by the end of the decade, reduced numbers to less than 50.[234]

After the end of the liberation war and the election of the ZANU-led government of Robert Mugabe in 1980 there were initially no radical changes in wildlife policy, although the wildlife department and some conservationists began to look at ways of bringing conservation and wildlife utilisation in line with development programmes to improve benefits for local people. New policies were designed to enable local communities, through district councils, to engage in regulated hunting for meat, hides and ivory and to tap into the lucrative but controversial safari hunting market. With hunting licences costing visiting hunters up to $3,000, the hunting of limited numbers of rhinos on private concessions supported efforts to restore rhino populations outside protected areas. This also created the opportunity to dovetail the economic and social development goals of the new government with the need to generate income for local communities and conservation, and to fund anti-poaching.[235]

Table 7.6 indicates Zimbabwe's rhino numbers between independence and 1997, as estimated by Emslie and Brooks.

Table 7.6: Zimbabwe rhino population[236]

Black rhino	1980	1984	1987	1991	1992	1993–4	1995	1997
	1,400	1,680	<1,775	1,400	425	381	315	339
Southern white rhino	1895	1948	1968	1984	1987	1991	1992	1993–4
	0	0	0	200	208	250	249	134
	1995	1997						
	138	167						

In contrast to the range states to the north, Zimbabwe's black rhino numbers increased during the 1970s, and by 1987 it was the only country with over 1,700 black rhinos, accounting for almost half the world's population at that time.

After 1980, the economy and wildlife sector initially continued to do well under the Mugabe government. Tourism to Hwange NP, Matusadona, Mana Pools and Victoria Falls increased, with European and North American tourists swelling numbers that under white rule had been largely limited to domestic white tourists and South Africans. It promised to bring in much-needed foreign exchange, and wildlife staff were relatively well paid and motivated – there was not at first the corruption and inefficiency seen in countries such as Kenya and Zambia, which had helped pave the way for rhino poaching. The continued growth in private ownership of wildlife on game farms and private safari areas (for photographic and hunting safaris) complemented state conservation efforts and enabled a steady increase in wildlife numbers. However, during the first decade of independence, the poaching of black rhinos grew, especially in the Zambezi Valley. White rhinos, originally brought in from South Africa, increased, especially in Hwange and on private game farms or conservancies such as in the Lowveld, Savé Valley and Bubye Valley. In many areas, regulated trophy hunting assisted in conservation efforts, by bringing in foreign currency income that helped to pay for reserve management and anti-poaching efforts.[237]

By 1993, Savé Valley Conservancy, Bubiana Conservancy and Bubye Valley were the leading private conservancies, established by joining marginal pastoral land and creating wildlife reserves on which a wide range of wildlife, including rhinos, was conserved but also legally hunted. The income from hunting and other sustainable-use methods enabled conservancies to increase their rhino numbers. Savé has 3,000 km² of rhino habitat with a potential carrying capacity of 600 rhino, while Bubiana at 1,300 km² can hold about 300.[238] Conservancy formation by local, regional and international investors coincided with black rhinos being moved from the Zambezi Valley where they were being heavily poached. Savé Valley Conservancy was formally established in 1991; all internal fences were removed from an area totalling 3,442 km² and a 350 km perimeter fence was constructed. Around 4,000 animals of 14 species were reintroduced, including elephants and rhinos.[239]

Large wildlife populations were concentrated around the Gonarezhou NP/Savé Valley area on the border with Mozambique, especially elephants but also rhinos and large herds of buffalos. Here efforts were made by conservationists in the late 1970s and early 1980s to gain the support of local people to stop the poaching of rhinos and elephants. Clive Stockil, founding chairman of the Savé Valley Conservancy and an honorary warden in Gonarezhou, said there was extensive poaching by the Shangaan community in the early 1980s, but by working with elders and with the development of community-based sustainable-use strategies, poaching was reduced.[240] However, illegal hunting was still occurring in central-southern Zimbabwe on state land, private reserves and conservancies, with 17 rhinos poached in the early 1990s on the Midlands Conservancy and at a private ranch in Karoi.[241] Organised gangs poached rhinos and elephants for horn and ivory, and hunting/snaring for bushmeat was carried on by local communities, many of which were aggrieved that they had lost land to the NPs and reserves under white rule.

While the conservancies began to help to increase wildlife numbers, Gonarezhou, during the liberation war and later the civil war in Mozambique, was plagued by poaching. Rhodesian special forces conducted operations into Mozambique from the park and the areas around it to assist the rebel Renamo and to fight Zimbabwean guerrillas. In the mid- to late 1980s, there was a major increase in poaching in Gonarezhou resulting from the conflict in Mozambique. Renamo, the Mozambican army, SADF/military intelligence and the Zimbabwean Army were all involved in poaching during the war against Renamo and the destabilisation of Mozambique by South Africa. Zimbabwean army units were based in Gonarezhou and clashes took place there between the army and Renamo. In 1981–2, one Mozambican gang operating in the park killed 80 elephants and an unknown number of black rhinos, and sold the ivory and horn to dealers inside Zimbabwe to buy food and other goods that were in short supply because of the war in Mozambique.[242] Zimbabwean troops poached in the park under cover of the counter-insurgency operations. When reports emerged in the press of a growth in poaching in Gonarezhou and the killing of 260 elephants and 32 rhinos there, the Zimbabwean government blamed Renamo, the SADF and poachers crossing from Mozambique, even though most of the killing was taking place in areas occupied by the Zimbabwe National Army (ZNA). Park staff tried to alert the government to the extent of poaching. Several conservation officers told the EIA that three reports had been compiled warning government ministers about the poaching, and implicating senior parks staff, army personnel, the SADF and organised smugglers. The central players in the smuggling operation were Gonarezhou warden Enoch Mkwebu, game scout Zephania Makatiwa, former Selous Scout and suspected ivory and horn smuggler 'Ant' White and Bill Taylor, an American-born dentist who had links with the Rhodesian Selous Scouts and the SADF, but also cooperated with poachers working for the ZNA. Mkwebu was moved from Gonarezhou and transferred to be acting provincial game warden at Gokwe, where there was a sudden and major rise in killing of elephants, allegedly for crop protection reasons.[243]

The start of the constitutional negotiations in South Africa, the scaling back of South African involvement in Mozambique and then the signing of the peace deal to end the civil war with Renamo that brought a decline in poaching and better enforcement by the NPs' APUs (now not obstructed by the ZNA or Renamo) deterred local poachers. Clive Stockil said that this produced a major improvement and a reduction in illegal killing. However, he warned that 'the main organisers of the poaching ring that operated in the 1980s have not disappeared and consistently try to find new ways to reopen poaching rings in the area', which they were to do in the late 1990s and the first decade and a half of the 2000s.[244] The Gonarezhou pattern was to be repeated in Zimbabwe and elsewhere in Africa, as it was a primary demonstration that illegal hunting by those from within the state itself presents a greater challenge to agencies responsible for wildlife protection than poaching by poor communities or criminal groups outside the networks of power and patronage. In these situations, the wildlife authorities struggle to combat illegal resource use because the 'illegitimate users' of that resource are from powerful state agencies or are highly placed army officers or government officials.[245]

The Zambezi Valley was spared extensive poaching of rhinos in the 1970s because of the Rhodesian military presence there to stop guerrilla incursions from Zambia. However, from about 1983 onwards, when rhinos were becoming scarce in Zambia, Zambian poaching gangs crossed into Mana Pools and other wildlife areas of northern Zimbabwe. Four rhino carcasses were found in late 1983 in Zimbabwe's Zambezi Valley with their horns hacked off; 12 were found in 1984, 68 in 1985, 149 in 1986 and 170 in 1987.[246] Zambians, and Angolan refugees in Zambia, were recruited to poach, and it was almost impossible for the Zimbabwean anti-poaching teams to police the whole 230 km stretch of the Zambezi that they crossed to do so.[247] Incursions became constant. The wildlife authorities started moving the remaining rhinos out, and between 1986 and 1989, 170 rhinos were relocated to southern Zimbabwe.[248] Blondie Leatham, one of the anti-poaching commanders, said that firefights were frequent between poachers armed with AK-47s and APUs. Many poachers died. He said, 'We had one period where we killed ten guys in fourteen days', and he believed that to get people to continue poaching, the middlemen who hired them 'simply upped the price per kilogram from $300 to $800'.[249]

The poaching took a heavy toll: in 1982, 1,171 live rhinos were seen alive and no carcasses; in 1984, only 746 were seen and 12 carcasses; in 1985, only 434 were seen and 149 carcasses; in 1986, only 359 were seen and 169 carcasses. 'By 1989, the Zambezi Valley population was put at 930, including 60 rhino on communal lands ... but in 1992 it was a disastrous 65 with none left on the communal lands,' according to Milliken, Nowell and Thomse.[250] In Sebungwe, Matusadona NP, Chizarira NP and neighbouring safari areas and communal lands, numbers had been 810, but were down to 60 in 1991, with none left in the communal lands. By November 1992, the AfRSG had been told that the Zimbabwean black rhino population had fallen to 420, with 623 black rhinos known to have been killed in the Zambezi Valley between 1984 and 1993 (393 in Sebungwe, 63 in Hwange and 51 in Gonarezhou); 1,130 rhino carcasses had been found across Zimbabwe in the same period, with the wildlife department fearing that a total of 1,500 had been killed.[251]

Although most of the poachers were Zambian, some with suspected links to senior politicians in the ruling Zambian Movement for Multiparty Democracy, there was evidence that Zimbabweans were assisting the poachers and were supplying them with food as Zimbabwe's economy declined and people in rural areas were hit the hardest, leading them to take money in return for assisting the poachers and guiding them to rhinos.[252] Some even started poaching as poverty bit deep and easy money beckoned. The Zimbabwean wildlife department admitted this in 1992 when it disclosed that 'during the past 18 months Zimbabwean locals have become significantly involved in both hunting horn and trafficking ... The current drought in the country is likely to lead to additional rhino mortality this year and involvement of Zimbabweans in illegal hunting and trafficking will probably increase as a result of economic problems in rural communities.'[253]

With the increased poaching and use of military-grade weapons, APUs developed military-style tactics that many senior rangers had learned as members of the Rhodesian security forces during the liberation war. If threatened or fired on, they fired back with the aim of killing armed poachers. The use of lethal force by APUs was backed by the government, with the Ministry of Home Affairs in 1985 granting them the right to use their weapons against poachers. Robert Mugabe intervened to stop a wildlife official from being charged with murder after the shooting of a suspected poacher.[254] In 1987, Mugabe approved a shoot-to-kill approach as part of an anti-poaching drive known as Operation Stronghold. Glenn Tatham was by then chief warden of Zimbabwe's NPs and he was quoted as stating: 'We knew we had to take the guys on and fight fire with fire. Our objective is to save animals; it's not to kill people. But we cannot afford the possible loss of life among our men by letting them walk into gangs of armed criminals without having the option of shooting first.'[255]

However, political pressure from the Zambian government, when the son of a senior Zambian policeman died on a poaching raid, led to Tatham and Blondie Leathem being arrested by the Zimbabwean police, hampering anti-poaching operations.[256] Although poaching was reduced through the use of uncompromising tactics, it did not end, and by the mid-1990s most of the surviving rhinos in the Zambezi Valley had been relocated to secure reserves and conservancies in the Zimbabwean Lowveld, away from the border with Zambia, in a programme led by Raoul du Toit, now the International Rhino Foundation's African Rhino Program Coordinator.[257]

Poaching continued in the mid-1990s. By 1993, there had been 873 known incursions by rhino poachers and 30% of them had led to contacts with game scouts; an estimated 167 suspected poachers had been killed – 68 in the Zambezi Valley, 51 in Sebungwe, 25 in Gonzarezhou and 23 in Hwange, while just 89 had been captured (28 in the Zambezi Valley, 32 in Sebungwe, 21 in Gonarezhou and 8 in Hwange) and convicted of poaching; more than 300 rhino horns had been recovered, about 15% of known rhinos poached.[258] This drew criticism by the domestic and international media, particularly in Zambia, as many of those killed were Zambians. However, anti-poaching personnel had to decide whether to shoot when faced by poachers armed with AK-47s who were unlikely to just surrender. As Milliken, Nowell and Thomsen put it, 'it is unrealistic to expect heavily-armed poachers to drop their weapons and engage in dialogue when detected by game scouts in the field'.[259] John Hanks, in his role as WWF International Africa Programme Director, worked to get funding for a helicopter to assist in Zimbabwean anti-poaching and wider rhino conservation work, but the funding was eventually withdrawn, despite indications that it was an important asset in combating poachers. The WWF, after criticism by the media, decided that its image was tarnished by use of the WWF-funded helicopter in the militarised Operation Stronghold, which had led to the shooting of poachers; it also became embroiled in the Operation Lock fiasco involving Hanks, as for a time that operation had funded the helicopter when WWF funding was withdrawn.[260]

In 1993, Operation Safeguard Our Heritage was launched with support from the army.[261] The plans involved setting up intensive protection zones (IPZs) in areas with surviving black rhino populations, such as Sinamatella Camp in Hwange. The approach also included allowing legal sport hunting of rhinos on private hunting concessions and trying (unsuccessfully) to change to CITES regulations to enable the sale of harvested rhino horn to bring in money to support conservation and anti-poaching.[262] One approach that was adopted alongside highly militarised anti-poaching and relocations to safe areas was dehorning.

In November 1992 it was reported that dehorning had started to deter poaching, with 117 black rhinos and 108 white rhinos dehorned in various parts of the country.[263] The CAMPFIRE sustainable-use scheme, introduced to boost local support for conservation in the former tribal trust areas and to boost rural livelihoods, was of benefit for a time in encouraging conservation and empowering communities to utilise wildlife resources, but had no great effect on rhino conservation, as rhinos had been poached out of many areas that came under the scheme, and the few left in unprotected rural areas could not be hunted as trophies. Like LIRDP and ADMADE in Zambia, CAMPFIRE started encouragingly, with local people buying into the scheme and receiving income, while schools and health clinics were improved thanks to money generated, but ZANU-run local councils began to take more and more of the income to fund their work, or it was skimmed off by corrupt officials.

One major conservation problem in the early 2000s, which affected conservancies and private reserves, was the government-backed process of land reform and farm invasions, as black Zimbabweans sought return of the land that had been taken from them under colonial and white minority rule. The fast-track land reform process, as it was termed by the Mugabe government, sanctioned occupation of white farms, which were turned over to black ownership. This affected some of the conservancies, which lost land or were invaded by those seeking land. Savé Valley originally covered an area of 3,490 km², but was reduced to 2,530 km² when an area measuring approximately 960 km² was reallocated to subsistence communal farmers and about 6,000 people settled there, resulting in the loss or dispersal of the wildlife.[264] In many other areas, white landowners were forcibly evicted from land that in some cases had provided wildlife conservation, and bushmeat hunting increased.[265]

Despite the land invasions and the effect on white-owned ranches and conservancies, poaching declined in the late 1990s and early 2000s, and in 2007 the Africa Rhino Specialist Group reported that there were in the wild in Zimbabwe 250 southern white rhinos and 535 black rhinos – up from 167 white rhinos and 339 black rhinos in 1997.[266]

CHAPTER 8

East and Central Africa, 2006–24

From the mid-1990s to 2006, poaching had declined. Numbers of black and white rhino began to increase, with black rhinos doubling by 2010 to 4,880 and southern white rhinos flourishing in southern Africa with increasing numbers (23,165);[1] there was a wider distribution, mainly through translocations, and income increased from tourism, live sales and trophy hunting.[2] The northern white rhino was functionally extinct in the wild,[3] as was the western black rhino. The number of rhinos poached increased from 2006 and rocketed from 2010 through to 2022. In 2006, 60 were poached and 62 in 2007. Thereafter poaching increased hugely, with a total of 11,703 rhinos known to have been illegally killed from 2006 to 2022. The peak year was 2015,

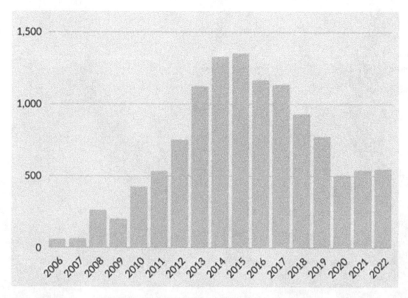

Numbers of rhinos poached in Africa, 2006–22.

when 1,352 rhinos were killed in Africa for their horns. And those figures may be an underestimate, as not all poached rhinos were discovered. South Africa, Botswana, Namibia and Zimbabwe were the countries hardest hit.

Table 8.1: Numbers of African rhinos poached annually[4]

2006	2007	2008	2009	2010	2011	2012	2013	2014
60	62	262	201	426	532	751	1,123	1,327

2015	2016	2017	2018	2019	2020	2021	2022
1,352	1,167	1,134	930	773	503	539	561

In its overview of the rhino populations in Africa at the end of 2021, CITES put the totals at 6,195 black rhino and 15,942 white rhino, giving a total of 22,137 in Africa. In a welcome development in October 2023, the IUCN's AfRSG updated the overall figure to a total of 23,171 rhinos, an increase of 5.2% on 2021, not including the nearly 1,300 captive in zoos and collections outside rhino range states. The total comprised 6,468 black rhinos (up 3.9% on 2021) and 16,801 white rhinos (up 5.6%), the latter registering the first increase in overall numbers since 2012.[5] This was more optimistic than the outlook at the CITES CoP in November 2022, which reported that African rhino numbers had fallen to 22,237 (not including 1,295 in zoos and collections) from 23,652 at the end of 2017 – the fall being mainly accounted for by white rhinos poached in South Africa; their numbers decreased by 3.01% p.a. between 2017 and 2021, while black rhinos increased by 3% p.a.[6] The distribution of black rhinos, as detailed at the 2022 CoP, showed that Namibia had 34.8%, South Africa 33.2%, Kenya 15.1% and Zimbabwe 9.9%, whereas for white rhinos, South Africa had 81.3%, Namibia 7.7%, Kenya 5.5% and Zimbabwe 2.6%. Botswana had both black and white rhinos, but they were in serious decline owing to poaching, with white down from 452 to 242 between 2017 and 2021 and black down from 50 to 23.[7]

Poaching and law enforcement

The AfRSG 2023 report noted that poaching across Africa 'generally declined from 2015 onwards', though with 561 rhinos poached in Africa in 2022, which was 2.49% of the estimated population – a lower level of poaching than the threshold of 3.4%, above which populations would fall (see Table 8.1).[8] However, it was not all good news, as poaching of rhinos had increased in some areas of South Africa, which still had the highest number of rhinos

Table 8.2: Rhinos poached in Africa, by country, up to the end of 2021

State	2006	2007	2008	2009	2010	2011	2012	2013	2014	2015	2016	2017	2018	2019	2020	2021
Botswana	0	0	0	0	0	0	2	2	1	0	1	9	18	31	55	n/a
Chad	-	-	-	-	-	-	-	-	-	-	-	-	0	0	0	0
DR Congo	0	0	2	2	-	-	-	-	-	-	-	-	2	0	0	0
Cote d-Ivoire	-	-	-	-	-	-	-	-	-	-	-	-	0	0	0	0
eSwatini	0	-	0	0	0	2	0	0	1	0	0	0	0	0	0	0
Kenya	3	1	6	21	22	27	29	59	35	11	10	9	4	4	0	6
Malawi	0	0	0	0	0	0	2	1	1	1	1	1	0	0	0	0
Mozambique	0	9	5	15	16	10	16	15	19	13	5	5	8	6	2	0
Namibia	0	0	0	2	2	1	1	4	30	97	61	44	93	56	40	40
Rwanda	-	-	-	-	-	-	-	-	-	-	-	0	0	0	0	0
Senegal	-	-	-	-	-	-	-	-	-	-	-	-	0	0	0	0
South Africa	36	13	83	122	333	448	668	1,004	1,215	1,175	1,054	1,028	769	594	394	451
Tanzania	0	0	2	0	1	2	2	0	5	5	0	2	0	0	0	0
Uganda	0	0	0	0	0	0	0	0	0	0	0	0	0	0	0	0
Zambia	0	1	0	0	0	0	0	0	0	0	0	0	2	0	0	0
Zimbabwe	21	38	164	39	52	43	42	38	20	50	35	16	34	82	12	4
Total	60	62	262	201	426	532	751	1,123	1,327	1,352	1,167	1,134	930	773	773	503

Table 8.3: CITES estimates of rhino populations in Africa at the end of 2021.[9] (Used with kind permission of CITES Secretariat)

Range state	Black rhinos				White rhinos		African rhinos	
	D.b. bicornis	D.b. michaeli	D.b. minor	Total for Diceros bicornis	C.s. cottoni	C.s. simum	Ceratotherium simum	Rhinocerotidae
Angola	-	-	-	-	-	3	3	3
Botswana	-	-	23	23	-	242	242	265
Chad	-	-	2	2	-	-	-	2
Côte d'Ivoire	-	-	-	-	-	0	0	0
DR Congo	-	-	-	-	-	20	20	20
eSwatini	-	-	48	48	-	98	98	146
Kenya	-	938	-	938	2	871	873	1,811
Malawi	-	-	56	56	-	-	-	56
Mozambique	-	-	2	2	-	14	14	16
Namibia	2,155	-	1	2,156	-	1,234	1,234	3,390
Rwanda	-	28	-	28	-	30	30	58
Senegal	-	-	-	-	-	0	0	0
South Africa	406	115	1,535	2,056	-	12,968	12,968	15,024
Tanzania	-	207	5	212	-	-	-	212
Uganda	-	-	-	-	-	35	35	35
Zambia	-	-	58	58	-	8	8	66
Zimbabwe	-	-	616	616	-	417	417	1,033
Africa	2,561	1,288	2,346	6,195	2	15,940	15,942	22,137

poached in 2022, with 448 illegally killed (451 in 2021).[10] This was up from 501 in 2021 and 503 in 2020, but both those years had lower than expected figures because of the effects of lockdowns, travel restrictions and other measures connected with the global COVID-19 pandemic. However, 561 is a vast improvement on the 1,352 African rhinos poached in 2015, the worst year in the twenty-first-century poaching crisis (see Table 8.2),[11] and Sam Ferreira, the Scientific Officer of the AfRSG, said that this was encouraging and the outlook was more positive than it had been for a decade.[12]

Between 2018 and 2021, 2,707 rhinos had been poached, with 90% killed in South Africa. The horns entered the illegal trade and made up the 'vast majority' of horns traded, with only a small proportion illegally taken from natural deaths, thefts from legal stockpiles or sold illegally from private stocks.[13] Between 2018 and 2020, CITES estimated that between 4,593 and 5,186 horns entered the illegal trade, of which between 2,418 and 2,869 were recovered through seizures, confiscation and arrests of smugglers and illegal traders. The COVID-19 pandemic and improved legal measures and law enforcement are believed by Knight, Mosweub and Fereira to have reduced the number of horns entering the illegal trade in 2020–1 from 1,531 to 1,729, compared with 2,378 in 2017. They also noted that the online trade in illegal rhino products was still operating, with a new product called rhino glue being advertised by dealers using the internet to spread their wares – it accounted for 27% of rhino products offered for sale online.[14]

What has become increasingly clear over the decade and a half of renewed poaching of rhinos is the multinational nature of the criminal gangs involved. As noted earlier, in Zambia there was evidence of Malian and Senegalese nationals acting as middlemen in the illegal trade, receiving horns from Zambian or Angolan poachers, while Somalis have always played a role in the trade in Kenya and Tanzania. In December 2022, further evidence emerged from the trial of the Guinean illegal wildlife trader Amara Cherif (also known as Bamba Issiaka) in the USA. He was sentenced to 57 months in prison for conspiring to traffic in rhino horns and ivory, and for the poaching of at least 35 rhinos. Cherif's co-conspirators, Moazu Kromah (also known as Ayoub, Ayuba or Kampala Man), a citizen of Liberia, and Mansur Mohamed Surur, a Kenyan, had already been sentenced to terms of 63 months and 54 months. They were part of a transnational illegal wildlife operation based in Uganda 'engaged in the large-scale trafficking and smuggling of rhinoceros horns and elephant ivory', and had also operated in the DRC, Guinea, Kenya, Mozambique, Senegal and Tanzania, obtaining horn and ivory and smuggling it to the USA and East Asia.[15] The court was told that the rhino horn they were caught with was worth $3.4m. As detailed in Chapters 5 and 7, Chinese, Taiwanese, Vietnamese and Middle Eastern middlemen and dealers selling to consumers were also involved in the trade, which spanned four continents. From 2018 to 2021 there were 1,588 arrests linked to rhino crimes, 751 prosecutions and 300 convictions in African rhino range states.[16]

Despite some improvement in law enforcement in some states, there remained a serious problem with the ability to use judicial systems, particularly in South Africa and Kenya, to convict those who commissioned poaching or other forms of illegal harvesting of horn, described as the kingpins of the illegal trade, who could string out court cases for years while out on bail and continuing their activities. Some evaded justice through bribery of magistrates and judges, while others, using skilled and expensive lawyers, found loopholes in poorly constructed prosecution cases and police evidence to gain acquittals. Those commissioned and armed to poach risked their lives and were the ones most likely to be arrested. For many years, they were given short sentences or fines that did not match the value of rhino horns, and the punishment did not match the crime or act as a deterrent. However, in South Africa, Namibia and Zimbabwe, sentencing began to be tightened up and the poachers received longer jail terms.

In South Africa, in the mid-2010s, there were advances in anti-poaching activity and increased arrests – rising from 67 in 2010 to 147 in 2014[17] – but with few convictions of major figures in the trade, a trend that continues. In Zimbabwe, research showed that by 2015, increased sentences and fines 'led to a reduction in the number of small-scale and subsistence poachers but had little impact on professional poachers who were financed by criminal networks'.[18] Kenya has a poor record, with only 4% of 743 cases involving wildlife between 2008 and 2013 resulting in jail sentences. Of those involving rhino or elephant poaching, 7% resulted in the jailing of offenders and 91% of fines imposed were below the maximum fine of KSh40,000 ($440).[19] In Tanzania, 'state corruption' enabled some poaching syndicates to operate with impunity, 'with members of parliament accusing a former donor of the ruling Chama Cha Mapinduzi party (Party of the Revolution) of using land allocated for legal hunting for the purpose of illegal poaching'.[20] Corruption in government, the judiciary and wildlife authorities in Tanzania has a long pedigree, and continues to be an obstacle to fighting wildlife crime.

Translocations

The period under review also saw an important series of translocations of both black and white rhinos to areas where rhinos had been rendered extinct by hunting and loss of habitat, including Malawi, Rwanda, Uganda, Chad and the DRC, and interest in plans for further translocations of both black and white rhinos to former ranges or to suitable areas not part of historic ranges, or where the northern white rhino had become extinct, for example to Uganda, Chad and the DRC; these included planned relocations of southern white rhino and eastern black rhino to Kidepo Valley NP in Uganda, southern white rhino to the NCA in Tanzania and the translocation of south-central black rhino to Karingani GR in Mozambique and both black and white rhino to Zinave NP there. Plans to translocate southern white rhino to Garamba NP in DRC, which had been the last stronghold of wild northern white rhino (details of the operation are noted later), came to fruition, with African Parks moving 16 white rhinos from private reserves in South Africa to Garamba, with more translocations planned.[21]

African Parks contracted with the respective governments to manage and develop parks works in Angola, Benin, CAR, Chad, the DRC, Malawi, Mozambique, the Republic of Congo, Rwanda, South Sudan, Zambia and Zimbabwe. Martin Rickleton, regional Operations Manager for African Parks, told me in July 2023 that translocations of black rhino to parks in Malawi, Rwanda and Chad had been successful.[22] The purchase in September 2023 by African Parks of Hume's Platinum Rhino operation will give the group a pool of white rhinos to use for future relocations of southern white rhino in former northern white rhino ranges, to run alongside restoring black rhinos to areas in which they have been depleted or rendered extinct by poaching.[23]

Another aspect of translocation has been moving rhinos within range states from vulnerable areas to more secure ones. In Kenya, this has taken the form of translocating rhinos to areas such as Tsavo, depleted of their rhinos by poaching and drought, but also to high-security sanctuaries where the level of protection is such that poaching is made very difficult, and where careful management of density and breeding is an important aspect of the 'national strategy to conserve the species as a meta-population'.[24] The dangers of trans-location as well as the benefits are examined in the following country sections. These were demonstrated during African Parks translocation of black rhino to Zakouma NP in Chad, when four of the first six rhinos relocated died, and the moving of black rhino from Lake Nakuru to Tsavo East NP, in which all of the 11 rhinos died – mainly from multiple stress syndrome, intensified by salt poisoning, dehydration, starvation and gastric issues, combined with 'clear professional negligence [which] took place at the release site, with poor communication between teams causing issues not to be acknowledged', and with one weakened rhino dying following an attack by lions.[25]

The ups and downs of private ownership: an overview

The 2000s saw an increase in the role of private safari reserves and game ranches in providing habitat and security for rhinos, continuing the trends started in South Africa, Zimbabwe and Namibia in previous decades. The private rhino sector in these countries has thrived for decades with income generated by tourism on some reserves, through trophy hunting and sales of live animals. South Africa permitted hunting of a sustainable offtake of 0.02–0.08% of its white and 0.13% of its black rhino populations per year, while Namibia allowed hunting of 0.37–1.78% of its white and up to 0.05% of its black rhinos annually between 2018 and 2021.[26] In 2018, across Africa, half of the white rhinos and a third of black rhinos were on privately owned land; 88% and 76% of Zimbabwe's black rhino and white rhino populations, respectively, were on private land, as were 27% and 75% of Namibia's black rhino and white rhino populations, respectively, with another 7% of Namibia's black rhino population on community conservancies, and 45% and 72% of Kenya's black rhino and white rhino populations, respectively, on privately owned conservancies, reserves or ranches.[27] In South Africa, the proportion of white rhino on private land increased from 25% in 2010 to 53% in 2021, but the number of black rhinos held privately increased from none before the 1990s to a peak of 27.4% in 2014, though this had declined in 2021 to 21.1%; custodian and community ownership contributed another 5.8%.[28] Pelham Jones, the chair of PROA, told me that according to data for 2022, out of about 13,500 white rhinos in South Africa, 8,000 were in private hands, while of 2,200 black rhinos, 750 were privately owned, meaning that over 60% of the national rhino herd was privately owned.[29]

Private ownership of rhinos, and other wildlife, has developed because of the creation in South Africa, Namibia, Zimbabwe and Botswana of enabling environments that permit sustainable use of wildlife and tourism as incentives for private owners to maintain wildlife habitat and a diversity of species. The threat to this sector of rhino conservation is poaching, with the changing focus of poaching from KNP and other NPs to privately owned rhinos entailing massive security costs.[30] Although tax incentives may be put in place in South Africa, there is no direct state support in any of the range states for management and anti-poaching costs for private owners, and a growing feeling that intensive captive rhino breeding, as at Platinum Rhino, is not the way forward.[31] The block to greater earnings for the state and private owners remains the CITES ban on trading internationally in rhino horn. The costs and obstacles to earning greater income from rhino ownership have led to a shrinking of the land set aside in South Africa and Kenya for private ownership.

The KWS, the County Government in Laikipia, the Ministry of Tourism, Wildlife and Heritage and the Association of Private and Community Land Rhino Sanctuaries (APLRS) 'have developed a draft strategy for rhino range expansion in Laikipia that could see up to 680,000 acres of land secured for rhinos during the next 15 years, including 'government, private and community sanctuaries under a collaborative and coordinated approach to conservation in a contiguous landscape'.[32] This is part of the Recovery and Action Plan for Black Rhinos in Kenya (2022–6), which aims to have 2,000 Eastern black rhino in suitable habitats by 2037, according to Jamie Gaymer, chair of the APLRS.[33] The key is securing appropriate habitat and funding for management and security, with Kenya more dependent than southern African range states on donor funding, while the latter utilise wildlife sales, tourism and trophy hunting, giving them more opportunities to generate income from rhino ownership.[34]

The ability of private owners to generate sufficient income could be threatened by pressure in Europe and North America to ban imports of trophies from hunting; in November 2023, Canada banned imports of hunting trophies that included rhino horn or ivory. CITES records that over 450 African elephant tusks, 16 rhino horns, 81 elephant trophies and 44 rhino trophies were legally imported into Canada between 2010 and 2021.[35] In July 2023, Belgium's government took the step of approving draft legislation to ban the import of hunting trophies of endangered animal species. Attempts by anti-hunting groups with the support of British MPs and the implicit backing of the Conservative government to ban the importing of a substantial range of hunting trophies, including rhino trophies and horns, failed to be passed into law but may be revived in the future. As Clements, Balfour and Di Minin cautioned, bans on trophy imports or trophy hunting could have serious effects on local livelihoods and rhino conservation in areas where regulated hunting brings in income for conservation and local community income.[36] In Chapter 10, the pros and cons of trophy hunting as part of a range of conservation measures and the southern African calls for the resumption of a legal rhino horn trade are examined in more detail.

East Africa

Kenya

Having been ravaged by poaching in the last decades of the twentieth century, with Kenya's rhino population (both black and white) falling from just over 15,100 in 1980 to 488 in 1992, numbers have climbed, and at the end of 2021 reached 1,811, with black rhino having grown from the low point of 414 to 938 and white rhino (including the two northern white females at Ol Pejeta) up from 33 in 1980 to 873 (see Table 8.3). In 2023, Save the Rhino reported the good news that in 2022 no rhinos were known to have been poached in Kenya and that numbers were estimated at just under 2,000, the world's third largest rhino population; this was only the second year, the other being 2020, in which no rhinos had been illegally killed there since 1999.[37] This was a vast improvement on the 2009–12 period when poachers still killed on average two rhinos each month, which was more than 2% p.a. of Kenya's rhinos since 2009. The government response then was to intensify anti-poaching, including converting rhino scouts on private rhino lands into Kenya Police reservists, and to relocate more rhinos from danger areas.[38]

The translocation of rhinos from areas where they were threatened by poaching to more secure sanctuaries and conservancies is an important part of Kenyan recovery plans. Following the relocations to sanctuaries in the 1990s and 2000s, NNP, Solio, Ol Pejeta and Ol Jogi remained important locations for both black and white rhino in the 2010s and early 2020s.[39] The Ngulia Rhino Sanctuary in Tsavo West continued to provide a secure habitat for black rhinos, following its expansion by 48% in 2008 to prevent too dense a rhino population from leading to lower breeding rates. The danger of over-crowding was also averted by relocating some rhinos into a surrounding IPZ in Tsavo West.[40] Relocation has security very much in mind, given that Kenya lost 145 rhinos to poachers between 2012 and 2016.[41] To further the national rhino recovery plans, in May 2020 the Kenyan government provided about $20m to support community wildlife conservancies and the KWS, which was intended, as Khayale et al. noted, to secure 'rhinos on community, private and State lands as well as sustaining suitable habitat for future rhino range expansion'.[42]

Kenya's rhinos were hard hit in 2013, one of the worst years so far this century for them, with 59 killed.[43] In 2014, improved security led to a 40% drop in poaching, with 35 rhinos illegally killed. Although this was an improvement, Ol Pejeta and Ol Jogi both experienced poaching of rhinos in 2013–14, despite extensive and expensive security measures.[44] In June 2014, for example, poachers entered Ol Jogi and killed four rhinos in one night.[45] In 2014, Ol Jogi lost nine rhinos, Solio lost seven and Lake Nakuru NP lost ten. In 2016–17, poaching had dropped, but Ol Pejeta still lost three rhinos to poachers, though stepped-up anti-poaching measures meant no rhinos were killed there in 2018. Ol Pejeta's website puts the population of rhinos there at over 165 black rhinos, 44 southern white rhinos and two northern white rhinos. Save the Rhino noted that the improved security came at a financial cost for Ol Pejeta, which now employs 150 rangers and a specially trained anti-poaching dog unit, which was a heavy financial burden especially given that Ol Pejeta, like other Kenyan tourism operations, lost significant income during the COVID-19 pandemic. However, this did have the effect of deterring poaching, with no poaching incidents reported in four consecutive years at the end of the 2010s.[46]

Jamie Gaymer, Chair of the APLRS, noted that at the end of 2023 Kenya was about to put into operation the Recovery and Action Plan for Black

Rhinos in Kenya (2022–6) to build on what had been achieved in reducing poaching in the preceding five years to less than 1% of the population per annum. However, he warned that Kenya was running out of secure areas to accommodate the growing rhino population, which was 'having negative consequences in the breeding performance of some populations and exacerbating negative rhino social dynamics' as the action plan worked to achieve its objective of having a population of 2,000 eastern black rhinos by 2037, with an initial aim of having 1,450 rhinos by 2030.[47] In an attempt to increase the land available, the Laikipia region was being viewed by KWS and APLRS as the most likely to increase the space for rhinos on privately and communally owned land. The plan suggested that 'There are significant opportunities for establishing new private and community sanctuaries in the Laikipia region … The longer-term strategy of merging the private and community sanctuaries that are in close proximity to each other should also be initiated during this Plan period.'[48] The objective is to increase the areas of contiguous land to allow expansion and dispersal of rhino populations there, with the hope that range expansion in Laikipia could see up to 680,000 acres secured for rhinos during the next 15 years.

In Laikipia, land is managed in a diversity of ways – wildlife conservancies, ecotourism ventures, sanctuaries for both black and white rhino, privately owned and government-owned livestock operations, group ranches that are communally owned and managed by pastoral communities, or smallholding agriculture. Much of Laikipia is unfenced, allowing wildlife to move freely between areas with different land uses, though in areas of Ol Pejeta, Solio and the other rhino sanctuaries there is some fencing and intensive protection of wildlife that is not found everywhere in Laikipia; the fencing was not designed to stop all movement of wildlife but to prevent entry and exit by rhinos.[49] Low walls with carefully spaced bollards have been built, which prevent rhinos from crossing but don't block elephants, antelopes or other mammals.[50] This method was used at Lewa Conservancy, which had a 142 km two-metre-high perimeter fence combining live electrical wires, but with a gap to allow movement of wildlife to and from the adjacent Borana Conservancy, which had wildlife but also cattle, and a gap leading to a 14 km elephant corridor to Mount Kenya National Forest Reserve. The walls and bollards exploit the inability of rhinos to climb over loose rocks or jump. 'Monitoring of the fences over three years showed that zebra, giraffe and elephant and smaller ungulates had no trouble in crossing but rhino could not. Predators like hyena, lion, leopard and jackal could cross; 37 species of mammal crossed the fence.'[51]

What is not clear in the planning for rhino range expansion in Laikipia is the extent to which there has been meaningful consultation with local

communities or whether there will be local community buy-in, given land
conflicts, grazing rights and questioning of the role and powers of privately
owned and NGO-linked conservancies in the region – which Mordecai
Ogada describes as 'a vehicle for elite capture of money, land and wildlife'
and what he and Mbaria labelled as the facilitation of 'a different form of
landgrab in which local people are led to falsely believe that they stand to
gain by setting up wildlife conservancies through a grand conservation
scheme that has no parallels'.[52] Kenya has more than 166 conservancies. They
cover an area larger than the country's NPs, and are home to more than 22%
of Kenya's ungulate wildlife biomass, with high densities of wildlife. Many,
such as Lewa, Ol Jogi and Ol Pejeta, are white-owned and were huge livestock
farms converted to high-cost safari tourism venues that include wildlife and
domestic livestock. The Northern Rangelands Trust (NRT) groups 43 conservancies, many of which are local institutions run for and by indigenous
people to support the management of community-owned land for the benefit
of improving livelihoods as well as conserving habitat and wildlife. The
NRT says that community conservancies give people a voice and provide a
platform for developing sustainable enterprise and livelihoods that are either
directly or indirectly related to conservation.[53]

There has been strong criticism of the NRT from within Kenya and by
conservationists at the US-based Oakland Institute,[54] who believe that conservancies have not served to empower local communities to both develop
their livelihoods and at the same time to further conservation goals in
ways that are sustainable. Ogada has focused much of his criticism on the
colonialist attitudes that he believes are an intrinsic part of the NRT and
much of the conservancy system in Kenya. In a paper for UNEP, he writes
that in Kenya lands utilised by pastoralists still had large wildlife populations
in comparison with other forms of agricultural land use, showing the compatibility of pastoralism with free-ranging wildlife, but that this compatibility
has not been at the centre of the conservancy movement in pastoral areas. In
Laikipia, the conservancy model represented by the NRT project, he argues,
has failed to include local communities as meaningful participants in the
development of conservation.[55] Mittal, Moloo and Mousseau have also argued
that many local communities in the Laikipia region do not believe that the
NRT and conservancies have served their best interests, but rather have 'dis-
possessed them of their lands and deployed armed security units that have
been responsible for serious human rights abuses', with NRT mobile APUs
accused of extrajudicial killings and disappearances.[56] The units received
paramilitary training and military-grade weapons from the KWS and from
a private security company, 51 Degrees. They have not only been engaged
in anti-poaching, but have also become embroiled in interethnic clashes.[57]

Whatever the precise roles played by APUs, their activities are viewed with suspicion and anger by many local people,[58] who are far from satisfied with what they have gained from the conservancy system and its model of conservation. This does not bode well for the aim in the rhino action plan to increase the area of land in Laikipia for rhino conservation.

In addition to the recovery and action plan for black rhinos, notably the eastern black rhino, Kenya has also adopted a White Rhino Conservation and Management Plan (2021–5) to encourage the growth of white rhino numbers, which have increased steadily since the first 51 animals were introduced between 1965 and 1994. In the first two decades of the twenty-first century, the population has grown at an average of 6.76% a year. By 2021, Kenya had white rhino resident at Il Ngwesi Community Conservancy Community, Lake Nakuru NP, Lewa–Borana Landscape, Meru NP, NNP, Nairobi Safari Walk, Ol Choro Oiroua Community Conservancy, Ol Jogi Conservancy/Pyramid, Ol Pejeta Conservancy, Ruma NP and Solio GR (Private) – with a national population of 873, including the two northern white rhino at Ol Pejeta.[59] The plan aims to keep poaching of white rhinos below 1% of the population and to monitor the populations across the parks, reserves and conservancies to ensure good biological management and prevent in-breeding.[60] The plan also notes the need to use assisted reproduction techniques, notably producing embryos *in vitro*, as the two remaining northern white rhino females are incapable of breeding even using sperm taken from the last living males.

One project designed to redistribute rhinos across secure areas and to re-establish them where they had become severely depleted ended in disaster, when all of the individuals translocated from Lake Nakuru NP to Tsavo East died very soon after release. All 11 perished – ten as a result of multiple stress syndrome, made worse by salt poisoning, dehydration, starvation and gastric issues, and the remaining one following an attack by lions within Tsavo East when it had been weakened by the stress of relocation and was very vulnerable. Save the Rhino reported that an official inquiry by Najib Balala, Cabinet Secretary for the Ministry for Tourism and Wildlife, noted that 'clear professional negligence took place at the release site, with poor communication between teams'.[61] The rhinos were supposed to be kept in a boma to acclimatise but were released early. Further planned relocations to Tsavo, with eight rhinos due to be moved from NNP, were then stopped. Former head of the KWS and former Chairman of the KWS Board, Dr Richard Leakey, issued a statement on 27 July saying that, prior to the end of his tenure as chair of KWS on 17 April 2018, the Board had noted that 'there was a deep concern about the lack of vegetation in the sanctuary that could sustain rhino and also, the real issue of available and safe water'. He said that his Board had not given approval for the translocation of rhinos into the Tsavo East sanctuary and that, to his knowledge, the new Board had not yet met; therefore, these decisions must have been made by the Ministry for Tourism and Wildlife, which 'is not provided for in the existing legislation governing KWS and its operations'.[62]

It is not only rhinos that have been in the crosshairs of the poaching gangs and the criminals who act as middlemen in wildlife crime, but also rangers and the commanders of APUs. The website of Saving Endangered Species Through Education and Justice (SEEJ), an organisation which was set up to monitor the conservation of Kenyan wildlife and report on the prosecution of those involved in the illegal wildlife trade, has highlighted the fact that rangers have died in firefights with well-armed poachers, and senior wardens and commanders of APUs have been killed in what appear to be targeted assassinations. On 31 August 2021, Warden II Banjila Obed Kofa, the commander of the rangers at Solio, had dropped his daughter at Nkuene Girls High School and was driving away when, on the outskirts of Meru, two Subaru Outback vehicles pulled alongside his car and sprayed it with bullets. 'It was hit at least 30 times … He was critically injured and immobile in the front seat. An occupant of the attack Subaru exited his vehicle. He was hooded and holding a semi-automatic handgun that was wrapped in hessian fabric … [He] shot Commander Kofa in the head three times.'[63] This was three days after the abduction and murder of Assistant Warden Francis Oyaro and a few months after the suspicious air crash in Nanyuki that killed KWS Pilot Ian Lemaiyan and Commander John Plimo. The Kenyan newspaper, *The Standard*, noted that Kofa was the fifth well-known 'environmental defender' to have been killed in a short period in Kenya.[64]

SEEJ said Kofa's killing 'was either a government-sanctioned execution, or an execution ordered by someone within government at a very high level', and that the motivation was that Kofa was 'either engaged in high level criminality, rhino horn trafficking for example, or he was preventing poachers and high level traffickers from easy access to the gold mine of rhinos within the Solio sanctuary. There is also a definite possibility that he had evidence as to the identity of the high level traffickers within government circles.'[65]

The SEEJ website noted that the Kenyan TV station, KTN, had broadcast an investigative report several years earlier providing evidence of a cartel of wildlife traffickers involving KWS rangers, police and government officials, and that the cartel was linked to the criminal syndicate of Moazu Kromah, known as Kampala Man, who was in prison in the USA for wildlife trafficking.[66] I have been unable to trace any reports relating to an investigation into Kofu's killing, and this once more demonstrates the combination of corruption, criminality and incompetence that remains a serious obstacle to conservation and the fight against wildlife crime in Kenya.

The inadequacy of the Kenyan judicial and law enforcement system was further demonstrated in August 2018, when a court ordered the release of wildlife trafficker Feisal Mohmed Ali, who had been jailed in July 2016 for trafficking in ivory and other wildlife commodities. He had fled Kenya in June 2014 after police seized 2,152 kg of ivory and issued a warrant for his arrest. With the help of Interpol he was arrested and extradited from Tanzania and jailed for 20 years. The 2018 judgment said that his trial was marked by serious irregularities in the original trial. The Mongabay website noted that Kenya's Office of the Director of Public Prosecutions and KWS were 'apoplectic with rage' at his freeing, as he had been seen as a driving force in Kenya's illegal wildlife trade.[67] Chris Morris, a former Canadian law enforcement officer and international criminal investigator, examined the case for Mongabay and concluded that the whole investigation and prosecution were deeply flawed – a senior police officer involved had raided the location where the ivory was found not to arrest wildlife traffickers but to extort bribes from them, leaving them with the ivory. The shakedown did not work and the police were then stuck with the ivory rather than a large cash bribe, and were forced to prosecute but without having gone through proper search, scene of crimes, forensic or evidence-logging procedures. There seems little doubt about Feisal Mohmed Ali's possession of illegal ivory or about the corruption and incompetence of the police, and the inadequacies of the prosecutors in bringing the case to trial in the way that they did.

Tanzania

The Tanzanian Minister for Natural Resources and Tourism, Pindi Chana, told the country's parliament in June 2022 that the Tanzanian black rhino population had been about 10,000 in the 1970s. It had declined to 3,795 by 1980 and dropped to 65 in the 1990s,[68] and then to 46 in 1997. Poaching in the SNP almost completely wiped out the black rhino there by 2007, but a few remained to form the basis of a hoped for recovery.[69] Tanzania's rhino numbers slowly climbed from that point in the 2000s to 161 in 2018 and 190 in 2020. At the end of 2021, the figure was put at 212.[70] The catastrophic decline was solely due to poaching.

While the Tanzanian wildlife authorities are publicly committed to conserving the remaining rhinos and trying to provide security and habitat for recovery in numbers, which has been progressing slowly, Minister Pindi Chana also said that they had plans to relocate a number of southern white rhinos to help to improve conservation and tourism in the NCA and in the Burigi-Chato and Mikumi NPs. She said that 30 white rhinos from South Africa would be translocated, but did not explain why stretched resources for conservation were being devoted to bringing in a species that had not been found in the areas of Tanzania into which they would be put. Sam Ferreira, the Scientific Officer of the AfRSG, told me that he could see some value in what he called 'novel ecosystem' approaches involving introducing rhinos into specific areas, but that in Tanzania the approach seems to be driven by tourism companies wishing to increase the appeal of their safaris – and so the motivation is commercial rather than conservation-driven.[71]

One area where efforts are being made to secure an existing but depleted rhino population and create conditions for its increase is the Grumeti safari concession area, adjacent to the western Serengeti, part of the wider Serengeti-Mara ecosystem. The reserve was created by the Tanzanian government in 1994 to protect the route of the wildebeest migration and the biodiversity of the whole ecosystem. In 2002, the Grumeti Community and Wildlife Conservation Fund (known as the Grumeti Fund), a not-for-profit organisation, was granted the right to manage and conserve the area, and in 2006 the Singita safari company took over management, with the aim of using high-cost, low-volume tourism to finance conservation, and bring in a profit for the company.

The Grumeti concession serves as a buffer between the NP and an estimated population of 90,000 on the boundary. The Grumeti Fund has set out to use the tourist potential and the conservation aims of the Fund to both restore a viable black rhino population and develop 'tangible benefits' for the communities neighbouring the concession.[72] The Fund operates APUs and tries to work with local communities to stop the poaching of rhinos and elephants and the snaring of antelopes, zebras and buffalos for meat. In the 2010s, some rhinos were poached, but increased security meant no rhinos or elephants were poached in 2019 and 246 poachers were arrested.[73] This was progress, which developed from the Black Rhino Re-Establishment Programme launched in 2017, which involved nine eastern black rhinos being translocated from South Africa to Grumeti with assistance from the Ministry of Natural Resources and Tourism and Tanzania Wildlife Management Authority and partners in South Africa. The presence of rhinos at Grumeti (along with elephants, lions, leopards and buffalos) enables Singita to offer tourists the chance to see the Big Five game species there. In addition to the black rhinos now in Grumeti, it was estimated in 2020 that about 40 eastern black rhinos were still surviving in Serengeti in the Moru region between the Serengeti plains to the east and wooded hills to the west.[74] In collaboration with the Frankfurt Zoological Society, the Tanzanian National Parks Authority (Tanapa) had established a rhino conservation project at Moru Kopjes, with a rhino protection post established to protect them. In September 2019, Tanzania reintroduced nine rhinos to boost the population of wild breeding rhinos around Moru and increase genetic diversity.

In August 2020, Tanzania celebrated the birth of a black rhino in the wild from animals reintroduced to the Serengeti.[75]

There are now black rhinos in a fenced rhino sanctuary in Mkomazi NP, bordering Kenya and the southern tip of Tsavo West NP, from which they had disappeared owing to heavy poaching in the 1970s and 1980s. The area was home to 200 eastern black rhinos in the 1960s, but by 1985 they had all been poached. The Tanzanian government began restoration efforts in the 1990s in collaboration with the George Adamson Wildlife Preservation Trust, with Tony Fitzjohn as field director.[76] The 54 km² Mkomazi Rhino Sanctuary was created as a fenced-off area of what was then the Mkomazi GR (it became an NP in 2008).[77] Four eastern black rhinos were translocated from Addo Elephant NP in South Africa to the Mkomazi sanctuary initially. By 2016, 11 black rhinos had been introduced from zoos across Europe to create a viable breeding population, and within a few years the population had risen through successful breeding to 35, with hopes that the number will increase to 50. The rhino sanctuary is open for tourists to visit, and some of the income is to be devoted to community projects in villages near the park, to incentivise local people to support conservation and so reduce the likelihood of people being drawn into poaching – with an encouraging factor being that Mkomazi has not experienced poaching for 22 years.[78] The Ministry of Natural Resources and Tourism has plans to reintroduce black rhinos to Arusha NP, where rhinos had been poached out by the mid-1980s, though no specific date has so far been given or the numbers to be translocated.[79]

Uganda

By 1984, Uganda's black rhino population had been rendered extinct through massive poaching over the previous decade and a half, and the northern white rhino had long since disappeared.[80] The southern white rhino had been introduced to the country and established within the Ziwa Rhino Sanctuary, a heavily protected and fenced 64.2 km² area in central Uganda, which had been established in 2005. Six white rhinos had been translocated there from Solio in Kenya and Disney Animal Kingdom in Florida.[81] At the end of 2013, the Ziwa population had risen to 13 – three adult males, three adult females each with a calf and four subadults.[82] Successful security and continued breeding increased the population to 21 by 2018.[83] In 2021, Ziwa's rhino warden, Moses Okello, said that the sanctuary, which had been increased in size to 70 km², now had 33 rhinos, and had plans to increase the population to 45 so that some could then be relocated to Uganda's NPs to create new populations of southern white rhino to replace the lost northern white rhino.[84] The first planned relocations are likely to be to Murchison Falls NP, where a feasibility study has established that there is suitable habitat. The long-term plan, according to Opio Patrick, a Uganda Wildlife Authority ranger at Ziwa, is to settle rhinos at 11 protected locations across suitable areas of the country.[85]

Alongside these positive developments, Uganda has worked to stop the smuggling of rhino horn through its airports and across the land borders. In May 2022, it was reported that a Yemeni national had been arrested at Entebbe International Airport trying to smuggle 16 kg of rhino horn worth Uganda Sh1.4bn ($368,284) to Yemen. The horns were detected hidden among a consignment of fruit by a scanner. Three years earlier, the notorious wildlife trafficker Moazu Kromah, Kampala Man, had been caught at Entebbe airport in possession of rhino horn and ivory.[86]

Central Africa

Chad once had thriving populations of northern white rhino and western black rhino, but they had been hunted to extinction by 1980 and 1990, respectively. Decades of civil war and incursions by Libya had rendered the country unstable and impeded conservation and anti-poaching work. The huge Zakouma NP in the south-east of the country, near the borders with Sudan and the CAR, was frequently raided by Baggara and Rizeigat (Janjaweed) poachers from Sudan, poachers from the CAR and Chadians made desperate by the ravages of war and poverty. The black rhinos and most of the elephants there had been poached for their horn and ivory, and bushmeat snaring was rampant. Zakouma's habitat mix of savanna, bush, forest and floodplains was ideal for rhinos, but insecurity and an inability to combat constant poaching had denuded it of wildlife. According to African Parks, who entered into a management agreement with the Chadian government in 2010, the park had lost all of its rhinos and 90% of its elephants to poachers. After the start of management by African Parks and 'with the overhaul of the park's conservation law enforcement and community engagement strategies', wildlife numbers began to increase, and by 2016 poaching was virtually eliminated.[87] Two years later, after careful planning, six black rhinos were translocated to Zakouma from South Africa. Unfortunately, four died within six months of arrival. None of the rhinos were poached, and post-mortem analysis did not indicate infectious disease or plant toxicity, but all four had low fat reserves; the statement by African Parks said this suggested that 'maladaptation by the rhinos to their new environment is the likely underlying cause, although tests to be undertaken on brain and spinal fluid may shed additional light on the exact cause of deaths'.[88]

The two remaining rhinos were carefully monitored and have survived. Martin Rickleton of African Parks told me in 2023 that 'they have successfully adapted to their new environment, marking the beginning of a thriving micro-population within the park. Building on this success, we are considering additional rhino relocations to Zakouma … A comprehensive assessment has been conducted, leading to the selection of an optimal release site that provides essential resources and diverse habitats. We have also taken seasonality into account.'[89] The intention to move more rhinos to Chad was confirmed to me by African Parks CEO Peter Fearnhead in June 2023.[90] The cost of translocations is huge as the rhino have to be captured, sedated, trucked to an airport and then flown 4,400 km from South Africa to Chad. Peter Fearnhead told the South African journalist Ed Stoddard that this cost in the region of $50,000 per rhino.[91] The cost did not deter African Parks from sticking to their plan, and in early December 2023 it was announced that five more black rhinos had been translocated to Zakouma in the hope of creating a breeding group that can be the nucleus of restocking this area with black rhinos.[92]

African Parks have also taken over management of the DRC's main rhino range park, Garamba, in the north-east of the country. The park had been the location of the last population of wild northern white rhino, declared extinct in the wild by the IUCN in 2008.[93] The IUCN decision followed last-ditch attempts to save the wild northern white rhino, beset as they were by a multiplicity of poachers, refugees from Sudan and the CAR, Congolese displaced by the wars in the east of the country, South Sudanese guerrillas from the SPLA, roving Janjaweed raiding and poaching parties and the remnants of the Ugandan LRA. The South African conservationist Lawrence Anthony attempted in 2006 to contact the LRA and meet its leaders, Vincent Otti and Joseph Kony, to persuade them to stop their followers poaching rhinos and elephants in Garamba and preventing the rangers from patrolling the park.[94] Talks with Otti had little effect, as in return for stopping poaching the LRA wanted South African intervention in the deadlocked peace talks and the dropping of cases against Kony, Otti and other leaders by the International Criminal Court.[95] The talks were dead in the water when in-fighting within the LRA led to Kony having Otti and other leaders killed. In January 2009 the LRA attacked the main ranger camp in Garamba, killing eight rangers before they abandoned the attack.

In 2005, African Parks signed an agreement with the Institut Congolais pour la Conservation de la Nature (ICCN) to manage the park. Not a great deal could be achieved beyond putting a management team in place and gradually restoring patrolling and conservation work, amid continuing insecurity resulting from LRA activities and other incursions into the park.[96] By 2016 the security situation had improved, and along with the ICCN, African Parks set about overhauling conservation laws related to the park and improving patrolling and law enforcement to stamp out poaching; it also began to work with local communities to convince them of the benefits of conservation and the potential income from future tourism, as well as from the stability that a better run park could bring to the region. African Parks says it was able to provide improved and expanded education and health facilities for local people.[97]

In terms of rhino conservation, the big breakthrough, building on the greater stability and security in the park, came in June 2023 when African Parks announced that southern white rhinos had been translocated to Garamba from private GRs in South Africa. In cooperation with the ICCN and the Canadian mining company Barrick Gold, which funded the rhino relocation, the rhinos were flown from South Africa to the DRC. The 16 rhinos were from &Beyond Phinda Private GR in KZN. They were airlifted from South Africa to Barrick's Kibali Mine airstrip in north-eastern DRC and then trucked to Garamba NP. African Parks's CEO, Peter Fearnhead, said 'This reintroduction is the start of a process whereby southern white rhino as the closest genetic alternative can fulfil the role of the northern white rhino in the landscape.' This viewpoint was supported by Sam Ferreira of the AfRSG, who told me that translocations like this should be seen in terms of what they could achieve functionally for rhino conservation rather than in strict taxonomic terms.[98] More southern white rhinos are expected to be sent to Garamba NP in the future. These, too, are likely to be sourced from South Africa, which, despite the heavy poaching there over the last decade, has the largest white rhino population at an estimated 12,968. The purchase in September 2023 by African Parks of Hume's Platinum Rhino operation has given the group access to over 2,000 white rhinos, which it now owns, and the possibility of large-scale translocations in the future to Garamba and other suitable locations, with Fearnhead describing the potential as exciting and globally strategic for the future of rhino conservation.[99] For the DRC and Garamba, the relocations are a big step forward in restoring white rhinos to the core of their former range – albeit a different subspecies. The head of the Congolese Institute for the Conservation of Nature, Yves Milan Ngangay, described the release of the rhinos and plans for future relocations as 'a testament to our country's commitment to biodiversity conservation.'[100]

DRC was not the only Central African former rhino range state to have rhinos translocated to establish new populations. Rwanda's Akagera NP had lost all of its black rhinos to poaching together with the effects of war and the aftermath of the 1994 genocide, with the last sightings of four rhinos occurring in 1998. As Rwanda started to rebuild its economy after the war and the killing of between 500,000 and a million people in the 1994 genocide, conservation and tourism became part of the recovery plan for the government of Paul Kagame. In 2010, with a view to rehabilitating Akagera and developing a safari tourism industry there to run alongside the lucrative gorilla tourism in the Parc des Volcans, the government signed a management agreement with, you guessed it, African Parks. The task was immense, as wildlife had been decimated or driven out by the influx of refugees fleeing the genocide and the final stages of the war that brought it to an end. At least 30,000 cattle had been brought into the park by displaced people, and poaching for food was rife. As poaching was brought under control and law enforcement improved, plans were put into effect to restore lions and black rhinos to the park between 2015 and 2017, and white rhinos were brought in during 2021.

Eastern black rhinos, for which Rwanda was part of the western limit of their range, were brought in from South Africa, with 18 being settled there, and with plans to expand the population with more translocations.[101] The programme was managed by African Parks and the Rwanda Development Board and funded by the Howard G. Buffett Foundation, the Dutch Government and the People's Postcode Lottery.[102] In 2021, 30 southern white rhinos were translocated from &Beyond Phinda in South Africa to Akagera, again with funding from the Howard G. Buffett Foundation. Ladis Ndahiriwe of African Parks told me that the translocation had gone well,[103] which is a good omen for further translocations if the black and white rhinos thrive. With Hume's rhinos now available, further relocations to countries such as Rwanda can be expected. Although southern white rhinos were never found there, as noted in previous chapters there are accounts of northern white rhinos there.

The translocations have been good publicity for both African Parks and the Rwandan government, the latter being keen to burnish its rather tarnished image arising from its incursions into DRC and support for rebels there, and its violent suppression of opposition politicians and dissent at home.[104] An expert on Rwandan politics, Michaela Wrong, told me that Kagame was very skilled at working out issues that would resonate with people in Europe and America – conservation being one that can make him 'look like a hero in the eyes of the Western world'.[105]

CHAPTER 9

Southern Africa's poaching storm

S outhern Africa became the eye of the poaching storm from 2009 onwards, but also saw translocations to repopulate parts of southern Africa depleted of rhinos and to bolster those in decline owing to renewed poaching. By 2015 there were 54 breeding populations of south-central black rhino in the region, with 1,580 animals, and further increases occurred over the next eight years. The eastern black rhino introduced, out of range, into South Africa had grown to 93, despite many being translocated to parts of its former range; by the end of 2015 there were 1,913 black rhinos in South Africa and eSwatini, In 2022, CITES estimated that there were 12,968 white rhinos and 2,056 black rhinos in South Africa at the end of 2021, 48 black and 90 white in eSwatini, 23 black and 242 white in Botswana, 56 black in Malawi, 2,156 black and 1,234 white in Namibia, 58 black and eight white in Zambia, and 616 black and 417 white in Zimbabwe.[1]

From 2009 to 2021, poaching in southern Africa increased, with South Africa hit the hardest and KNP bearing the brunt, most rhinos poached in South Africa being killed there in the mid- to late 2010s. Poaching led to a fall in white rhino numbers from 10,621 in KNP in 2011 to a maximum of 1,988 at the end of 2022; black rhinos were down from 415 in 2013 to 208 at the end of 2022.[2] Poaching levels in South Africa fell in the latter part of the 2010s, as did KNP's losses and its proportion of rhinos killed, with poaching falling from 504 in 2017 to 247 in 2020, 195 in 2021 and 124 in 2022.[3] By the early 2020s, the focus of poaching was moving from KNP, where 124 were killed in 2022, to KZN, where 244 were poached. In 2023, 499 rhinos were killed by poachers in South Africa. This disappointing figure is 51 (37%) up on the poaching numbers for 2022. It is a chilling reminder that the South African government and wildlife authorities have not got poaching under control. Releasing the annual poaching figures, the Minister of Forestry, Fisheries and the Environment, Barbara Creecy, said that while poaching had dropped in KNP, previously the focal point of rhino crime, 'pressure again has been felt in the KwaZulu-Natal (KZN) province with Hluhluwe-iMfolozi Park facing the brunt of poaching cases,' with 307 of the 499 rhinos killed there – 244 rhinos were poached in the whole of KZN in 2022.[4] In KNP, 78 rhinos were killed in 2023, compared with 124 in 2022 – a decrease brought about by better anti-poaching measures, but also a much smaller rhino population owing to over a decade of heavy poaching.

Despite these depressing figures, black rhinos increased in numbers regionally and, despite a very bad year in 2022, Namibia's rhinos were not as badly affected by poaching, while Botswana experienced severe losses in the latter half of the 2010s owing to poaching, internal conflict and poor management following the transition from Ian Khama's presidency to Mokgweetsi Masisi's. In Mozambique there was some light at the end of the tunnel with translocations of rhinos to Zhinave NP and other parks, while Malawi benefited from improved management of its parks and transloca-tions of black rhino from South Africa. Zambia was able to continue with the project to build on introductions of black rhino to North Luangwa NP, and Zimbabwe, despite several years of quite high levels of poaching, saw its overall rhino population rise from 785 in 2007 to 1,033 in 2021, perhaps partly because the focus of rhino poaching switched to South Africa. However, poaching had increased in Namibia by 93% to 87 in 2022, up from 47 in 2021 – most of them killed in Etosha NP.[5]

Despite the rise in poaching in Namibia in 2022, at the CITES CoP 19 in November that year, Namibia had its white rhinos transferred to Appendix II of CITES trade regulations, thus allowing international trade in live animals for conservation. South Africa and eSwatini had already been granted this dispensation at CoPs in 1995 and 2005, respectively. The deliberations at CoP 19 said that Namibia's white rhino population was growing at 6.7% p.a. The same CoP rejected another attempt by eSwatini to allow it to trade in rhino horn; it had also been rejected by the 17th CoP in 2016.[6] The southern African rhino range states, all members of the SADC, are united in trying to get the restoration of a legal, controlled international commercial trade in rhino horn, something that would bring in significant income for conserva-tion and rural development, but which is unlikely in the foreseeable future to garner sufficient support to lift the CITES ban. Pelham Jones of PROA, who is lobbying for a legal trade, is hopeful that a united SADC stand could make progress in getting a legal trade in the long term.[7]

Angola

In Angola there were estimated to be three white rhinos in 2021. The large black rhino population in south-eastern Angola was wiped out during the mass harvesting of rhino horn and ivory by the South African army and UNITA rebels in the 1970s and 1980s. The IUCN Red List for the south-western black rhino suggests that some survived, but doesn't give an estimate of numbers.[8] The presence of landmines in large areas of the south-east, including Mavinga NP and Luengue-Luiana NP, left from decades of civil war and South African destabilisation, has made surveys almost impossible, and some black rhinos may survive in areas that have not been surveyed for decades. Work has been underway since 2002 to clear landmines in southern Angola, with $60m committed by the government in 2019 for a five-year clearance project. As the mine clearance NGO HALO has said of wildlife in southern Angola, the presence of landmines 'makes it almost impossible to apply the conservation measures needed to protect this vital resource', with anti-poaching action and the monitoring of species very hazardous.[9] The availability of military-grade weapons from the civil war and the extreme poverty in many areas make poaching an inescapable resort for many who are trying to survive with little economic support from a massively corrupt, oil-rich government.

There is the hope that with African Parks taking over management of Angola's Iona NP in December 2019 (on the southern coast bordering Namibia), rhinos may be reintroduced there, as the organisation is committed to repopulating the parks it manages with rhinos.[10] Desert-adapted black rhinos are present over the border in Namibia's Kunene region, and the habitat is suitable. On its web pages about Iona NP, African Parks states that it is hoped that Iona 'will also be the recipient of reintroductions of black rhino, lion, and possibly even, elephant, to restore this spectacular desert park to its historic glory'.[11]

Angola plays a negative role in large mammal conservation, being a hub for the illegal wildlife trade on the Atlantic coast of southern Africa. There are many Chinese resident in Angola as diplomats and managerial staff or as labourers on projects being built with Chinese funding. They are a ready market for wildlife commodities, which are smuggled back to China.[12] Angola has a ban on rhino horn and ivory trade domestically, and as a CITES member on paper adheres to the 1977 ban on international trade in rhino horn. The CITES CoP in Panama in November 2022 noted that Angola was a player in the illegal rhino horn trade and had made 13 seizures of horn between 2014 and 2021. It was a hub for trade in horns 'sourced from countries other than South Africa, including Kenya, Botswana, and Namibia'.[13]

Botswana

The black and white rhinos that survived the poaching onslaughts at the end of the twentieth century were augmented by the translocation of 33 white and five black rhinos from South Africa between 2001 and 2003, bringing Botswana's population to an estimated 80, distributed across a wide area.[14] After 2009, as regional rhino poaching increased rapidly, Rhino Without Borders, a South African NGO, received permission from the Botswana government, despite misgivings within the DWNP,[15] to relocate 87 white rhinos from South Africa to the Okavango Delta. Conservation efforts and lower levels of poaching in the 2010s helped Botswana's rhinos to recover in number, and they were boosted by further translocations from South Africa. In late 2019 it was estimated that Botswana's rhino population exceeded 500, following the reintroduction of 215 white rhinos since 2015 and population growth among survivors of previous waves of poaching. Wilderness Safaris relocated a significant number of black rhinos to Mombo from South Africa and Zimbabwe in 2015.[16] Emslie et al. believed that Botswana had the fourth largest white rhino population on the continent.[17]

However, rhino poaching was increasing throughout southern Africa, and elephant poaching was increasing in northern Botswana following the decision by President Ian Khama to suspend trophy and commercial hunting.[18] Amos Ramokati, the regional wildlife officer for the DWNP in Maun, and Michael Flyman, the DWNP's wildlife census head, both told me that following the 2014 hunting suspension there was an increase in the number of local people assisting poachers, and that this was partly to blame for an increase in poaching.[19] The implicit social contract between rural communities – who benefited from selling hunting quotas, getting employment with hunting operators and receiving meat from the hunts – and the wildlife and law enforcement agencies had broken down, and people were hunting for meat and helping poachers coming into northern Botswana from Zambia via Namibia.[20] Trophy hunting, though controversial and opposed by many, including President Khama and animal rights NGOs, was an important source of income for rural communities living alongside Botswana's wildlife. The Sankuyo community, on the edge of the Okavango, earned significant income from selling the hunting quota for its land, receiving meat from animals hunted and employment with the hunting operators. The income provided cash dividends for every household, 101 jobs, help with funeral costs for residents, educational scholarships, social housing and old age pensions.[21] When Khama suspended hunting, communities such as Sankuyo lost income and jobs. Chief Timex Moalosi of Sankuyo told me in May 2018 that the hunting suspension had cost the Sankuyo community trust $600,000 a year from the sale of hunting quotas, and they also lost meat supplies. The grievance felt by rural communities that lost income and jobs may have been a motivation for those who engaged in poaching or helped poaching gangs to find rhinos and elephants.

The rhino relocation programme from South Africa to Botswana was initially linked to the formation of an elite rhino protection squad by the government, but it came at a time of increased poaching in the region as a whole. By 2017, the Khama government had been forced by economic circumstances to cut funding for the protection squad, which had no fuel for its patrol vehicles. The Environment Minister Tshekedi Khama II, the president's brother, made clear that Botswana's reputation for shooting to kill poachers was now the main defence for the rhinos.[22] He bemoaned the poor response from donor countries and agencies who encouraged the setting up of the specialised paramilitary protection and intelligence-gathering unit to protect the rhinos being relocated from poaching hotspots in South Africa and Zimbabwe.[23] The relocated rhinos became targets for poachers infiltrating Mombo and the Delta. Declining rhino numbers and intensified anti-poaching measures in South Africa led poachers to seek new sources of horn, infiltrating northern Botswana and taking advantage of the local effects of the hunting ban and the growth in factionalism in the wildlife and law enforcement sectors following Mokgweetsi Masisi's assumption of the presidency in 2018. Ian Khama rapidly fell out with his successor, partly because Masisi responded to popular demands for the lifting of the hunting suspension, and Khama worked to undermine him. There was conflict between the BDF, police and wildlife authorities over failures to protect rhinos or catch poachers, and there was suspicion that wildlife officials were giving rhino locations and anti-poaching patrol details to poachers.[24] Factionalism between and within APUs, military and police hindered anti-poaching efforts. The BDF was deemed to be very ineffective at anti-poaching.[25] Botswana became a soft target for Zambian poaching gangs, who crossed the river and swamps along the Namibia–Botswana border. In northern Botswana they were sometimes helped by local Batswana who guided them to the rhinos. Poached horns were taken out through Namibia into Zambia.[26]

Botswana's rhino numbers fell substantially, with the 452 white and 50 black rhinos in Botswana at the end of 2017 declining to 242 white and 23 black rhinos by the end of 2021. Poaching had started to rise in 2017, with nine poached in that year; 18 were poached in 2018, 31 in 2019 and 55 in 2020.[27] At least five poaching gangs were operating in and around Wilderness's Mombo reserve – seven poachers had been killed by the BDF or APUs but no major poachers were brought to trial.[28] The Zimbabwe-based Bhejane Trust, which monitors rhinos and poaching, wrote in March 2020 that 'relentless slaughter of rhino continues unabated in the Okavango Delta and at the current rate there will be no rhino left in the Delta within a few months'.[29] The Trust believed that protection was sadly inadequate and that more than two rhinos a week were being killed by poachers. It stated that

official estimates of 47 having been killed in a few months were too conservative, and that the population was probably down to about 195 white rhinos
and six black rhinos in the Okavango.[30]

The problems in deterring or catching poachers became obvious as the
wildlife authorities became ever more sensitive and splits between the army,
police and DWNP came out into the open. In March 2021, the Director of the
DWNP, Kabelo Senyatso, demanded that the media verify anti-poaching data
with them before publishing anything, to avoid misleading statements 'that may
have a negative implication on Botswana';[31] it was feared that rhino poaching
was tarnishing Botswana's reputation as a safari destination. In August 2020,
Cyril Taolo, Acting Director of the DWNP, admitted that 56 rhinos had been
killed between August 2018 and August 2020. However, Simon Espley of Africa
Geographic said his sources believed that between April 2018 and early 2021
between 100 and 140 had been killed, while some sources said that another
60 rhinos were unaccounted for and were 'almost certainly poached', which
could mean up to 200 had been poached in total, and there were reports of a
BDF helicopter pilot guiding poaching gangs to the rhinos.[32]

On 11 March, the Botswana newspaper *Mmegi* reported that a soldier
and a poacher had been killed in a firefight on Chief's Island in the Moremi
reserve, noting that 11 poachers had been killed by the BDF in 2019.[33] When
five poachers escaped a BDF patrol in Chobe NP soon after, there were re-
criminations between the army, police and DWNP, with the BDF fiercely
criticising the others for releasing information to the press.[34] The row fuelled
speculation that there had been insider involvement in rhino poaching,[35] and
local sources told me that it was suspected that there was involvement in
poaching of a special commando unit that was aggrieved at not getting a pay
increase. The lack of cooperation between the BDF and the police and the
conflict between Masisi and Ian Khama meant that anti-poaching was poorly
managed and had lost direction.[36] Khama accused the Masisi government of
disarming APUs, which was untrue. What had happened was that Ian and
Tshekedi Khama had armed units with military-grade weapons, against the
country's firearms regulations; these weapons had been withdrawn by Masisi
and replaced with the semi-automatic weapons normally given to the units.
The Botswana-based wildlife conservationist and veterinarian Erik Verreyne
argued much the same in his perceptive analysis published in the South
African *Daily Maverick*.[37] He added that security for the rhinos could only
be achieved by concentrating the population in smaller areas, 'where we can
concentrate our defences optimally'.[38]

In 2020, the Botswana government revealed that 92 rhinos were killed
in 2019 and 2020, but it did not release figures in 2021 or 2022. Zambian
poaching gangs were blamed for most killings, but there was little cooperation

between the Zambian authorities and Botswana because dozens of Zambians had been shot by the BDF in anti-poaching operations, many of whom were claimed by the Zambian government to be fishermen and not poachers.[39] The failure to stop the poaching in the Okavango, Mombo and surrounding areas led to the capture and relocation of all the rhinos that could be found and the dehorning of rhinos to make them less of a target for poachers. This was not wholly successful, and in August 2022 a white rhino that had been moved to the Khama Rhino Sanctuary near Serowe was killed by poachers and had its horn removed; the DWNP denied this, despite local journalists citing sources within the DWNP about the incident.[40] In February 2023, the Assistant Minister of Local Government and Rural Development, Mabuse Pule, admitted that two rhinos had been poached in the sanctuary in October and November 2022, despite the supposedly tight security.[41] The Botswana-based wildlife vet Erik Verreyne told me in September 2023 that 12 rhinos had been poached in the Khama sanctuary since July 2022, and that two or three black rhinos were still at Mombo; 30 white rhinos had been moved to Mathlophuduhudu in the west of Botswana, and most remaining black rhinos were moved to Orapa.[42] Botswana's Director of Wildlife and NPs, Kabelo Senyatso, said at the CITES CoP in November 2022 that after the relocation and heightening of security, rhinos were very safe and poaching had been massively reduced.[43] However, the poaching at the Khama sanctuary is a worrying sign that despite increased security and translocation away from the northern borders, poaching is not over despite government optimism, and that Batswana connected with conservation may be involved.

At the CITES CoP in November 2022, the Botswana government disclosed that a total of 138 rhinos had been poached between 2018 and November 2022. Botswana reported to the CoP that the DWNP said there were 285 white rhinos and 23 black rhinos secured on several well-protected reserves and sanctuaries away from the Okavango, and that over 100 white rhinos had been dehorned.[44] CITES gave a lower estimate of 242 white and 23 black rhinos.[45]

eSwatini

After the poaching of the late 1980s and early 1990s had subsided, better security and law enforcement enabled an increase in the national population of black and white rhinos to 90 in 2015, though it fell to 66 in 2017 as a result of several years of drought. After the drought the population increased to 98 animals by early 2022, and it is currently growing at an annual rate of 8%.[46] The kingdom's rhino population is in two locations – Hlane Royal NP and Mkaya GR – but some animals may be translocated to the Mlilwane Wildlife

Sanctuary. There is a strong commitment by King Mswati III to conservation of the rhino as a national symbol of wildlife protection.[47] Only three rhinos have been poached in the last 12 years – two in 2011 and one in 2014.[48] The costs of protection are high for a small and relatively poor country, and where the monarch has a lavish lifestyle at the expense of the rest of the population.[49] The need for funding for security and rhino management led the eSwatini government to request that CITES downlist its white rhino, enabling horn obtained from dehorning and natural mortality to be legally exported to bring in income for conservation.

The Swazi proposal was considered at the 17th CITES CoP in Johannesburg in September and early October 2016. It led to fierce debate, with most southern African rhino range states (excepting Botswana) supporting the Swazi position but with the European Union, the USA, Kenya and other CITES members opposed; they defeated the proposal by 100 votes to 27 with 17 abstentions.[50] The pro-trade group had argued that horn from Swazi rhinos could be used sustainably to fund conservation and local economic development around the park and reserve. Debate centred on the critical need for sustainable funding for rhino conservation, the potential impact on demand-reduction efforts should trade be legalised, whether or not CITES and its trade ban has failed or worked for rhino conservation, the importance of community livelihoods and concerns about the risk of illegal laundering of horn if a legal trade was not properly controlled.[51] The head of eSwatini's park and reserves, Ted Reilly, said that Swaziland would have used the funds from the sale to increase protection and conservation measures, and provide incentives for local people to support their efforts. Addressing the CoP, he reminded delegates of the financial and human costs of protecting rhinos, and said that the ban was not working and a regulated trade was the only answer.[52] Reilly had hoped that a legal trade would have raised approximately $9.9m at a wholesale price of $30,000 per kilogram, which would have been placed in an endowment fund to yield approximately $600,000 annually and made a huge difference to resources available for protection, as well as allowing the development of wider conservation programmes in protected areas and benefiting staff and local communities.[53]

Western conservation and animal welfare NGOs were jubilant about the vote against Swaziland's proposals. Kelvin Alie, director of wildlife trade at IFAW, said: 'At a time when rhinoceros are more under threat than ever from poachers due to rapidly increasing black market prices in their horn, this decision by parties to deny Swaziland's request to trade in white rhino horn is to be applauded.' His view was echoed by the head of Born Free, Will Travers. Shortly before the CoP, Travers had debated the pros and cons of a legal wildlife trade with John Hume in London, and made it clear that he and many other NGO activists in the conservation field would never support a legal trade – even in natural mortality horn or horn removed without injury to the rhino to deter poachers from killing the animals.[54] I attended the debate, and it was clear that there was total opposition of animal rights groups to trade or sustainable-use approaches to rhino conservation under any circumstances.

Malawi and Mozambique

In 2000, Malawi had six black rhinos in a carefully guarded sanctuary in Liwonde NP. After African Parks took over management of Majete NP in 2003, two rhino bulls were taken from Liwonde to Majete;[55] five females and another bull were moved there in 2007.[56] Through protection, successful breeding and translocations of rhinos from South Africa, by the end of 2021 Malawi had 56 south-central black rhino.[57] African Parks, in collaboration with the Malawian wildlife authorities, managed to massively reduce poaching (whether for horn, ivory or bushmeat) and successfully reintroduced 3,000 mammals of 15 species to the depleted NPs. The rhinos in Liwonde have been breeding successfully and the first calf was born in Majete in 2008.[58] In 2015, the organisation contracted to manage Liwonde. In November 2019, 17 black rhinos were flown from iMfolozi GR in South Africa to diversify the Liwonde rhino population, while one rhino was taken from Majete to Liwonde and two from Liwonde to Majete. African Parks announced that no rhinos or elephants had been lost to poaching in its parks between 2016 and 2019.[59]

In the early 2000s, Mozambique had fewer than 20 black rhinos and no white rhinos, following the decades of war and heavy poaching. Poaching continued in the 2000s, and the Sabie GR in Mozambique, bordering KNP, lost at least 20 rhinos between 2005 and 2010, some of them having dispersed into Sabie from KNP.[60] By the end of 2021, through translocations of white rhino there were 14 white rhinos but only two black rhinos in the country.[61] Poaching was still a problem in many areas of Mozambique, though there were so few rhinos that illegal killing concentrated on elephants and ungulates for meat. The poaching of rhinos involving Mozambicans took place in KNP, with poachers living on the edge of the Mozambican section of the Greater Limpopo Transfrontier Conservation Area crossing into KNP to poach rhinos.[62]

In his study of the rhino horn trade, Rademeyer identified villages near the transfrontier park boundaries that were centres for recruiting poachers, notably Canhane. There are few job or other income-generating opportunities there, and many villagers try to get into South Africa through KNP to find work; others are recruited to poach rhinos. One villager cited by Rademeyer earned R3,000 a month in the village, but would get between R15,000 and

R80,000 in 2012 for a set of horns, depending on their size.[63] Poaching in Kruger was risky, and became more so as anti-poaching operations improved and the South African military became involved. There was also poaching on the Mozambican side of the transfrontier park. Twelve white rhinos had been moved from KNP into the Mozambican side between 2001 and 2008, and became targets for the poachers; camera trapping from 2019 to 2021 did not reveal any white rhinos remaining in the LNP section of the transfrontier park.[64]

Between March 2010 and March 2012, at least 12 Mozambican men were shot in KNP after having been tracked down and identified as suspected poachers – out of 21 poachers shot in that period. The head of security at KNP, Ken Maggs, said that the high price of horn, cash paid for it by middlemen in the trade and unemployment in Mozambique meant that middlemen had no trouble finding men who were willing to risk their lives poaching – the lure of cash income meant that shooting poachers had no real deterrent effect.[65] Rademeyer wrote that in Massangir in Mozambique, adjacent to the southern end of KNP, a shop and taxi business owner, Jonas Mongwe, linked to the poaching gangs said that the poaching syndicates would visit the town and provide guns for men to poach in KNP, and there was a string of villages along the park boundary from which poachers were recruited and sent into KNP.[66]

Hübschle carried out a study of the origins and motivations of poachers from Mozambique, which is worth reading in full to get a detailed picture of who is recruited, how, why, and how the gangs operate.[67] She found that in 2013–14 about 70% of rhino poachers operating in KNP were Mozambican and entered along the eastern boundary. By 2016, because of increased security at crossing points, they entered through South Africa having crossed the border outside the park. They were motivated by a number of factors. One was community grievance over expulsions from the LNP when the transfrontier park was formed, with over 7,000 people deprived of land and much of their livelihood from raising cattle, with a consequent loss of food security. This led to a feeling that for the Mozambican and South African authorities wildlife was more important than people, and there was no incentive for the affected people to support conservation or refrain from poaching, bushmeat hunting and gathering firewood within the park. Rhino horn brought in more money than a Mozambican villager from these communities could earn in a year.[68] To add to this, rhino poaching kingpins who lived in or visited the villages along the park border were seen as Robin Hoods, stealing rhinos from the rich and distributing benefits to the poor villagers through payments to poachers, injections of cash into communities, payments to local chiefs or political officeholders, and providing cows to slaughter and beer to celebrate the return of poachers with horns. Despite the lure of poaching and the inability of the Mozambican authorities to raise living standards, provide incentives to support conservation or smash the poaching gangs, there have been some positive developments, with white rhinos being translocated to an 18,500 ha fenced sanctuary in Zinave NP, east of Gonarezhou NP, which straddles the Zimbabwe–Mozambique border. In July 2022, 19 out of a planned 40 white rhinos arrived in Zinave from South Africa – the first rhinos in Zinave for 40 years.[69]

Namibia

In Namibia, the white rhino population grew at an average annual rate of 6.7% between 2002 and 2021. The population started with 16 animals imported from South Africa in 1975 and now numbers between 1,234 and 1,237, alongside a population of south-western black rhino of 2,156 in 2021, over three times the population of 707 in 1997.[70] However, poaching is a threat. For many years it has seemed to be falling, despite occasional spikes. In 2012, only two rhinos had been poached; this rose to 25 in 2014 and jumped to 101 in 2015,[71] then numbers fell back to 55 in 2017 but rose again to 83 in 2018, and fell to 57 in 2019, 42 in 2020 and 44 in 2021.[72] However, 2022 became a disastrous year for poaching, with indications that many of the rhinos poached that year in Etosha NP were killed with the collusion of park staff.

In June 2022, the first warning signs appeared when 11 black rhinos were killed in two weeks in Etosha. The PR head for the Ministry of Environment, Forestry and Tourism (MEFT), Romeo Muyunda, said that ranger patrols had found 11 dead rhinos with their horns removed and that 22 rhinos had been poached in the country since the beginning of the year.[73] The year got worse, and 93 rhinos were killed in total. Only five horns from the 93 dead rhinos were recovered and only 80 arrests made.[74] Forty-six of the rhinos were killed in Etosha and about 61 black rhinos were killed across Namibia. Of the rhinos killed outside Etosha, 15 were poached on custodianship farms and 25 white rhinos were killed on private farms.[75] To the relief of the conservancies, no rhinos were killed on communal conservancies in 2022.[76] By May 2023, 16 rhinos were reported to have been poached in the year's first five months, half of them in Etosha and the other half on private farms, but none on conservancies. Muyunda, after announcing the killings, said that eight suspects had been arrested over poaching in Etosha, though no details were given of their relationship to the wildlife authorities in the park.[77]

One of the most worrying aspects was that Minister of Environment, Tourism and Forestry Pohamba Shifeta announced that his ministry was investigating some of its own employees working in Etosha, as the poaching was suspected of being 'an inside job', with staff working for poaching syndicates.[78]

A new anti-poaching head was appointed for Etosha with the mandate to crack down on poaching and investigate staff involvement. After the announcement of the overall rise in rhino poaching, Kenneth Uiseb, Deputy Director of Wildlife Research and Monitoring at MEFT, told me that 'it is a real pity that we continue to see poaching of rhino in the flagship national park … I believe a change in strategy and involving more the communities neighbouring the park holds the solution to poaching.'[79] Another major problem in combating poaching has been the backlogs of cases and slow judicial processes that mean those arrested may be on bail and able to continue poaching for a considerable time. MEFT said in its 2023 annual report that of 682 suspects arrested since the start of 2015, only 40 had been convicted by 15 May 2023, and that 'businessmen and prominent members of society who have been arrested for rhino-horn trafficking are able to secure excellent legal representation that is able to block rapid convictions'.[80]

MEFT continues to work closely with the SRT, which is playing a major role in rhino monitoring, generating funds from rhino tourism, ensuring involvement of local communities and maintaining the presence of its rangers in the rhino-inhabited areas of Damaraland/Kunene, with patrolling effort up 360% from 2012 when poaching started to increase.[81] The cooperation between SRT, MEFT and the conservancies has maintained the successful rhino tourism system, with SRT guides taking small groups of tourists to track rhinos on foot, especially from the Wilderness Safaris Desert Rhino camp in Damaraland.[82] In the six years up to 2023, the rhino encounters have earned over $1m, which led to a 340% increase in the employment of local rhino rangers; Knight, Mosweub and Ferreira believe that this has contributed positively to the reduction in poaching in the conservancies where SRT operates.[83] The conservancy system plays a major role in providing a safe habitat and in lessening human–wildlife conflict in Damaraland and Kunene, to the benefit of the rhino. By 2023 there were 86 conservancies registered, covering 20.2% of Namibia's land (166,045 km²) and 238,701 inhabitants, with streams of income ranging from rhino tourism and joint-venture tourist lodges to trophy hunting, regulated harvesting of medicinal plants, such as devil's claw, and livestock husbandry. Access 'to alternative income streams will become increasingly important' for people living in these areas, given the marginal land that cannot support large numbers of livestock or crop production, as climate change endangers some livelihoods.[84] Some academics, such as Sullivan, have criticised the conservancy system as being driven by expatriates and not necessarily in the interests of Namibians, effectively a continuation of old conservation policies with a façade of community ownership, and have strongly attacked the use of trophy hunting as a conservation tool.[85] Muntifering et al. believe that 'these conservancies have probably contributed to a decrease in poaching and a general widespread increase in wildlife on communal land, including threatened mammals such as the black rhinoceros'.[86] They note that since 'the adoption and expansion of joint-venture tourism enterprises, the rhinoceros population has more than doubled and sustained consistent positive growth rates despite persisting almost entirely on formally unprotected lands'.[87] Torra Conservancy, which houses Wilderness's Damaraland Camp and has a share, with Palmwag Concession, in Wilderness's Desert Rhino Camp, gains significant income from rhino tourism in addition to hunting – it earns about $3,542 from trophy hunting and $326 from hunting for meat annually.[88] Desert-adapted rhinos in Damaraland/Kunene are not hunted. The conservancy employs 243 local people, some of them game guards. Derek de la Harpe, Corporate Affairs head at Wilderness Safaris, told me that before COVID-19 reduced tourist income, Torra earned $225,442 p.a. from the payments from Wilderness, including wages for staff, most of whom come from the region.[89]

In addition to rhino-viewing tourist income, Namibia's rhino conservation policies include limited and regulated hunting of rhino on private reserves, according to quotas agreed with CITES. The legal trophy hunting of Namibian rhinos has proved sustainable, with small proportions of populations hunted each year – both black and white rhino.[90] In 2004, CITES allotted Namibia a quota of five black rhinos that could be hunted annually on special permits as trophy animals. The animals hunted are all males past breeding age, 'the rationale being that these animals may in any case be terminally injured as they are displaced by younger bulls who take over their territory',[91] and 'Removing old bulls from the population also increases the rhino population growth rate', as old bulls can obstruct breeding by younger males, which might be killed or injured in territorial fights.[92] Brown and Potgeiter concluded in 2020 that trophy hunting of rhinos (in addition to hunting on a sustainable basis of a wider variety of antelopes and some predators) brings local communities and the Namibian conservation sector 'direct benefits' that include 'income and improved food security from photographic and hunting tourism, which operate within the same areas in Namibia without negatively affecting each other ... the two industries are complementary, photographic tourism could not fully replace hunting if the latter were banned'.[93]

Substantial funding is needed to protect rhinos, some of which comes from hunting and tourism income, some of which comes from the auctioning of the black rhino hunting quota. The permits are sold to the highest bidder and usually raise $250,000–400,000, not including the cost of the services of a professional hunter, accommodation, transport and taxidermy fees. A hunt in 2020 raised $400,000, which was paid to the national Game Products Trust Fund (GPTF), which is the link between trophy quota fees, live animal sales and other wildlife income and the funding of on-the-ground conservation and sustainable development projects in rural areas.[94] A central aim of the fund is to manage endangered species such as black rhino and to reduce human–wildlife conflict to benefit people and wildlife.[95] Between 2012 and 2018, 61% of the funds generated were spent on anti-poaching and wildlife management, 15% on mitigating human–wildlife conflict and 13% on protected area management, with the remainder spent on supporting conservancies, managing wildlife sales and hunting and research.[96]

When the auctions take place there is always an upsurge of protests by animal rights and anti-hunting groups. In 2015, for example, a hunter, Cory Knowlton, paid $350,000 for the permit to hunt a male black rhino. He paid for and took part in a three-day hunt before shooting the selected rhino, while being filmed by CNN. The hunt and the hunter's willingness to be filmed on the hunt led to social media and other criticism, and Knowlton received death threats.[97] There was increased criticism in the USA as the auction was conducted by the Dallas Safari Club in Texas rather than by MEFT in Namibia.[98] John Jackson III of Conservation Force, a charitable foundation that helps with marketing and acts as a conduit for the 'charitable part of the donation' from the auctions, told me that the auctions may raise $250,000 but more generally raise nearer $500,000, which goes to the GPTF, while the hunter pays the safari operator/landowner separately for the conduct of the hunt and the government for a licence to hunt in Namibia.[99]

Namibia also allows regulated hunting of the less endangered white rhino on private land. From 2018 to 2021, hunting of white rhino increased, with 11 shot in 2016 but 17 in 2019 and 22 in 2021; hunting fell away in 2020 because of COVID-19 restrictions on travel.[100] The legal hunting of white rhino in Namibia ranged from 0.37% to 1.78% of the population, which is lower than the natural mortality rate. In the same period, Namibia reported to CITES the hunting of three black rhinos, substantially lower than the full CITES quota of 20 over that period.[101] Sales of live rhinos also bring in income through the game fund for the state and for private owners – there are regular transfers of animals from private custodian properties to other private owners and also to suitable recipients outside Namibia.[102] By 2020, 25 commercial game farms/ranches and 10 communal conservancies were part of Namibia's custodianship programme, with 572 black rhinos on private land and 150 in conservancies.[103]

With rhino tourism being a key part of conservation, the SRT and wildlife authorities have carefully monitored how tourists affect the rhino. Jeff Muntifering of the SRT carried out a survey that found black rhinos were susceptible to disturbance by tourists viewing rhinos on foot or in vehicles in the conservancies, basing the study at Wilderness's Desert Rhino Camp.[104] Noting that the camp, established in 2003, had successfully deterred poaching and raised significant income through rhino tourism, including tracking on foot, he found that rhinos reduced use of vital waterholes where the proximity and usage of roads by safari vehicles was higher, and that they generally avoided areas around an airstrip and lodge roads. Some 52.7% of the rhino range around the camp was 'made unavailable to the rhinoceros population by frequent vehicle movements'.[105] As a result the SRT and Wilderness developed a rotation system around the camp and game-viewing roads, which led to a 61% reduction in rhino avoidance of high-value habitat. They also monitored the effects of on-foot tracking and viewing of rhinos; there were 112 observed encounters with 33 different 'known' rhinos in the study period, of which 45 'of the encounters (37%) resulted in the rhino remaining unaware, 45 (37%) were disturbed but not displaced, and 33 (26%) were displaced'.[106] New measures were put in place to avoid displacements and keep tourists at a reasonable distance from the rhinos, for the comfort of the rhinos and the safety of the tourists.

When I visited Desert Rhino and tracked rhinos on foot with SRT guides, they were careful to keep at least 200 m from the rhinos and to approach on foot downwind of them from behind a low ridge. The rhinos did become aware of us, but were not unduly bothered by our presence and stayed where they were for the 20 minutes for which we watched them. In another encounter that I had, the cow and calf seen from a vehicle in the Torra Community Conservancy were not alarmed, perhaps being habituated to vehicles.

Near to where I encountered the rhino cow and calf at Torra, I met a Herero pastoralist with a small cattle post, who was a member of the Torra Conservancy. He told me that income from rhino tourism had enabled the conservancy to build a wire-fenced boma surrounded by shadecloth to keep his livestock safe from predators at night. Before he had the boma, lions, hyenas and leopards would get into thornbush enclosures or stampede the livestock, killing cattle or goats. Since the new, costly boma had been built he had not lost a single animal. He showed me the solar-powered water pump that pumped water into a trough: he did not have to buy diesel for the pump and it stopped elephants damaging his pump and pipes, as there was always water in the trough. This was paid for with conservancy income from wildlife tourism.[107] The conservancy system has served Namibia's communities and rhinos well. In its Strategic Plan for 2023–8, the SRT in Namibia stressed that successes in rhino conservation 'can be attributed to local communities' ... pride for their rhino, a strong level of trust and partnership with the Namibian government, partnerships with various civil society organisations, and most importantly, our grassroots approach.[108] The Trust emphasised that:

> 100% of rhinos living on community lands in the Kunene
> Region are being protected by the people that live alongside
> them. They are incentivised to protect the rhinos because
> of a sense of ownership and responsibility, and because
> they receive income from rhino monitoring – through
> the provision of incentives to rangers – and rhino tourism
> which provides employment and direct cash payments to
> conservancies.[109]

The Trust works with 14 conservancies in Damaraland/Kunene (and Nyae Nyae in north-east Namibia south of Khaudum NP, on the border with Botswana) and is endeavouring to support conservancy management:

> One of SRT's biggest threats is poor conservancy
> governance. Without functioning conservancies, SRT
> cannot adequately monitor rhinos as their work heavily
> relies on conservancies reinvesting into conservation
> through the provision of rhino rangers for patrols and
> keeping rhino areas free from settlement and unsustain-
> able tourism. Conservancies are also custodians of wildlife
> and have a responsibility to ensure transparency and good
> governance in their activities.[110]

A cloud on the horizon for Namibian rhino conservation, beyond the Etosha poaching and the need to strengthen conservancy governance, is the danger of habitat being lost to human activity, notably mining. Although no rhinos were poached on conservancy land in recent years, some have been displaced from habitat on the //Huab Conservancy, east of Torra in Kunene, by blasting and other activity at a copper mine whose mining was approved by the MEFT minister, leading to a dispute between the conservancy and the government.[111] In an open letter to the minister, Pohamba Shifeta, //Huab Conservancy's Management Committee said that 'rhinos, local jobs, and our conservancy have been imperilled by this groundless, uninformed, and reckless decision' to permit mining on conservancy land. Emma Gomes, chairperson of the conservancy, said that the mine had prompted black rhinos to migrate, reducing the community's income as well as damaging the conservation of rhinos and other wildlife on the conservancy. Conservancy members held a meeting to demand that mining be stopped.[112] The income from rhinos is derived mainly from rhino viewing and tracking on foot through Ultimate Safaris, which had invested $470,000 in a camp at //Huab but had been forced to close the camp permanently because of the effects of the mining. Tristan Cowley, managing director of Ultimate Safaris, told me on 28 December 2023 that the camp closure meant he was

> still considering litigating against the Minister of
> Environment, but with all lost it may just be throwing good
> money after bad in an environment with very unpredictable
> courts. The //Huab Conservancy has lost their entire income
> as a result and all the support we offered as a result of our
> presence in the area … The impact is massive and to date we
> are yet to receive a verdict from the Minister on our lodged
> appeal![113]

Kenneth Uiseb of MEFT told me that 'the Minister didn't allow the mining. The office of the environmental commissioner granted the environmental clearance based on an EIA [Environmental Impact Assessment] report which was flawed. The tourism operator appealed to the Minister but no decision was taken on the appeal since the tourism operator ceased operations and moved out of the conservancy.' He said that the impact report was flawed as 'It was done at the time of Covid lockdowns, and also didn't consider the wildlife related impacts and overlooked the rhino and how that would impact the local livelihoods and rhino conservation.'[114] The failures in this process that have affected the desert-adapted rhinos, and the conservancy's income, reek of incompetence at best and the favouring of mining interests over conservation and local people.

South Africa: weathering the poaching storm

The years under review here were the toughest for the South African rhino population and for the wildlife authorities, conservationists and private owners trying to protect them. Poaching was at its height in the mid-2010s, having started to rise in 2006 and taking off in 2009. In 2014 and 2015, a total of 2,390 rhinos were poached in South Africa. At the end of 2010, South Africa had an estimated 18,796 white and 1,916 black rhinos – about 93% of the world's white rhino population and 40% of its black rhino. White rhinos were the hardest hit during the poaching crisis; surprisingly, south-western black rhino numbers rose from 198 to 242 between 2011 and 2014, but south-central black rhino numbers fell by 6% to 1,522 from about 1,619 in 2014 – mainly owing to poaching in KNP.[115] The estimate from CITES for the end of 2021 put the white rhino population at 12,968 and black rhino at 2,056,[116] giving a total of 15,024, according to the 2022 State of the Rhino Report,[117] which represents a massive fall of 5,828 in white rhino numbers and a small increase of 140 in black rhino numbers (see Table 9.3).

Table 9.1: South African rhino poaching statistics[118]

Year	2007	2008	2009	2010	2011	2012	2013	2014
National tally	13	83	122	33	448	668	1,004	1,215

Year	2015	2016	2017	2018	2019	2020	2021	2022
National tally	1,175	1,054	1,028	769	594	394	451	448

Table 9.2: Poaching statistics for recent years broken down by SANParks reserves and South Africa's provinces

	2020	2021	2022
SANParks	247	209	124
Gauteng	2	2	2
Limpopo	18	38	25
Mpumalanga	13	39	21
North-West	19	32	24
Eastern Cape	0	0	0
Western Cape	0	4	0
Northern Cape	1	1	4
Free State	1	24	4
KZN	93	102	244

Poaching led to a fall in white rhino numbers from 10,621 in KNP in 2011 to 1,988 at the end of 2022; black rhinos were down from 415 in 2013 to 208 at the end of 2022.[119] By the early 2020s, poaching in South Africa was moving from KNP, which had been the hotspot, where 124 rhinos were killed in 2022, to KZN (Table 9.1). The numbers poached there went up from 93 in 2020 to 244 in 2022 (228 on reserves and 16 privately owned) (Table 9.2). A poaching level of 282 was predicted for 2023 in the province's reserves (especially Hluhluwe-iMfolozi),[120] a prediction that turned out to be too optimistic, as 307 rhinos were killed in 2023 in Hluhluwe-iMfolozi alone – the highest poaching level ever recorded in KZN, according to Environment Minister Barbara Creecy.[121]

Table 9.3: Kruger rhino numbers[122]

	2011	2012	2013	2014	2015	2016
White rhino	10,621	10,495	8,968	8,619	8,875	7,235
Black rhino			415	310	383	407

	2017	2018	2019	2020	2021
White rhino	5,142	4,360	3,549	2,607	2,250
Black rhino	507	296	268	202	208

The rise of poaching in KZN in place of KNP was because numbers there had fallen dramatically, which meant rhinos were harder to find and kill. Cedric Coetzee, head of rhino protection at Hluhluwe-iMfolozi in KZN, told me in 2016 that whereas it might take poachers days to track a rhino in KNP, the high density of animals in the KZN reserves meant that poachers might only spend two to three hours there before killing a rhino and escaping with its horns.[123] The upsurge in poaching in KZN reserves in the early 2020s demonstrated how the accessibility of rhinos in reserves such as Hluluwe-iMfolozi was exposing them to poaching as gangs moved in, having shifted the focus of their activities from KNP.

Corruption and mismanagement in the Hluhluwe-iMfolozi Reserve in KZN have been partly blamed for the failure to contain poaching there. Sources within the park and conservation NGOs told me in 2018 that APUs were being diverted from patrolling poaching hotspots at particular times by corrupt employees; South African conservation NGOs have raised concerns about financial irregularities in the management of the reserve, which have led to senior figures being moved to other KZN reserves.[124] Dr John Ledger, a past director of the EWT, looked into the problems at KZN following the sharp rise in poaching, noting that poaching gangs were well organised but the management of Ezemvelo KZN's reserves was not, which had a negative effect on anti-poaching effectiveness. He found that poor management and low morale were reducing the ability to fight poaching, and that reserve managers not living on reserves reduced oversight, all of which created 'fertile grounds for disaffected employees who will pass on information to the poaching syndicates, sweetened by some welcome cash, perhaps'. To cap it all, the major rhino crime investigator there was moved from his post, which made life easier for poachers.[125]

Across South Africa during these years there has been a persistent failure to deal with corruption within the national and provincial wildlife authorities, the police and the judiciary. While arrests have gone up, convictions have not followed suit and suspects are often able to endlessly delay trials, remain on bail and continue to poach or commission poaching. Tom Milliken of TRAFFIC pointed out that the National Prosecuting Authority and the Crime Intelligence Division of South Africa's police were 'dysfunctional',[126] often taking years to bring suspects to court, only for judges to acquit them on technicalities (often because of incompetent prosecution cases and poor evidence gathering by the police). Poachers may be caught, but the middlemen or kingpins who organise poaching and smuggling are arrested less often, or if there are prosecutions they are slow, often dismissed by magistrates or judges of dubious integrity or the sentences handed down are light. Cathy Dean of the British-based Save the Rhino NGO told me in 2018 that she was concerned about the KZN poaching crisis, failures in the conservation system there, and the failure of the justice system to convict poaching middlemen and kingpins.[127] The failure to speed up the prosecution of alleged poaching or illegal trade kingpins who have been charged but not stood trial, such as Dawie Groenewald, Hugo Ras, Big Joe Nyalunga and Dumisani Gwala (until his acquittal as a result of the police evidence being deemed inadmissible), and whose trials have been repeatedly delayed or derailed by incompetence in the law enforcement and justice systems, is a long-term problem. The South African government's own statistics bear out the delays and incompetence in the judicial system when it comes to prosecuting and convicting poachers. Environment Minister Creecy announced in February 2022 that 189 people had been arrested for poaching and illegal wildlife trading but only 38 had actually progressed through the justice system, with 37 of the cases resulting in the conviction of 61 poachers, which meant that less than a third of those arrested had been convicted and that a large number of cases had not made it to court.[128]

In her report, which mentioned the small drop in poaching (mainly due to COVID-19 movement restrictions), Creecy adopted a self-congratulatory tone, but this masked the terrible statistic that poaching may be decreasing because there are fewer rhinos to kill. In 2009, there were 21,087 rhinos in South Africa, and 122 were poached – 0.57% of the population. This means that reproduction outweighed poaching, and so numbers increased. In 2021, in KNP, which had 2,250 white and 208 black rhinos, the poaching figure was 195 or 8.63%. Rhino cows do not breed until they are about four to six years old, and females usually have calves every two and a half to three years. This means that a high poaching rate prevents recovery in numbers. The 2022 poaching figure of 448 was still very high in comparison with the numbers of living rhinos in the country, amounting to 3% of the national population being poached; if the prediction of 600 poached in 2023 is true, that would be about 4.1% of the population. The AfRSG 2023 report noted that a level of poaching above 3.6% would not enable a population to recover through breeding.[129] Sam Ferreira, the scientific officer for the AfRSG, told me in December 2023 that KNP in 2023 had the lowest rate of poaching in a decade, though final figures were not available at the time, but the loss of rhinos as a percentage of the total KNP population was down to about 2.2%. And so for the first time in years the outlook was, Ferreira said, positive as far as KNP was concerned.[130] However, the increased poaching level for South Africa as a whole in 2023 does not suggest that the poaching scourge has been seriously reduced.

The episodes of illegal dehorning, killing of rhinos by owners and poaching that have been responsible for South African horn reaching the illegal trade will now be described.

Pseudo-hunts and the Boere Mafia

As noted in Chapter 5, the first signs of the Vietnamese entering the illegal market for rhino horn in a substantial way was in around 2003, when Vietnamese nationals started visiting South Africa for the purpose of white rhino trophy hunting, with the horns being legally exported to Vietnam as personal trophies but once there illegally sold to dealers or consumers for a huge profit. The East Asian smugglers and dealers developed business relationships with private rhino owners and professional hunters, many of them Afrikaners who became known as the Boere Mafia. They identified privately owned rhinos to be hunted and carried out the hunts or hired professional hunters to do so. The Asian dealers arranged for Vietnamese or Thais, often prostitutes or young women with no obvious hunting background or interest, to enter South Africa to hunt rhinos. Often, as detailed in previous chapters, the women did not shoot the rhinos; instead this was frequently done illegally by the professional hunter or rhino owner, though the name of the non-shooting client would be on the permit, making the hunt appear legal. The horn was then exported with full CITES papers. The dealers who commissioned the process would sell on the horn.[131] One thing that suggested the illegality of the hunts and the role of the Vietnamese or other nationals on the permit as mere conduits for horn and not genuine hunters was the number of 'incidents where rhinos have fallen from single, well-placed "kill shots", which indicates a highly-skilled or professional hunter ... such individuals, through participation in illegal rhino horn sales or pseudo-hunting scams with Asian clients, developed business relationships with foreign criminal syndicates to export supposedly legal horn for sale in Asia'.[132]

Starting as a small-scale rhino horn scam, the pseudo-hunts multiplied to feed mounting demand, especially from Vietnam. One hunter in Free State

Province, Randy Westraadt, was involved in 34 hunts involving Vietnamese between September 2009 and November 2010. Rademeyer said that in those hunts the rhino horn cost $4,000 an inch, increasing to $4,500 if over 25 inches.[133] The infamous poacher and horn trafficker Dawie Groenewald was involved in running pseudo-hunts, often using rhinos he had bought from SANParks at auctions. He told Rademeyer he was charging the Vietnamese R50,000–70,000 (about $5,680–8,000) per kilogram for horn trophies when the Vietnamese hunts started. Groenewald admitted to Rademeyer and to John Hume that he had killed rhinos without permits, and added that he killed rhinos in these so-called hunts to be able to effectively sell the horn, as he was opposed to the ban on the international trade in rhino horn.[134]

In his ground-breaking exposé of the pseudo-hunts, Rademeyer gave the numbers of trophy horns leaving South Africa for Vietnam or other destinations in East Asia (Table 9.4).

Table 9.4: Numbers of trophy horns that left South Africa for East Asia[135]

2005	2006	2007	2008	2009	2010
24	98	146	98	136	131

This amounted to about 316 rhinos killed for the export of their horn for illegal commercial trade. By the time that the South African government caught on to the illegal nature of the whole arrangement, 659 horns from 329 rhinos had been exported, raising $200–300m at a cost of about $20m in permit and trophy fees to those running the illegal trade.[136]

Even before investigative journalists such as Rademeyer and the trade researchers Milliken and Shaw had started to dig into the trade, the Professional Hunters' Association of South Africa (PHASA), representing legal, professional hunters, had registered concerns in 2005 about rhino trophy hunting involving visiting 'hunters' from Asia, and required members to supply a written account of their rhino hunting activities to ensure compliance with national and association rules.[137] At first this had no effect on government regulation of rhino hunting to prevent circumvention of rules on exports and the nature of legal hunts. However, in April 2012, by which time the investigations by journalists and trade experts had started to reveal the extent of the illegal trade derived from pseudo-hunts, the government amended the Norms and Standards of trophy hunting, and hunters from Asia were effectively blocked from hunting rhinos to gain access to horn.[138] The Department of Environmental Affairs advised 'provincial authorities not to issue hunting permits to Vietnamese citizens due to various concerns regarding illegal hunting practices'.[139] Other measures to prevent legal hunts from being used to feed the illegal trade included limiting permits to one hunt per hunter in any year and the need for hunts to be officially witnessed. This effectively put an end to the pseudo-hunts, though not to other illegal methods of obtaining and selling rhino horn from privately owned rhinos killed or dehorned to supply the trade. After the restrictions stopped the use of Vietnamese as supposed hunters, there was a short-lived increase in the number of Polish and Czech hunters killing rhinos on legal hunts, but this did not survive as a method to get horn, as the Czech police investigated some of those applying to hunt rhinos in South Africa and found they were not legitimate hunters.[140] After the cessation of the Vietnamese and East European pseudo-hunts, there was a fall in the number of white rhinos hunted annually in 2012–15 compared with 2004–6, a trend that Emslie et al. observed to have continued through 2016 and 2017. They suspected that 'some pseudo-hunting has continued since 2012, although at a much lower level as a result of improved regulation and some hunting applications being rejected in South Africa (a total of eight in 2016–17, usually where the applicant had no prior hunting experience)'.[141]

Groenwald, Ras and the illegal trade kingpins

From the late 2000s to the present, it has become clear that groups of private rhino owners, professional hunters, veterinarians, game capture specialists and some safari tour operators were involved in the illegal trade. Drawn mainly from the Afrikaner community, 'members of these groups had been called the "Boere mafia" or "khaki-collar criminals" in local parlance', and many had the skills and experience as wildlife professionals to transport, immobilise and kill rhinos.[142] The most notorious is Dawie Gronewald, a safari tour operator, who worked closely with a group of professionals including former KNP wildlife capture head Dr Douw Grobler, vets Johannes Gerhardus Kruger, Karel Toet and Manie Du Plessis, and professional hunter Hugo Ras, who were all involved in various forms of illegal killing or dehorning of rhinos.[143] The emergence of information and the arrests and charges levelled against them for illegal dealing in rhino horn created a new picture of poaching beyond the stereotype of poor poachers or criminal poaching gangs killing rhinos in parks, reserves and on game farms. These people were capable of mounting delaying tactics and employing good lawyers to keep them on bail and out of prison. Of 29 white suspects arrested for rhino-related crimes between 2006 and 2012, 90% were granted bail and only two were convicted in that period; 20% of those arrested were professional hunters, 20% were vets, 17% were safari operators and 72% were employed by or connected with private game farms.[144]

Groenewald, the only high-profile tour operator or hunter to be expelled by PHASA, was first arrested in 2010, accused of breaking the laws on the trade in rhino horn. The police found 20 rhinos that had been killed and buried on Groenewald's land, all without horns. Douw Grobler was linked with this and with the suspected use of etorphine (also known as M99), the drug used in legal dehornings, illegally supplied to the Afrikaner rhino horn networks, so that rhinos could be killed silently.[145] In March 2012, three vets, including Grobler, appeared in court on charges relating to the supply of drugs for knocking out or killing rhinos and the illegal use of a tranquillising drug favoured by rhino poachers.[146] Tom Milliken of TRAFFIC was concerned that the role of these wildlife professionals and people in the private game industry had a corrupting influence on others, including employees of KNP and other parks or reserves.[147]

Despite his arrest and the charges against him, Groenewald did nothing to disguise his desire to make money from killing rhinos. He told Rademeyer that he would shoot 100 rhinos a year as 'It's good business', so it is no wonder that the long police investigation into his activities linked him to hundreds of rhino poaching incidents, including on a game farm where rotting carcasses were exhumed from mass graves. Altogether by 2012 he was facing 1,736 counts of racketeering, money laundering, fraud, intimidation, illegal hunting and dealing in rhino horns, including the killing of 59 of his own rhinos for their horns, burying or burning the carcasses and selling the meat to a local butchery, and the illegal dehorning of rhinos and sale of 394 horns over a four-year period.[148] He is thought to have bought rhinos at auctions for the purpose of covertly killing them and selling their horns to make a huge profit. Originally a drug-squad police officer, he had been forced to leave after allegations that he and his brother were linked to a syndicate smuggling stolen cars to Zimbabwe,[149] following the pattern where many involved in the illegal rhino horn trade are also involved in other illegal activities including car theft, cash-in-transit robberies and drug dealing. Groenewald was arrested and convicted in the USA in 2010 for trying to sell an illegal leopard hunt to a prospective American client. At the same time he was being watched and investigated by the police special investigations unit, the Hawks, who were also investigating Ras and Deon van Deventer, the latter being one of the few khaki-collar criminals to be convicted and jailed, in his case for killing 22 rhinos and removing their horns.[150]

Hugo Ras was a former safari operator and part owner of a bar in Pretoria, notorious for drug dealing and other criminal activity. He was arrested in 2011 along with Grobler, the game capture vet. Ras's trial on a variety of charges connected with dealing in rhino horn was continually delayed by his lawyers and by the slow and corruptible court processes in South Africa. By December 2018 he had still not had to appear in court and got another year-long delay of the trial in which he, his wife and brother, a brother-in-law, former Hawks officer Willie Oosthuizen, former attorney Joseph Wilkinson, game capture pilot Bonnie Steyn, Willie van Jaarsveld and Matthys Scheepers were charged with a range of criminal charges involving rhino poaching, alleged theft and illegal possession, transport and sale of rhino horn.[151] In May 2021, with the rhino horn charges still outstanding, Hugo Ras and Van Jaarsveld were sentenced for fraud relating to the sale of a boat, and Ras also for possession of an illegal firearm and ammunition, with a sentence of 29 years' imprisonment.[152]

However, Groenewald was able to continually avoid trial, which was delayed until February 2021, after his lawyers appealed to the High Court in Pretoria. In July 2021, while out on bail, Groenewald was arrested again and charged with illegal possession and sale of rhino horn, having been found, along with co-accused Schalk Steyn, in possession of 19 rhino horns.[153] The COVID-19 pandemic led to the suspension of courts and a further delay. Groenewald and Steyn made a brief appearance in court on 11 April 2022 at which further investigations were deemed necessary, and there was another year's delay as Groenewald and his co-accused appealed to the Constitutional Court to have the charges dropped.[154] Watch this space for further delays.

As the focus in the mid-2010s was on the poaching in KNP, Hluhluwe-iMfolozi and on private ranches, less attention was paid to the killings by the khaki-collar criminals, but these hit the news again in December 2023 when an American, Derek Lewitton, was arrested by the Hawks following the unearthing of 26 rhino carcasses on his game farm near Gravelotte. The Hawks discovered ten unmarked rhino horns in a safe on the farm – the dead rhinos and horns did not have the necessary documentation, which implicated Lewitton in the illegal horn trade.[155] The Provincial Commissioner of Police, Major General Jan Scheepers, said, 'Everywhere you looked, rhinos were lying there dead', adding that there are likely to be 'many more carcasses on the property'.[156] Lewitton has been charged with wildlife trafficking and firearms offences, which he denies. On his website he claims to be saving rhinos and engaging in captive breeding to increase the population of an endangered species.[157]

Thefts from stores and the Rathkeale Rovers

Some of the Boere Mafia, along with corrupt SANParks employees, were suspected of involvement along with other criminal syndicates in robberies of horn from NP horn stocks and those held privately.[158] In 2012 it was reported that at least 37 horns had been stolen from private game ranches, a taxidermist legally preparing trophies was robbed of 18 horns and there was an attempted armed robbery of rhino horn stocks from the government store within Addo NP in Eastern Cape province in June 2009; given the high security provided for rhino horn stocks it is generally suspected that robberies involve inside information or direct participation of insiders, including NP staff.[159] Outside South Africa there have been thefts of rhino horn from museums, often connected with an Irish traveller gang known as the Rathkeale Rovers, who are also involved in robbery, money laundering, drug dealing, airport heists and trading in counterfeit goods.[160] There were at least 20 thefts of horns from museums and art dealers in a short space of time blamed on the Rovers in 2012, and after an international police operation most of the gang were arrested after a theft of valuable jade artefacts from museums and then imprisoned, after which the robberies largely stopped. In 2017, criminals went as far as to kill a white rhino and saw off its horn at a zoo in Thoiry, west of Paris. The rhino was shot and had its front horn removed with a chainsaw, but the criminals left before they could remove the back horn.[161] As a result, Dvůr Králové Zoo in the Czech Republic, which has a herd of 21 black and southern white rhinos, decided to dehorn all its adult rhinos. The zoo was famous for having been home to the last captive northern white rhino.

The most recent horn theft occurred at the headquarters of the North West Parks Board in June 2023. Thieves broke in and stole 51 rhino horns, in what was described by police as 'a well-planned heist executed with military precision'. Pieter Nel, the North West Parks Board's acting chief conservation officer, described the incident as being 'like a kick in the gut'.[162] It took the police eight hours to arrive after the robbery was reported. After the robbery, the Hawks remained silent on what had happened to the horns, with a value estimated at R9m, The Hawks' head, Godfrey Lebeya, told Parliament that they were unable to disclose where the 51 rhino horns were sold, and 'due to the sensitivity of the ongoing investigations revealing the destiny of the commodities will jeopardise the probe'.[163] In early July it was announced that a 40-year-old man had been arrested in Rustenberg by the Hawks and charged with 'business burglary' – his name was not released. Martin Ewi, the Institute for Security Studies specialist in southern Africa's organised crime, told News24 that the robbery was most likely carried out by 'foot soldiers', with a middleman having meticulously planned the robbery and escape after an order for rhino horns from a rhino trade kingpin in a market country.[164]

From pseudo-hunts to a tsunami of poaching

Between 1990 and 2006, rhino poaching in South Africa had been at a low level with an average of 15 rhinos poached annually, but with a jump to 36 in 2006, falling again to 13 in 2007 and then starting to take off alarmingly, with

83 poached in 2008 and 122 in 2009 before continuously climbing to 448 in 2011 and 1,215 in 2014 (Table 9.1). After that it slowly declined as rhinos became scarcer in KNP, IPZs were established to improve security, and the poaching focus moved to KZN and to private reserves in some provinces (Table 9.2). The steep rise from 2009 onwards was a result of increasing demand that was not being met by the pseudo-hunts, illegal killing by professional hunters and rhino owners on private land and the existing low level of what could be called conventional poaching. The ending of pseudo-hunts, as 't Sas Rolfes suggests (see Chapter 5), could also have been a factor in increasing poaching, as one source of large numbers of rhino horns to meet demand in Vietnam was removed. Most illegal hunting was carried out by poaching gangs armed with rifles, but some was carried out by wildlife professionals using helicopters and tranquillisers that could be fired from silent dart guns – enabling poachers to track rhinos, sedate them, remove the horns and leave them to suffer long and painful deaths.[165]

High levels of poaching remove breeding animals and so reduce natural recruitment through breeding. As previously mentioned, once poaching exceeds 3.6% of the population, the effect is to lead to a sustained fall in numbers as reproduction fails to keep pace with deaths. Emslie et al. record that from 2014 to 2017, reported minimum poaching levels in KNP remained high, with a geometric mean of 8.1% per year; this, combined with periods of drought, which accounted for nearly 20% of total rhino deaths in that period, took a heavy toll, with numbers falling with little short-term chance of recovery. They also note that because of ranger shortages, carcass detection was not very accurate, with as many as 20% undetected, so there could be an even higher death rate. White rhino numbers declined in this period by 51% in KNP and 26% in other state parks and reserves.[166] Poaching, drought and natural mortality reduced KNP's white and black rhino populations. This concerning situation in KNP and elsewhere led to the search for solutions; one solution opted for in KNP was a more militarised and technological approach.

Jooste's war on poaching and other measures

In December 2012, as poaching increased and there was evidence of corruption and poor management of anti-poaching in KNP, retired SADF Major General Johan Jooste was appointed as Head of Special Projects with SANParks, which involved rhino protection, including anti-poaching and intelligence gathering, utilising SANParks anti-poaching capability, and working with the South African National Defence Force (SANDF) and the police. He immediately started a programme to transform and operate the anti-poaching teams on a paramilitary basis and to root out corrupt staff, 'including rangers, who were either poaching or supplying information to criminals'.[167] Jooste found the park poorly managed and riddled with corruption. Funds for conservation and anti-poaching were being skimmed off, leading to poor working and living conditions for the rangers, and shortages of food and basic equipment – factors that often caused staff to turn to poaching.

Aware that poaching in KNP was being carried out by Mozambicans as well as locally recruited South Africans, Jooste sought to gain the cooperation of the Mozambican government and police, but lamented that there was little or no enforcement of wildlife laws in the Mozambican section of the combined park or along its boundaries, while South African APUs could not pursue poachers into the Mozambican section of the park.[168] Intelligence reports that were reaching Jooste indicated what has already been described earlier in this chapter – that Mozambican poaching kingpins were driving poachers right up to the border running through the park and dropping them off to poach in KNP unmolested by Mozambican rangers, border forces or the police. Jooste did not feel he could trust his Mozambican counterparts with details of his operations and strategy.[169]

Jooste was appointed by the SANParks CEO David Mabunda with the mandate to turn the ranger force into a hard-hitting paramilitary force able to tackle the poachers head on.[170] Mabunda publicly warned poachers that their 'days are numbered', and declared 'we are on their trail and closing quickly on them'.[171] As Humphrys and Smith commented, 'The intention was to send signals, particularly for international consumption, that conservation was being toughened up.'[172] However, it was an uphill task, and during Jooste's first full month in post, January 2013, 42 rhinos were killed in KNP. The general's initial strategy was to train the rangers as paramilitary units with strong discipline and efficient organisation, and to supply them with better equipment. He wanted a helicopter, sophisticated surveillance and sensor equipment to detect and track poachers, and drones and microwave communications equipment. This could not be funded by the meagre resources available from the government and SANParks revenue, but he was able to get private funding from the Howard Buffett Foundation in the USA, which has been involved in rhino conservation, including funding translocations by African Parks. The Foundation promised $17m and then upped this figure in stages to $23m, which enabled the purchase of a helicopter.

With improved training for APUs, better intelligence gathering and more sophisticated equipment, Jooste, with the backing of Mabunda, took an aggressive approach and set up an IPZ in the southern part of the park, where most of the rhinos were situated, encompassing an area of 4,000 km² between the Sabie and Crocodile rivers. It was created in 2014 to concentrate anti-poaching resources to improve the security of rhinos in high-density areas using advanced technology, equipment and infrastructure. With funding from the Peace Parks Foundation and the Dutch lottery, Jooste established a joint operations centre and started installing sensory equipment along the border between the IPZ and Mozambican territory.[173] He also needed to work with the SAP and the SANDF, with army units being seconded to anti-poaching work.

Working with the SANDF proved far from easy, despite Jooste's military background. One of his anti-poaching commanders, David English, told him that the army units were useless, and Jooste wrote that they were motivated to control territory rather than seek out mobile poaching gangs. They used ten-man patrols that were easily spotted or heard, and established noisy camps, with lights on at night, which were easily avoided by poachers.[174] These conclusions were corroborated by Xolani Nicholus Funda, head ranger at KNP, whom I spoke to in August 2016 at a meeting in KNP. Funda said that despite some problems of corruption and incompetence in the police, they worked well with rangers, but neither the police nor the army were experienced or trained in how to operate in the bush, and there had been minor clashes between rangers and soldiers when the SANDF was first deployed.[175] Despite Jooste's need for intelligence on poachers and the poaching middlemen, Funda admitted that intelligence gathering was not good around the boundaries of KNP inside South Africa, and even poorer when it came to Mozambique. However, Funda did say that the establishment of the IPZ had borne fruit by concentrating anti-poaching resources, catching poachers and deterring incursions.

This slightly optimistic note jars with South Africa's overall poaching figures (Table 9.1). Jooste was responsible for security in all the NPs and was expected to be able to bring down overall poaching figures, but they rose from 668 in 2012, the year of his appointment, to 1,004 in 2013; between 2013 and 2018, when they started to decline substantially, 6,245 rhinos were killed, and in that period KNP's rhino population dropped from 8,968 white and 415 black rhinos to 4,360 white and 296 black rhinos – dropping further to 1,988 and 208 respectively, by 2022. KNP was worst hit in the mid-2010s, with over 4,300 poached there between 2013 and 2021; 2015 (891) and 2016 (767) had been the worst years, and it was only in 2017 (504) that KNP losses fell appreciably. It was only in 2019, the year when Jooste parted company with SANParks, that losses dropped to 499 and then to 247 in 2020, 195 in 2021 and 124 in 2022.[176] This did not suggest success for the militarised and more technologically based approach, elements of which had been adopted (unsuccessfully) by

Ezemvelo KZN, which runs Hluhluwe-iMfolozi and other provincial reserves in KZN, nor did it suggest success in rooting out corruption.

In his own account of the fight against poaching, Jooste blamed some of the failure to do more to reduce or even contain poaching on jealousies and in-fighting within SANParks, and the closing down of his intelligence gathering operation, run by a private company outside the control of SANParks, by the Environment Minister Edna Molewa.[177] Jooste also lamented the problems with law enforcement, with poor follow-up of intelligence leads and incompetent development of prosecution cases, caused by the jealousies evident in the senior ranks of the police, their delays in finding or processing evidence and the suspicions about their corruption, something also noted by Funda when I interviewed him. It is clear that institutional rivalries and corruption obstructed the work to reduce poaching, and also that the paramilitary strategy failed as there was no support for anti-poaching in local communities, which felt harassed. The situation was exacerbated by shortages of rangers and tracker dog units and evidence of continuing corruption among SANParks staff, including rangers.[178] In the late 2010s, several rangers, other park staff and police officers were arrested for engaging in or assisting with the poaching of rhinos.[179]

Jooste's attempts to develop integrity testing for rangers and other staff to prevent the recruitment of potentially corrupt people and weed out those suspected of criminality were obstructed. He was not able to sack employees who failed these tests and, as Vigne and Martin concluded, his anti-poaching operation could not deal with the underlying causes of poaching, 'the criminal gangs and corrupt government people on the outside',[180] or the massive pool of potential poachers in South Africa and Mozambique; all of these were a result of underlying socio-economic issues, such as unemployment, poor chances of making a decent livelihood and the existence of rich, clearly corrupt individuals in the ruling ANC and among government and provincial office holders, the police, judiciary and the business sector. Pinnock notes that Jooste's attempts to root out corruption and take the fight to the poachers were undercut by the institutional conflicts and poor management of conservation, which led to the loss of half of the grant from the Buffett Foundation and then the virtual collapse of the IPZs. Jooste and SANParks parted company in December 2019.[181] Jooste's book contains no clear explanation of why he resigned, and he has remained tight-lipped since.

Three years after his departure, and with continuing signs of rangers and other SANParks staff helping poachers, in December 2022, SANParks announced that it would introduce lie detector (polygraph) tests as part of integrity testing of its employees. Testing would be on a voluntary basis at first, but a SANParks press release said 'the intention is ultimately to make polygraph testing compulsory to certain job categories'. In a written statement in answer to a question in Parliament, Environment Affairs Minister Barbara Creecy said the tests would start in early 2023 to ensure that staff involvement in poaching and other crime was eliminated from the 4,000+ workforce.[182]

The result of the decade of heavy poaching and its continuation, albeit now at a much reduced rate, has been to reduce the population and its ability to increase through reproduction. Nhleko et al. recorded in 2022 that current 'poaching levels have resulted in a reduction to the life-time reproductive output per cow from approximately 6 to 0.7 calves: a compound effect of 5.3 future offspring. Under current levels of poaching, we project a 35% decline in the Kruger rhino population in the next 10 years', but said that this could be stopped and the population enabled to increase substantially if the rate of poaching was halved.[183] They wrote that 'maintaining and improving the lifetime reproductive output of rhino cows should thus be the highest management priority', along with improved anti-poaching measures, higher sentences for poaching rhino cows, efforts to dehorn as many rhino cows as possible and measures to keep cows away from areas of heavy poaching.[184] Selier and Di Minin endorsed these as short-term fixes, but stressed that other issues needed to be grasped, such as whether a legal and regulated trade in rhino horn through CITES would weaken the economic forces that drive poaching, adding that the CITES ban, 'in place now for more than 40 years, has failed to effectively provide strict protection to the species, despite the numerous anti-poaching measures implemented in South Africa'.[185] They also, in the view of this author quite rightly, noted that motivations to poach were wholly bound up with poor life chances, unemployment, and unachievable status and livelihood aspirations among young men and their communities given current socio-economic conditions in rhino range states: 'the incentives gained from the illegal killing and trade in rhino horn far outweigh the risks (e.g., being fined, incarcerated, or even killed by the authorities or by the animals they target)'.[186]

Above all, community incentives to support rhino conservation and community empowerment were needed, given an unemployment rate of 46.5% in the areas adjacent to KNP. Economic upliftment, improved employment and income prospects, and developing community intolerance of poaching are the only ways in which young men can be dissuaded from being drawn into poaching by the kingpins who arm, commission and reward rhino poaching while law enforcement and the courts have failed to deal with those running poaching. A militarised response could catch, kill or sometimes deter individual poachers, but would not address the factors that motivate people to poach – and clearly would not deal with the overall problem of demand and high prices for horn in East Asia that provided the monetary incentive to organise poaching by middlemen and the cash to recruit the foot soldiers of poaching.

Nyalunga and the big kingpins

While the Boere Mafia were heavily involved in some aspects of the illegal killing of rhinos and sale of rhino horn, other criminal syndicates and major players were involved in commissioning poaching, trafficking the horn and supplying the international syndicates that smuggled it to markets in East Asia. It would be impossible in the space available to detail all of those involved, but there are a few major criminals whose names are frequently in the media as poaching kingpins. One, who became notorious not just for his criminal activities but for his ability to avoid trial and conviction, is Big Joe Nyalunga (often rendered as Nyalungu), a former policeman who turned to rhino poaching and drug smuggling. He was arrested at Hazyview, Mpumalanga, a few kilometres from KNP in 2011 and then again in 2012; he was charged with possession of four rhino horns and a large quantity of cannabis, 60 hunting knives and pangas, silencers for .375 and .458 rifles and stolen laptops, all at a time when he was on bail for suspected money laundering.[187] Investigations over a long period, amid repeated attempts to get him in front of a judge, were frustrated by endless delays engineered by Nyalunga's lawyers, but they led to the discovery of large amounts of cash that he was suspected of using to pay for rhino poaching, and the charging of six Mozambicans suspected of having poached rhinos to order for Nyalunga.[188] As the case dragged on with continual postponements, and with several of his co-accused disappearing or being deported, Nyalunga was arrested again in May 2023 on suspicion of rhino poaching, after a high-speed car chase following a poaching incident at Lydenburg.[189] Initially remanded in custody, he was eventually granted bail and is due in court at some unspecified date in the future. On 30 July 2024, Nyalunga and his wife were arrested and charged with tax evasion – an odd parallel with Al Capone if this gangster gets away with poaching but is jailed for fiddling his tax.

Another known criminal and rhino horn dealer, Clyde Mnisi, was shot dead in a suspected gangland killing in March 2023. He was due to appear in court in April 2023 for his role as a key player in the rhino horn trade as part of a syndicate including policemen and employees at KNP. He was also believed to be involved in cash-in-transit robberies. Mnisi's wife was killed in another gang-style shooting soon after.[190] He had close connections with Nyalunga and with another rhino horn kingpin, Petros Mabuza, who had also been killed in what appeared to be a criminal turf war. Despite Mabuza's reputation as a poacher and criminal, his funeral was attended by leading members of the ANC in Mpumalanga.

The criminal syndicates involved in poaching often have close political connections with corrupt police, judges and members of the governing ANC, as was demonstrated in Mnisi's case. The diverse links with corrupt individuals in key state institutions have enhanced their ability to recruit NP and provincial reserve employees and use them to aid poaching operations. As Rademeyer acutely observes,

> Toxic politics, deep-seated inequality, corruption and embedded organised criminality have profoundly affected the park and surrounding communities. Crime and corruption in the Kruger National Park should not be viewed in isolation without taking the impact of organized crime in Mpumalanga, including kidnappings, cash-in-transit heists, ATM bombings, illegal mining, extortion and corruption, into account.[191]

Ending corruption within KNP, Hluhluwe-iMfolozi and other parks and reserves is crucial to reducing poaching in the long term, as 40–70% of KNP's anti-poaching and law enforcement staff are thought 'to be aiding poaching networks or involved in corrupt or criminal activities in some way including high levels of fuel theft'.[192] Police and judicial corruption and the role of provincial ANC postholders as godfathers of a diversity of crimes makes this criminal nut a very hard one to crack. The ability of the criminal syndicates not just to frustrate investigations but also to kill lead investigators was demonstrated by the assassination of Hawks investigator Lieutenant Colonel Leroy Brewer in March 2020 – he was a key figure in fighting organised crime.[193] Another key figure in rhino protection, Anton Mzimba, a ranger at Timbavati Reserve adjacent to KNP, was murdered in July 2022 in a suspected gangland killing to remove a thorn in the side of poachers. Mzimba was renowned as a committed defender of rhinos and as being incorruptible.[194] The murder was another blow to SANParks' attempts to stop employees from helping poachers and the kingpins, as it suggested that if they opposed the poachers their lives would be at risk, and not just in clashes with armed poachers. The extent of the involvement of SANParks rangers in poaching was revealed by arrests carried out between April 2018 and February 2023 of a series of experienced rangers from KNP who were subsequently charged with corruption, money laundering, fraud and involvement in poaching. The Hawks organised crime unit said they were supplying 'tactical information to rhino-poaching syndicates for large sums of money'.[195] Nine relatives of the rangers were arrested in December 2022 for involvement in bribing rangers. By February, 14 people had been arrested for aiding poaching in KNP.[196] Against this background, in April 2023, Environment Minister Creecy told parliament that KNP and SANParks had only been able to fill five of 87 vacant ranger posts in KNP; 51 of the posts had been vacant for over five years.[197]

The judiciary has also been under the spotlight for both incompetence and corruption when dealing with rhino poaching and linked cases – often enabling the endless delays in cases coming to trial, and with magistrates and judges dismissing cases on questionable grounds. The prime example of this has been the investigation and suspension of KZN Regional Court President Eric Nzimande following accusations of corrupt payments, racketeering and receiving bribes to appoint unsuitable candidates as attorneys in cases. The Save the Wild NGO had campaigned for his suspension, believing him to have been key in obstructive cases against suspected poaching kingpins.[198] Nzimande was then charged with these offences and is awaiting trial. Implicated in his crimes is Z.W. Ngwenya, who represented Dumisani Gwala in his court appearances. Gwala was arrested in 2014 on rhino poaching charges along with Wiseman Mageba. Over a period of nine years there were 30 postponements of his trial after a series of objections by his defence team. In July 2023, a magistrate dismissed the case against him, declaring him not guilty of the poaching charges on the basis that evidence presented by the prosecutors was for some reason not admissible, though he did receive a suspended sentence and a small fine for resisting arrest.[199]

Private ownership and the demands for legal rhino horn sales

In 2009, the growing number of private ranch and reserve owners who had bought rhinos at auctions and established over several decades a growing herd of mainly white rhinos came together to form PROA to represent their interests, lobby government over establishing a legal trade in rhino horn, and try to improve coordination and cooperation between private owners of rhinos in South Africa in response to the growing poaching threat – this followed the government's suspension of the legal domestic trade in rhino horn. At the time of its formation, private owners had under a quarter of South Africa's rhinos in their possession, mainly white rhino but a growing number of black rhino. There were about 395 private owners. The year after the association was formed, ownership had risen to 25% of white rhinos and an increasing number of black rhinos (which reached a peak in 2014 with 27.4% privately owned, after which there was a gradual decline to 21.1% in 2021, with some owners of a large number of rhinos, such as John Hume, divesting themselves of black rhinos in favour of expanding white rhino numbers). Pelham Jones, chair of PROA, told me in January 2024 that according to data for 2022, out of about 13,500 white rhinos in South Africa, 8,000 were in private hands, while of 2,200 black rhinos, 750 were privately owned, meaning over 60% of the national rhino herd was privately owned.[200] Private ownership has been pivotal in increasing rhino numbers, distributing animals widely across the country and providing protection for them outside the state or provincial-run system. Pelham Jones told me that while private owners have 60% of rhinos, their losses are under 20% of the annual national poaching figures. This was corroborated by research by the wildlife journalist Ed Stoddard in 2023, who found that private owners had been successful in deterring poaching – they had lost 37 out of 394 rhinos killed in 2020, a worrying 124 of 451 in 2021, but 86 of 448 in 2022.[201] Out of 282 rhinos believed to have been poached in KZN in 2023, only 11 were privately owned.[202]

By 2016, the number of private owners was declining as poaching peaked, with 300 still operating, compared with 395 in 2008. Knight noted that despite the fall in the number of owners by about 15%, more white rhinos than ever were in private hands – up from 4,950 at the start of 2014 to 6,014 by the start of 2016.[203] However, there were more signs by 2020 that enthusiasm for rhino ownership in South Africa was waning because of the costs and dangers of security provision, with 28% of private owners divesting from rhinos, 57% holding on to their rhinos and around 15% buying rhinos from those who were getting out of the business, so owner numbers fell but not rhino numbers in private hands.[204] Costs of monitoring and security had doubled between 2010 and 2016 and were still rising, with a conservative estimate being $5 per hectare per year. The rise in poaching on private reserves in the late 2010s and the start of the 2020s, now slowing, put pressure on owners through mounting costs and dangers to owners and their employees from poaching gangs. John Hume told me in August 2016 that security at his Buffalo Dreams ranch cost him R3m monthly (about $229,745 in current values).

The very high costs of protection and management, and the failure of the lifting by the Constitutional Court of the moratorium on domestic trading in horn in South Africa in April 2017 to provide significant income for private owners with stocks of horn, led eventually to John Hume's sale of his Platinum Rhino/Buffalo Dreams ranch and rhinos. He simply could not sustain the mounting costs with no prospect of income through sales of horn from dehorned rhinos. The Constitutional Court decision, the result of a long-running legal action by PROA to prevent the government from maintaining the moratorium, forced the government to cobble together regulations for the sale of horn and the circumstances under which horn could be exported. Bizarrely, given the international trade ban, a foreign citizen visiting South Africa could get a permit to export a maximum of two rhinos per year (or their horns), meaning the already overstretched South African wildlife authorities would be required to police both legal and illegal trade. The rules have huge potential for laundering poached horns and perhaps a new version of the notorious pseudo-hunts.[205] The lifting of the moratorium did not lead to a bounty for private rhino owners with stocks of horn, as there was no real domestic market for horn, which still could not be exported legally to the main markets in East Asia. The auction, held in August 2017 with about 500 kg of horn in 250 lots for sale, was a flop – with few bids and very low sales bringing in little money. After the auction, the South African government issued 21 permits for the domestic sale of 574.6 kg of horn, 689 pieces of horn, between 2018 and 2021, but it is not known how many permits actually led to sales. Domestic sales have not become a major source of income from rhino or legal horn stockholders.[206]

Despite this setback for owners, private ownership and management of rhinos remains an important aspect of overall rhino population conservation and expansion. Low income, corruption, poor management and frequently poor relations with local communities can 'impede the expansion of state-run parks and can result in existing parks being poorly managed and ineffective', with conservancies, private safari reserves and game ranches becoming 'a key aspect of conservation (and development) in several countries', financing themselves through limited, regulated trophy hunting, live sales and some wildlife-based tourism. About 25% of South African private rhino owners offer trophy hunting of rhino, 45% sell live animals and 62% have wildlife tourist businesses.[207] However, the attitude of the Environment Ministry towards the future of private rhino ownership is hard to assess. In December 2020, a government-commissioned panel of experts published their study of the phasing out of breeding and keeping of rhinos on private land and the return of rhinos to what was termed 'wilder land', warning of the dangers of in-breeding by private owners.[208] Environment Minister Creecy said after reading the report that the government would issue a position paper regarding what to do about the recommendations in the report, but there has been no clear signal from the government about private ownership of rhinos,[209] though she has arranged some tax incentives to offset their costs.

Against this far from rosy outlook, many private owners are sticking with their rhinos and absorbing the costs, despite them rising to above $170,000 a month for some, without significant income beyond some trophy hunting and occasional live sales.[210] Hunting and tourism income dropped in 2020–2 during the COVID-19 pandemic and is taking time to recover, and with some owners getting out of the business, live sales between private owners are not booming and bringing income at a level that rewards owners; prices have dropped by 75% over the last decade.[211] Rubino and Pienaar looked at why owners continued to conserve rhinos despite the costs, and found that while some were hoping for future profits from exports if CITES removed its ban, others seemed driven by environmental knowledge or concerns, and by simple 'emotional attachment' to rhinos.[212] Owners who responded to their questions spoke of the vital importance of good security (some using private security companies, some having their own anti-poaching teams and some using a mixture – 'one jackal watching the other'), well-maintained fences and expensive surveillance equipment.[213] All of the owners contacted wanted the international trade in horn legalised so that they could sell horn to pay for the security measures. This corroborated the results obtained by Wright, Cundill and Biggs in their study of owners' attitudes to legalising the trade in rhino horn, with the majority favouring legal trade, though they had little confidence that CITES would lift the ban.[214]

Hume's story

John Hume is a very wealthy former property developer who started keeping rhinos on land he owned, buying six animals from Ezemvelo KZN Wildlife in 1996. He established a ranch in Mpumalanga and by 2008 had bought rhinos from owners across South Africa, making his growing herd genetically diverse, with only KNP and Hluhluwe-iMfolozi having greater genetic diversity among their rhinos. Between 27 and 157 white rhinos were introduced annually from 2008 to 2016, totalling 957 rhinos, sourced from 98 different sites in South Africa.[215] He told me that he moved his rhinos to Buffalo Dreams ranch in North West Province because the Mpumalanga one was not suitable as it provided too much cover for poachers.[216] He financed the ranch and the management and security of the rhinos from his own fortune and through occasional sales of rhinos to other private owners. Between 1996 and 2023, Hume said that his rhinos produced 1,870 calves, and by the time of the sale of the rhinos and ranch to African Parks in September 2023, the rhinos totalled over 2,000 – all white, as he had sold all his black rhinos several years earlier. In 2023 he owned over 13% of the entire global population of white rhinos.

The cost of security and providing supplementary feed for his rhinos in the dry season and at times of drought was huge. In 2016, he told me it was costing him R3m a month just for security. He was strict in choosing and monitoring his security team – they all had to surrender their mobile phones while on duty. This paid dividends, as he said he had not lost a rhino to poachers since 2011, and this clean scoresheet was maintained until he sold the rhinos. Prior to that, the ranch had lost 32 rhinos to poachers between 2008 and 2011.

However, the costs were exhausting his funds, and in June 2022 he announced plans to give 100 rhinos a year to owners or the state for rewilding on suitable ranches, reserves or parks in southern Africa.[217] Although it appeared that no one was willing to take him up on the offer between him making it and the sale of the rhinos in 2023, it was not an outlandish or unviable offer. Mike Knight of the AfRSG wrote that Buffalo Dreams was a professionally run, registered captive breeding organisation with significant biological assets, because of the diversity of origins of the rhinos, and that the development of a population of over 2,000 was 'a major conservation achievement', with great potential for translocations and rewilding of white rhinos.[218]

With no takers for rewilding, Hume announced in early 2023 that he would have to sell the rhinos and the ranch, with an auction scheduled for 26 April to 1 May. Hume set the opening bid price at $10m for over 2,000 rhinos, the 8,500 ha ranch and all the security and surveillance equipment. His large stocks of rhino horn, obtained by dehorning all of his rhinos on a regular basis, were not included in the sale; if there was a legal trade internationally, they would be worth considerably more than the asking price for the rhinos and the ranch.[219] However, even though some pre-auction interest was expressed, no bids were made at the auction, although Hume said some had been received offline and offers were being considered privately.[220]

The piecemeal sale of rhinos and land was averted when, on 4 September 2023, African Parks announced that it had agreed to buy the ranch and rhinos from Hume for a significant but undisclosed sum, including 213 buffalos, 11 giraffes, seven zebras and five hippos. In correspondence with African Parks CEO Peter Fernhead and senior members of his staff, I was told that they intended to rewild all of the 2,000+ rhinos over the next ten years, which if achieved would represent a new chapter in the restoration of white rhino, albeit only southern white rhino, in Central, East and southern Africa. The organisation said that 'with the support of the South African Government, as well as having secured emergency funding to make the transaction possible, African Parks agreed to purchase the farm and all 2,000 rhino. African Parks has one clear objective: to rewild these rhino over the next 10 years to well-managed and secure areas'.[221] The source of funding was not revealed, but in the past African Parks has obtained funding for relocations in Malawi, Chad, Rwanda and DRC from groups such as the Howard Buffett Foundation (which was involved with the Garamba relocations), the Acacia Conservation Foundation and the Rob Walton Foundation, and it has at times received funding from the European Union, USAID, the German Deutsche Gesellschaft für Internationale Zusammenarbeit and the Dutch National Lottery. The sale and African Parks plans were welcomed by South African Environment Minister Creecy and by Mike Knight and Richar Emslie of the AfRSG.[222]

Having bought the ranch, African Parks will need funds to maintain it and to provide security and supplementary feed during lean times. However, the big cost will be moving the rhinos when suitable locations and deals for translocation are worked out. African Parks is experienced in this, as has been detailed in earlier chapters, but costs are high and, as the loss of four of the first six black rhinos relocated to Chad demonstrated, there are risks attached. Just to dart and capture a rhino costs about $1,500, while moving a rhino overland in southern Africa costs around $5,250, and sending one to Chad or DRC can cost $50,000, as they have to be flown there.[223]

After the announcement of the sale, I asked Pelham Jones of PROA whether this would prompt other private owners to sell, hoping for a good deal from an organisation such as African Parks. He said that despite the costs and some private owners dropping out of ownership, private reserves that were still operating had suffered much lower losses than NPs or provincial reserves in recent years. He did not think that the sale of Hume's operation was a sign that other private owners would follow the same route, as 'the JH deal was unique' and until African Parks came along there were 'no buyers'.[224]

Zambia

By the first decade of the new century, Zambia's previously substantial black rhino population had been exterminated, and the only hope of recovery was through the 25 black rhinos in North Luangwa NP, the result of a carefully planned and executed translocation programme. Five black rhinos were translocated to North Luangwa NP in 2003 and further rhinos were brought in to bring the population up to 25 in May 2010 in a fenced sanctuary of 220 km² within the park.[225] There had been deaths following translocation, with four rhinos dying between 2009 and 2010 as a result of fights between males and the effects of trypanosomiasis, bringing the numbers down to 21. Successful breeding of the rhinos increased the total to 30, but the deaths from conflict were a reminder that careful management would be needed to avoid overstocking of the fenced area, and that release into the wider park was a necessity.[226] In 2018, two more bulls were introduced to the park to improve genetic diversity among the breeding adults. Save the Rhino reported in 2023 that no poaching had been reported in the park since the translocations, and that the numbers had risen to between 50 and 100.[227]

White rhino numbers, following their out-of-range translocation from South Africa to Mosi-oa-Tunya NP on the northern bank of the Zambezi Valley starting in 1964 and carried out again in 1994 to replace poached rhinos, had been hit hard by poaching, and by 2008 only a single male remained, so another male and three females were introduced, which produced three calves. Plans were drawn up to expand the park by adding to it the Dambwa Forest Reserve.[228] A pair of adult white rhinos was introduced to Lusaka NP in 2011. By 2021, CITES estimated that Zambia's total rhino population was 58 black and eight white rhinos.[229]

Zimbabwe

In 2007, the AfRSG estimated that there were 250 southern white rhinos and 535 black rhinos in Zimbabwe, an increase of 83 white and 196 black rhinos in ten years since the reduction in poaching and the relocation of rhinos away from poaching hotspots in the Zambezi Valley and adjacent safari and hunting areas.[230] By 2021 it was estimated by CITES that the population was 616 white and 417 black rhinos,[231] an increase but one that was limited by outbreaks of heavy poaching across southern Africa in the first 20 years of the twenty-first century. Zimbabwe was hit hard in 2008 when 123 rhinos were poached, three times the 2007 number and the highest since 1987. The Bubiana Conservancy, an area of south-western Zimbabwe, bore the brunt of the attacks, and 71 rhinos were killed by poachers in the Lowveld region in 2008, with 24 more killed in the first five months of 2009.[232] The remaining rhinos in Bubiana were moved out to other locations, with help from rangers of Bubye Valley Conservancy (BVC), which was then hit, with 38 rhinos lost in 2009, only four through natural mortality, with much of the poaching by Harare-based gangs armed with AK-47s and axes, some of whom were former or serving soldiers using military tactics on poaching raids.[233] Others poaching in the Lowveld were local men, who had long poached zebras, antelopes and buffalos for meat for their communities. The local poachers, nicknamed zebra gangs, sold poached horn to South African middlemen, many of whom were connected with what Rademeyer called the Musina Mafia, based in the South African town just south of the Beitbridge border crossing into Zimbabwe. An Afrikaner, Johan Roos, was a leading member of the poaching set-up, and bought horns and supplied silenced rifles to poachers.[234] He was well known to APUs and conservancy heads in the Lowveld as a poacher and a buyer of skins and horns. Blondie Leatham, the manager of BVC, told Rademeyer he had caught Roos poaching on the road from Beitbridge to Bulawayo:

'That fat-arsed bastard was klapping animals with three
locals and selling the meat in Beitbridge,' he says. 'I had
him on the ground with an FN [rifle] at his head, and he
kept whining, "Ag, meneer, meneer, ek is jammer. Meneer,
asseblief, ek het 'n klein kind. Ek sal dit nooit weer doen

nie." [Sir, sir, I'm sorry. Sir, please, I have a small child. I'll never do it again.]'[235]

At the Savé Valley Conservancy, on the border with Mozambique north of Gonarezhou, despite the loss of land during the land reform process the black rhino population had grown from 20 relocated there from areas of high poaching to 120 by 2014.[236] Efficient anti-poaching, funded by combined tourism and hunting income, had enabled the increase in numbers – even though in 2010 the conservancy lost 16 rhinos, three of which had been dehorned; in 2011, six rhinos were poached. The worrying aspect of the 2011 poaching was that the six rhinos that were poached between January and August 2011 had all been dehorned within 19 months (one was killed within 24 hours of being dehorned, and another within five days).[237] Raoul du Toit of the Lowveld Rhino Trust concluded that 'dehorning can be effective in reducing poaching of rhinos unless the risk to poachers of being detected is so low that it is still worth the poachers obtaining horn stubs'.[238] This is demonstrated by continued poaching in reserves where all the rhinos have been dehorned. Rhino populations in Hwange NP, Matobo NP, Matusadona NP and Chipinge Safari Area 'have been almost completely dehorned in the last two years and yet have suffered severe poaching'.[239] This is an indication that dehorning, while useful in deterring poaching when combined with effective anti-poaching measures, is not in itself a solution to poaching, because in the absence of rhinos with horns, poachers will clearly kill for the small disc of horn left after careful dehorning – or may even kill to demonstrate that they are not deterred by dehorning.

In the 2010s, poaching dropped from the disastrous level of 164 in 2008 to between 20 and 42 annually until 2015, when poaching was at its height in southern Africa, and Zimbabwe lost 50 rhinos to poaching (Table 9.5), up from 20 in 2014.[240] The numbers fell dramatically in 2020 and remained low for three years because of the national and international restrictions during the COVID-19 pandemic. Six of the seven rhinos poached in Zimbabwe in 2022 were in Matobo NP.

Table 9.5: Zimbabwe rhino poaching figures[241]

2006	2007	2008	2009	2010	2011	2012	2013	2014
21	38	164	39	52	42	31	38	20

2015	2016	2017	2018	2019	2020	2021	2022
50	35	37	34	82	12	4	7

The reduction of poaching across the 2010s and into the 2020s was partly thanks to better anti-poaching on conservancies and in NPs, partly because of a switch of poaching to South Africa, and also due to the effect of better policies regarding human/wildlife conflict outside protected areas and efforts to improve relationships with, incentivise and involve communities in conservation. One example of success in doing this was at Malilangwe Wildlife Reserve at the southern end of the Savé Valley Conservancy. There, in 2008, a gang of five poachers led by a former game scout from the reserve killed a rhino for its horn, and the following year poachers led by a former employee badly wounded but did not succeed in killing another rhino.[242] Ball et al., who researched the poaching at the reserve, noted that 'Analysis of the mechanics of poaching incidents in Zimbabwe has shown that in most cases poachers were assisted by corrupt employees of PAs, and that reserves with high levels of poaching had poor or non-existent intelligence systems.'[243] The managers and security team at Malilangwe responded by developing more rigorous selection and training procedures for game scouts, fencing the reserve and working to improve relations with the local community, including intelligence gathering, which they believe prevented 19 poaching attempts that were intercepted, with only one rhino killed.[244] Informers in local communities were nurtured and well paid, which alerted the security team to imminent threats. The financial cost is high, with Malilangwe spending about $750,000 p.a., and scouts and other employees being paid good wages; this enables a recruitment process that gets good applicants, weeds out any potential threats, and reduces staff turnover. Departing staff with any sense of grievance are clear threats, as they have the knowledge and perhaps the motivation to help poachers.[245] The boundary fences are patrolled daily and patrols are conducted at night too, with night observation posts manned when there is a full moon. In addition, the reserve has provided boreholes for safe water supplies, support for health clinics and schools, and supplementary feeding programmes for children in local communities to improve relations with communities from which poachers might be drawn.[246]

With poaching reduced, partly due to COVID-19 restrictions, it was felt secure enough for the wildlife authorities to start reintroducing rhinos to Gonarezhou for the first time in nearly 30 years. In May 2021 the Zimbabwe Parks and Wildlife Management Authority announced that it was starting the translocation of both black and white rhinos to the park, which is part of the Great Limpopo Transfrontier Park with South Africa's KNP and Mozambique's LNP.[247] Translocations started in 2022, with black rhinos taken from BVC and Bubiana to establish a new breeding population with the help of the Gonarezhou Conservation Trust.[248] This demonstrated the value that private ownership and conservancies have had for the conservation and restocking of depleted areas with rhinos. By the end of the 2010s, 88% of black rhinos and 76% of white rhinos were being conserved on private land and conservancies in Zimbabwe.[249]

Many conservancies had been hit by poaching, but had still been able to increase rhino numbers. The Lowveld conservancies, notably BVC, are good examples. In 2019, 60 rhinos were poached there, compared with 37 in 2017 and 48 in 2018.[250] Increased BVC anti-poaching efforts and better monitoring of the rhinos with the help of the Lowveld Rhino Trust 'ended well with rhino poaching incursions dramatically reduced, and those that have occurred, were intercepted before any rhino could be killed'.[251] As a bonus, ten black rhino calves were born within nine months in 2019 and 2020. In September 2023, Raoul du Toit told me that there were 285 rhinos (black and white) at BVC and 1,000 in the Lowveld as a whole, and that with improved security since 2019, only eight rhinos were lost between 2019 and mid-2023. Blondie Leatham of BVC said that in that period poaching had been kept down in the Lowveld region as a whole.[252] The limiting of poaching was assisted by the arrests of poachers in Bulawayo, including two BVC scouts suspected of involvement in poaching in 2022, who were linked to poachers who were active in Matobo NP and BVC.[253] Leathem and Du Toit told me that some of the poachers were local people who had been recruited by poaching syndicates active in many areas of Zimbabwe, that Zambian poachers had been caught at BVC and that a Mozambican was shot in a firefight with APUs; on occasion, poachers had been assisted by corrupt BVC staff members, who were being 'weeded out'.[254] Many of the poachers came from other areas of Zimbabwe. Du Toit added that whenever possible rhinos on BVC were dehorned. BVC and other conservancies support themselves through either tourism or trophy hunting, or through a mixture of both. He told me that hunting was an important source of income but rhinos were not hunted, even though that would be technically legal. As far as he knew, no rhinos had been trophy hunted in Zimbabwe for 30 years.[255]

CHAPTER 10

What next?

t is clear from the narrative so far that despite strenuous efforts to reduce poaching, the improvements in conservation methods and the varied toolkit of measures to combat poaching, reduce demand for rhino horn, translocate rhinos to augment depleted populations or restore them to ranges where they have been rendered extinct by poaching, black and white rhino are under threat and some subspecies have been exterminated or brought to the brink of extinction. Poaching has gone down, but the signs are that it could be rising again in southern Africa.

This chapter surveys the tools being used in attempts to halt the decline in numbers and revive or restore populations. The spotlight is on the attempts to recreate northern white rhino from stored eggs and sperm, the translocation of rhinos from southern Africa to East and Central Africa, dehorning to deter poaching and experiments in poisoning horn, demand reduction in the main markets in Asia, the bitter debates over whether a legal trade would cut poaching, the role of hunting in rhino conservation, funding experiments and the key role of community empowerment. Each of these would warrant a chapter or even a book in its own right, but I summarise the arguments and possible outcomes for the sake of brevity and offer my opinions on the way forward.

Breeding northern white rhino

With just two female northern white rhinos surviving at Ol Pejeta, the only way to produce new generations is through lab-based artificial methods of creating embryos from stored eggs from females and sperm harvested from the last living males, which have since died. As long ago as 1986 the northern white rhino expert Kes Hillman-Smith and colleagues lamented the decline of the subspecies and suggested that the score or so of them in captivity could be used for captive breeding and release, and hormone and other treatments could be tried to encourage the production of calves.[1] Particular hope was pinned on the northern white rhino at Dvůr Králové Zoo in the Czech Republic (then still Czechoslovakia). Even at this stage artificial insemination, *in vitro* fertilisation and implanting northern white rhino embryos in white rhino females were being suggested.[2] However, the prospects of recovery were viewed as slim, being based on a very small number of captive animals that could breed, and requiring significant human intervention and funding. There were also hopes that with sufficient protection from poaching and habitat loss the remaining wild rhinos in Sudan and Zaire (DRC) could recover. As detailed in previous chapters, this hope was an illusory one.

Breeding in captivity proved difficult. It utilised animals that were captured in southern Sudan in the 1970s and had been transferred to zoos, and those already in Dvůr Králové, San Diego, Khartoum, London Zoo and Marwell (in the UK). Dvůr Králové and San Diego became the centres for the breeding programme, but one female died and a male had to be euthanised, and although four calves were produced, numbers in captivity were not increasing enough for relocation to be considered.[3] In 1999, after over two decades of captive breeding efforts, Emslie and Brooks said: 'Captive reproduction of northern white rhinos has not yet been successful despite the efforts of the zoos involved. Currently, there are just nine northern white rhino left in two zoos: at Dvůr Králové in the Czech Republic and at San Diego in the USA.' This was down on the 14 in captivity in 1984; the last birth in captivity was in 1989.[4]

By 2009 there were four breeding-age northern white rhinos (two male, two female) at Dvůr Králové, and with support from Fauna & Flora International these last four breeding individuals were flown from the zoo to Kenya's Ol Pejeta Conservancy in a final attempt to save the subspecies from extinction. It was hoped that a more natural environment would stimulate them to breed. One of the males died in 2014 and the other, known as Sudan, in 2018, leaving two females. It was hoped that using frozen sperm and *in vitro* techniques they could still produce calves, given that natural reproduction had not occurred. Further consideration was also given to southern white rhino surrogates as a way to produce northern white rhino calves and maintain the latter's genes into the future, even though the subspecies was functionally extinct. At Oxford University, a Rhino Fertility Project was established; this grew eggs from ovary tissue of deceased females, which could then be inseminated with stored sperm.[5] This supplemented work led by Thomas Hildebrandt at the Leibniz Institute for Zoo and Wildlife Research in Germany, where they had managed to collect eggs from the last remaining northern white rhino females at Ol Pejeta. The immature eggs then matured in lab conditions and were fertilised with frozen sperm at the Avantea laboratory in Italy to produce embryos. So far, 22 quality embryos have been created by the lab in Italy.[6] They are frozen and awaiting implantation in a surrogate female southern white rhino, as the two remaining northern white rhino females are viewed as incapable of nurturing embryos to full term to produce calves.[7] Two of the eggs fertilised by the Leibniz team have grown to early-stage embryos, described by the International Rhino Foundation as 'a phenomenal achievement'.[8] Transferring embryos into surrogates to produce baby animals is a process that has been well established for lots of species, including horses and cows, though it is in the development phase for rhinos. In July 2023, the Leibniz Institute reported that five new embryos had been created with frozen sperm from different now deceased northern white rhino bulls, and two southern white rhino females had been chosen as potential surrogates.[9] However, despite the decades of research and the efforts to produce viable embryos, so far none have been implanted into southern white rhinos, and the prospect of producing sufficient calves to regenerate a northern white rhino population with sufficient genetic diversity and numbers to breed and eventually restore a wild population in Central Africa seems distant if not illusory. One has to question whether the funds and scientific effort being invested in this would be better employed in the conservation and recovery of functional rhino populations.

Translocations and establishing new populations

As the preceding chapters have shown, translocations have been underway ever since Ian Player and his team at Hluhluwe-iMfolozi successful bred and developed capture techniques to relocate rhinos from KZN to other locations in South Africa and then to Zimbabwe, Botswana, Namibia, Zambia and later Kenya. Over the following decades up to the present, translocation has been refined and improved, and both black and white rhinos have been sent to parts of their ranges where numbers had been depleted by poaching and to areas, some outside the historical ranges, where populations had disappeared completely, notably Malawi, Rwanda, Chad and the DRC. As Emslie and Brooks warned back in 1999, 'Establishing new discrete populations is both risky (for the new arrivals) and expensive.'[10] The deaths of rhinos during or soon after translocations, as happened in Kenya and most recently in Chad, demonstrate the risks to the rhinos involved and demand ever more stringent efforts to improve safety and acclimatisation. However, when successful, the process can relieve pressure on smaller reserves or private ranches where rhinos are reaching carrying capacity or where poaching has become a serious threat to the survival of a population. It can also work well to distribute populations to restore those that have been lost or are in danger of this, and can even create new populations. Locations to which rhinos are to be sent need to have sufficient carrying capacity in terms of area, browse and access to water, and the management capacity to acclimatise, monitor and protect translocated animals, including monitoring of the density of the population and ensuring genetic diversity over time. There is a requirement to select animals for translocation on the basis of gender and age structure that will ensure good breeding prospects and avoid conflict between males, while locations safe from civil or interstate conflict, insurgency and poaching by armed militias have to be chosen. There also has to be an absence of human–wildlife conflict with local communities, or processes for mitigating this.[11]

South Africa is the main but not sole source of both black and white rhinos that are translocated to other rhino range states; Zimbabwe and Namibia have contributed some. The South African environment ministry works with the WWF Black Rhino Range Expansion Project to contribute to the expansion of black rhino numbers and ranges, and has assisted with translocations to Botswana and Malawi, as well as the monitoring of black rhinos in Kenya.[12]

African Parks, as detailed in preceding chapters, has played a major role in recent years, with significant funding from donors such as the Howard Buffett Foundation, in translocating both black and white rhinos from South Africa to Malawi, Rwanda, Chad and the DRC; other successful, though more limited, translocations of southern white rhinos have taken place from South Africa to Ziwa reserve in Uganda. The purchase by African Parks of John Hume's Platinum Rhino operation in South Africa holds out significant potential for the translocation of over 2,000 southern white rhinos to locations elsewhere in Africa – whether to parks and reserves already being restocked to recover depleted populations, or to introduce southern white rhino to former northern white rhino ranges that are now devoid of their original rhino populations. In carrying out translocations the key, I was told by Sam Ferreira, the scientific officer of the AfRSG, is to plan for functional introductions that create viable populations. This is much more important for the future of rhino conservation and recovery than rigid adherence to taxonomy and therefore only restoring rhino subspecies that were originally found in the locations chosen for translocations.[13] The first 40 rhinos from Hume's ranch were relocated to Munywana Conservancy, a 300-km^2 private game reserve in Northern KwaZulu Natal, in May 2024, and by 7 June another 120 rhinos had been translocated to member reserves of the Greater Kruger Environmental Protection Foundation (GKEPF) in Mpumalanga and Limpopo provinces, South Africa, to rebuild rhino numbers in the Kruger ecosystem at a time when poaching rates there had declined. Peter Fernhead, the Africa Parks CEO, told me that 'We are not specifically focussing on either public or private reserves per se, but rather those that are ecologically suitable, and which pass a security test as well as management competency'. In theory this could be either public or private reserves, but in practice it is likely to be the latter, at least for the early years. He added that more white rhinos would be translocated to Garamba in the DRC later in 2024.[14]

Dehorning and other measures to combat poaching

Some of the first attempts to use dehorning to make rhinos less attractive to poachers took place in Namibia after the loss to poachers of desert-adapted black rhino in Damaraland, Kaokoveld and Etosha in the late 1980s; they were successful as not a single dehorned rhino was poached.[15] In November 1992, dehorning had started in Zimbabwe to deter poaching, with 117 black rhinos and 108 white rhinos dehorned at various locations across the country.[16] Dehorning was also used on the rhinos that survived in Botswana after the poaching wave that hit the country in the late 2010s and early 2020s. A large proportion of private rhino owners – whether on ranches breeding them or private reserves keeping rhinos as a tourist attraction – have used dehorning to deter poachers, in combination with strenuous attempts to protect rhinos from poaching with ranger patrols and surveillance equipment. Dehorning creates large stocks of horn that have to be securely stored because of the ever-present danger of robberies from stocks.

Dehorning has worked, or seemed to in Zimbabwe in the 1990s,[17] to cut poaching levels, but it is not on its own a solution to poaching. Raoul du Toit of the Lowveld Rhino Trust and International Rhino Foundation believes dehorning can be effective in reducing poaching of rhinos, but if security is poor and there is little risk of poachers being caught, the removal even of the stub of the horn may still make it worthwhile for poachers to kill a rhino.[18] This was demonstrated by continued poaching in Hwange NP, Matobo NP, Matusadona NP and Chipinge Safari Area, where most of the rhinos had been dehorned but protection measures were insufficient and a few rhinos still had horns; poachers entered these areas seeking rhinos with horns, but killed dehorned rhinos if they had a sufficient stub of horn to make them worth poaching. There have also been cases of dehorned rhinos being killed in South Africa since the upsurge in poaching from 2007 onwards.[19] This suggests that dehorning is only a viable deterrent in combination with effective anti-poaching measures, such as creating IPZs where rhinos are being dehorned. Dehorning must also be carried out on a regular basis, as the horn regrows constantly. Dehorning is costly, requiring the services of a veterinarian and a team of rangers to dart and secure the rhino, in addition to the drugs needed to sedate and revive it, and sufficient vehicles to carry the team and locate the rhino. In KNP in 2011 it was estimated that it cost $620 per rhino for each dehorning and that it was more expensive on private reserves, costing about $1,000 for each one.[20]

There are also questions concerning impacts on the behavioural ecology of rhinos,[21] the ability of females to protect their young from lions and hyenas, the consequences for male territorial and mating competition, and the wider territorial behaviour of male rhinos. In addition, there is a need for research into the long-term effects of administering drugs during the repeated dehornings of rhinos, especially pregnant cows. Recent studies have suggested that frequently immobilising free-ranging black rhinos for dehorning, ear-notching or to attach radio collars led to longer intervals between producing calves than for rhinos not immobilised. Other studies found no difference in intercalf intervals. Penny et al. looked at the intercalf intervals for white rhinos on a private reserve in South Africa's Northern province, and found that the intervals decreased after dehorning rather than increasing, suggesting there is no negative impact on reproductive health in the population examined.[22]

A study by Chimes et al. of Namibian black rhinos found that there was no appreciable difference in calf survival rates between horned and dehorned black rhinos,[23] contradicting a 1994 study that found that in an area with a hyena population all calves of dehorned mothers died within a year of birth.[24] What was not clear from that account was how large the hyena population was and the effect that a severe drought at the time of the study had on mortality of all rhino calves, but it did suggest (as studies in Ngorongoro Crater revealed; see Chapter 6) that the presence of substantial numbers of large hyena clans reduces the survivability of rhino calves – horned or dehorned. Chimes et al. also recorded that deaths from rhino fights were not reported among dehorned rhinos and there was no observable decline in reproductive productivity.[25] However, one effect that has been recorded is that dehorned rhinos restricted their movements more, had smaller territories and did not interact as much as horned rhinos with others of their species, which raises a question about whether in the long term dehorning will affect social behaviour and breeding.[26] The study found that some dehorned rhinos lost 80% of their territory on one reserve, and data from nine other South African fenced reserves showed that over 15 years of recorded observations, dehorning led to a loss of about 45% of territory size, with a decrease of 53% for females and 38% for males. Duthé, who led the study, suggested that the lack of a large horn perhaps affected rhinos' confidence in interactions with others of the species.[27] This lack of confidence may have resulted in defensive aggression in dehorned rhinos, and Penny et al. reported an increase in aggressive challenges in this group, which perhaps suggests dehorning leads to alterations in dominance hierarchies – though they noted that agonistic behaviour did not lead to fights with serious injuries but to avoidance of appeasement behaviour.[28]

The jury is still out on the biological, reproductive and behavioural effects of dehorning, though it does seem clear that the defensive capability of cows is reduced in the presence of predators, especially large hyena clans.

However, it is worth concluding this section with a brief account of my experience of taking a small part in and observing two dehornings on John Hume's Buffalo Dreams ranch in August 2016:

> The large bull rhino, accompanied by a couple of rhino cows, was about a hundred metres away. The jeep carrying the darting team moved closer, there was a popping sound and the bull moved off with a dart clearly visible in his upper leg. Within two minutes he was down on his knees. The dehorning team was out of the jeep, attaching blinkers to cover his eyes and a group of ranch hands held him down and attached a rope to his back leg. The vet monitored the rhino's vital signs – it was sedated but not unconscious and not obviously alarmed or in pain. The dehorners measured and meticulously recorded the circumference and height of the horn and calculated how much to remove. All the while the rhino was breathing loudly but steadily and made no attempt to get up or even shake off attention. A battery driven saw was then used to cut through the horn, which

Dehorning a white rhino on John Hume's Buffalo Dreams Ranch, South Africa, 29 August 2016. (Keith Somerville)

Dehorned white rhino on John Hume's Buffalo Dreams Ranch, South Africa, 29 August 2016. (Keith Somerville)

took little longer than a minute – all the time someone was spraying cold water on to the horn to prevent over-heating and burn injuries.

The team cleaned up the edges of the horn stump and brushed off any shaving or horn dust – which all went on to a big plastic sheet under the rhino and was gathered up in sealed and marked bags. The two horns were measured, weighed and marked with indelible ink and their specifications recorded. When a rhino, all of whom are tagged and ID chipped, is first dehorned, DNA samples are taken so in future any horn from that rhino can be clearly identified. From the first rhino dehorned, the main horn weighed 565g, the smaller horn 67g and the shavings 45g. The horns and shavings from this rhino would be stored in a safe in a bank or secure depository in South Africa.[29]

Horns from privately owned rhinos are routinely removed, and the stored horns have built up into a huge stockpile that if the CITES ban was lifted would bring in tens of millions of dollars for private owners and for the South African government, not to mention other countries such as Namibia, Botswana and Zimbabwe. When I interviewed him after seeing the dehornings at his ranch, John Hume was categorical that while he wanted the ban lifted and to be able to legally sell his stockpiled horn, this was not his motivation for breeding rhinos, something that has often been said about private owners who dehorn their animals, which is practically all of them. Mike Knight of the AfRSG estimated in 2019 that the legal stockpiles of horn and regular dehornings of thousands of rhinos in South Africa could sustainably produce from 5.3 to 13.4 tonnes of horn a year legally, and that a conservative estimate of the weight of the African rhino horn stockpile was 52.16 tonnes – the majority of this being legal stocks.[30]

Related to dehorning have been experiments in injecting dyes or toxic substances into horns to render them unusable or unpleasant to ingest, an approach which developed as the poaching wave hit southern Africa in the 2010s. Those in favour of this method claimed that it rendered horns useless and that consumers of rhino horn who ingested horn treated in this way would suffer ill effects, and thus be deterred from using rhino horn as a medicine, to prevent hangovers or as a prestige substance to mix with wine; also dyed horn would be useless for artistic or ornamental purposes. Proponents say that treated horn would become unsaleable and would reduce demand by deterring consumers. Sam Ferreira and colleagues in the SANParks scientific team examined horns treated with poison and the literature relating to this deterrent method, and 'found the information on which the assumptions are based to be weak, and refute claims that discolouring horns is a viable method … We argue that conservationists should not use this technique to deal with the rhino poaching threat.'[31] They added that if dealers could not sell treated horn and if treating horn reduced the quantity of untreated horn on the market, it would push up prices and 'simultaneously increase poaching incentives'. Poaching would not be reduced partly because poachers would not be aware from the appearance of the horn that it had been treated with poison, and partly because poachers are not the end users; they added that fake horns were in circulation on the market, and this had not served to affect demand for poached horn.[32] Poison treatment was used by the Rhino Rescue Project in South Africa on 230 rhinos. Four were subsequently poached, with 37 untreated rhinos poached in the same area, which indicates its questionable role as a deterrent.[33]

Demand-reduction campaigns

Poaching is essentially driven by demand. In the main markets for rhino horn in Vietnam, China and among the Chinese diaspora across south-east Asia, high demand and inconsistent, lax or corrupted law enforcement have created conditions for poached horn to meet demand. To overcome these problems, efforts have been made by an array of conservation NGOs, such as Save the Rhino, Wild Aid, IFAW, Born Free, Education for Nature Vietnam and Humane Society International, both international and domestic in consuming countries, to reduce demand through billboard, poster, TV, radio and social media campaigns, usually involving prominent celebrities, leading national and international business leaders and Buddhist priests. Save the Rhino used social marketing techniques to try to persuade Vietnamese horn consumers/buyers with conservation messages about the effects of poaching on the remaining rhino species and highlighting the illusory nature of the health and other benefits said to come from consuming powdered rhino horn or medical compounds, including rhino horn. They argued against the oft-repeated claims that it can cure cancer, prevent hangovers, reduce fevers and have other benefits, saying instead that it 'lacks any scientific backing as rhino horn is made from keratin: the same material as your nails and hair'.[34] However, in trying to reduce demand, the campaigns came up against strong cultural norms concerning the efficacy and power of rhino horns and the millennia-old traditional medical practices in China, Vietnam and other parts of south-east Asia. There was some evidence that demand reduction, combined with government efforts to reduce the use of rhino horn-based medicines, did work to reduce consumption in Japan and South Korea,[35] but demand reduction has not had an observably significant effect in China or Vietnam.

One campaign in Vietnam, beyond the rather crude one of telling people that rhino horn as a medical treatment is of no more use than biting one's nails, has been the so-called Chi campaign; this was launched by TRAFFIC to focus on the concept of an individual's inner strength, emphasising that consuming rhino horn does not make you a stronger person. The campaign was aimed at the business community, as market research 'identified the key user group as wealthy middle-aged businessmen, keen to show off their new-found wealth' and to impress their peers or encourage prospective business partners by plying them with rhino horn wine or gifts of horn. The campaign used Vietnamese businessmen to put across the message on billboards in the country's biggest cities: 'A successful businessman relies on his will and strength of mind. Success comes from opportunities you create, not from a piece of horn … [and] masculinity comes from within.'[36] The efficacy of this approach has not been scientifically tested, but demand has not been noticeably dropping in Vietnam. Wild Aid used rather cruder campaigns, notably the nail-biting one mentioned above. This used celebrities including the actress Maggie Q and international businessman Richard Branson. Posters, billboards and social media showed them biting their nails, with the message that this is the same medically as taking rhino horn.[37] A total of over 30 celebrities were used in these campaigns in China and Vietnam.

Hoai Nam Dang Vua and Nielsen examined the nature and effects of demand-reduction campaigns, and were critical of the claims by campaigning NGOs that there was no scientific evidence of rhino horn efficacy and the way the message was presented, noting that many of the campaigns were culturally insensitive, crude and might backfire – thus having the opposite effect to that intended.[38] In her report for CITES, Nowell found that there was a short-lived therapeutic effect of using rhino horn to reduce fever in children, though it was not as effective as acetaminophen (a common non-steroidal anti-inflammatory drug), but it could have enough of an effect to reinforce beliefs in rhino horn's efficacy as a fever cure.[39] Nowell noted that use of rhino horn as a medicine was promoted in China and Vietnam through well-publicised claims that prominent celebrities had been cured of ailments and that it could help in the treatment of cancer, bolstering popular belief in its curative properties and its detox value in preventing or curing hangovers.[40] Another factor was that families would use any possible cure to save a relative with cancer. Adherence to these newly acquired beliefs, which were supported by claims of cures by sellers of rhino horn powder or medicines, combined with the long-held cultural beliefs in rhino horn power and traditional Chinese and Vietnamese medical uses of horn, has proved a serious obstacle to demand reduction. The campaigns have often failed to appreciate the motivations for use – whether curative, derived from traditional beliefs or as elite gifts. Hoai Nam Dang Vua and Nielsen found this in a survey conducted in Vietnam in 2015–16 at the height of poaching in Africa. There was very little concern among users about the danger of rhino extinction, but rather 'pride or joy experienced by using, possessing and, in several cases, also sharing rhino horn due to its rarity and preciousness'; most expressed a preference for wild rhino horn rather than horn taken from farmed rhinos.[41]

The survey did not find fertile ground for demand reduction, and this was corroborated by Anh Ngoc Vu's study of this subject, which concluded that such 'ungrounded environmentalism … typical of these campaigns is problematic and risks deepening historical stereotypes and cultural mis-representations'.[42] Vietnamese NGOs carrying out campaigns do so mainly in large cities and in collaboration with or funded by international animal rights NGOs, whose approach can be crude, as in the nail-biting examples, or appear to demonise a whole country because of consumption of rhino horn. Anh Ngoc Vu emphasised that Western-inspired campaigns were undermined by their cultural insensitivity and crass rejection of 'longstanding beliefs in the efficacy of traditional medicines', giving the example of a TRAFFIC representative saying that 'Buying rhino means buying human fingernails with high price.'[43] Such campaigns not only went against beliefs and cultural values but also reeked of a superior, colonialist attitude that alleged Vietnamese use of wildlife products was 'evil, wrong, uncivilised, and dumb'; this created resistance and did not gain social traction to make use of rhino horn unacceptable.[44] Another study concluded that the nail-biting and similar campaigns got feedback from Vietnamese surveyed that the methods and slogan used 'were untrue and disrespectful of Vietnamese culture.'[45] While it is possible that more carefully targeted and culturally sensitive campaigns might have an effect in the future, there is no evidence that demand-reduction campaigns are having a substantial effect in reducing demand, and therefore in helping to stop poaching.

Legalising the trade in rhino horn: the controversy continues

As noted in the preceding chapters, one solution put forward to reduce poaching and combat the illegal trade in rhino horn is to have a legal trade in horn that is regulated and secure, bringing in income for conservation and for the communities who live with or alongside rhinos, giving those communities direct benefits and a sense of ownership to counteract the monetary lure of poaching. The international trade in rhino horn was banned by CITES members in 1977, but ever since its imposition the efficacy of the ban has been questioned, and some conservationists, including leading members of the AfRSG, wildlife economists and private rhino owners, have advocated looking at a legal trade in horn from legally held stockpiles, dehorned rhinos, natural mortality and horn seized from smugglers. At the eighth CITES CoP in Kyoto in 1992, Zimbabwe and South Africa submitted proposals for a controlled legal trade in rhino horn. The Zimbabwean delegation argued that income from sales would be used 'to augment national expenditure on rhino protection. Zimbabwe estimated that it requires double its current wildlife protection budget to conserve remaining populations effectively', while the South Africans said that it would raise funding for rhino conservation and help to fund intelligence gathering and anti-poaching work.[46] The proposals were deemed premature by TRAFFIC and fiercely opposed by animal rights NGOs such as the US-based AWF, IFAW and Born Free. With opposition from the USA, Britain and other EU member states and African states such as Kenya, the proposal did not succeed and the ban remained. At the tenth CoP in Harare in 1997, South Africa put forward a legal trade proposal, which again was resoundingly rejected by the same coalition of member states backed vocally by anti-trade NGOs.[47] In South Africa, private rhino owners, provincial wildlife organisations such as Ezemvelo KZN Wildlife

and conservation economists all supported moves to reinstitute legal international trade, as did the governments of Namibia, Zimbabwe and Botswana
(the latter until Ian Khama became president and opposed trade).[48]

The arguments for a legal trade, as put forward by 't Sas Rolfes, Wiltshire,
Eustace and others, have a logic and seek to establish strong economic and
market arguments for developing a carefully controlled legal trade.[49] 't Sas
Rolfes concluded from his research into the rhino horn market that the
1977 CITES ban had 'no discernible positive effect on poaching', and that an
alternative was needed that would meet demand legally and without killing
rhinos, especially in the face of the very limited effect of demand reduction.
What was needed was a legal rhino horn trade regime that would remove the
current monopoly enjoyed by the illegal trade, establish transparency and
have visible market prices for horn, and allow monitoring of demand through
recording of legal sales – so that 'the incentives for speculative stockpiling by
criminals would be greatly reduced, if not altogether removed'.[50] By having
regular sales and reducing the speculation that drives up prices, the price
of horn could be brought down, reducing the income that could be illegally
garnered from poached horn. Income would also be generated for state
and private rhino owners to invest in anti-poaching, habitat conservation
and breeding programmes to enable the recovery of numbers.[51] However,
attempts to form a legal trade would be opposed by those who had a strong,
one might even say intransigent, ethical stance against it, and would not agree
to it under any circumstances, a view put forward strongly at a debate with
John Hume in London in August 2016 by Will Travers, the head of Born Free.
Travers also argued that a legal trade would undermine demand reduction
and could provide a conduit for laundering poached horn.[52] Opponents of
trade also argue that legal sales would encourage the use of horn for a variety
of purposes, and with increased demand poaching would persist as legal sales
would not meet demand.[53] Ferreira and Okita-Ouma cautioned in 2012 that
'the provision of rhino horn may stimulate dormant markets because of af-
fordability to a larger fraction of potential consumers', and this had to be
considered when discussing the pros and cons of a legal trade.[54]

Over the last two decades the debates over legal trade have continued,
and at times have become extremely bitter and polarised, 'with few objective
considerations of the costs and benefits of market approaches under different
trade regimes and timeframes', as Milliken and Saw noted as early as 2012.[55]
This polarisation and the vociferous anti-trade campaigns have ensured that
successive southern African proposals to allow a legal regulated trade have
consistently been voted down without getting near the two-thirds majority
of CITES voting members needed to effect a change in the trade regime. This
absolute obstacle to reversing the ban has often led South Africa and other

states to refrain from putting forward further proposals for trade, the failure of CoP17 in 2016 to approve the eSwatini proposal for a legal trade to raise money for rhino conservation being a good example of the problem faced by pro-trade countries and proponents. Between 2009 and 2017, South Africa suspended the legal domestic trade in rhino horn, with part of the motivation for this being to avoid bad publicity about rhino horn at the 2010 football World Cup in South Africa, which was expected to promote tourism. The Environment Minister at the time, Edna Molewa, argued that there were too many obstacles to a legal trade through CITES, and there was no likelihood that South African withdrawal for CITES would work, as the likely horn trading partners – China and Vietnam – would remain CITES members, and so would be prohibited from engaging in legal trade.[56]

Those who favour a legal trade have tried to overcome the objections concerning the possible laundering of poached ivory through a legal trade by arguing that DNA analysis could be used to distinguish between legal and illegal horn, an argument put to me by Hume, Eustace and Wiltshire when I discussed the proposals for a legal trade with them in South Africa in August 2016.[57] Wiltshire and Eustace advocated adopting a structure like the De Beers diamond company's Central Selling Organization, which auctioned diamonds in a tightly controlled process. A legal trade would generate funds for conservation and anti-poaching, and could be used to provide rhino-related income for local communities. Biggs et al. said that evidence from other legal wildlife product trades, such as that in crocodile skin, suggests that legality can reduce incentives for poaching as long as poached horn cannot enter the legal trade (with DNA identification of legal horn the likely answer), a legal trade can deliver sufficient quantities of horn and at a cost-effective price through something like a central selling organisation, and demand does not escalate to dangerous levels that encourage poaching to top up the legal supply.[58] Legal stockpiles held across southern Africa and the regular provision of horn from dehorned rhinos would supply a legal market. In 2016, Hume told me that South Africa had at least 32 tons of horn stockpiled, at least 22 tons of that in government hands. Governments in southern Africa tend to be cagey for obvious reasons (because of thefts) about releasing exact details of their horn stocks. Eustace, who favours a central selling organisation, wrote in 2015 that a legal trade could sell at least 400 horns annually from legal stocks, 300 from natural deaths and 500 from dehornings, which would meet the majority of demand, though he conceded that poaching would continue but at a much lower level. The central selling organisation would set a quota for sales and tight regulation of who could sell, thus helping to prevent corruption.[59] A big fly in the ointment is the history of poor management and corruption in

South Africa's wildlife, government, law enforcement and judicial systems, already detailed in Chapter 9.

I strongly believe that the trade ban has not worked and will not work. Rhino horn is a resource that can be harvested non-lethally and sustainably. It can earn income to encourage breeders, pay rangers and anti-poaching teams realistic salaries, and provide sophisticated surveillance and supply benefits that will gain the support of people around parks, reserves and ranches. However, opposition is strong and there is no chance in the short or medium term that CITES members will vote to lift the ban, while leaving CITES is not really an option for southern African rhino range states that favour a legal trade, as their most likely trading partners in Asia are unlikely to favour leaving CITES. Much more work needs to be done on the nature of a legal trade, safeguards against corruption and laundering of poached horn, and developing a very positive but realistic picture of how rhinos and people would benefit from a legal trade, because the opponents will not give up. Pelham Jones, representing the pro-trade private owners, said he was an optimist in the long term, but that major political changes were needed before South Africa would push for a legal trade and get sufficient support to achieve it, noting that the legal domestic trade in South Africa had not materially benefited rhino owners.[60]

Trophy hunting as a conservation strategy for rhinos

Throughout the narrative concerning rhino conservation in southern Africa, trophy hunting has been referred to as a source of income for private rhino owners, conservancies and governments. It is clearly used as part of a toolkit for conservation that includes tourism, funding from NGOs, governments and philanthropists, and ideas of conservation or rhino bonds to raise finance for rhino conservation, recovery and translocation programmes. Like the efforts to establish a legal trade in rhino horn, trophy hunting is very controversial, with some conservationists and academics publishing on conservation and many animal rights/welfare NGOs – IFAW, AWF, Born Free, Lion Aid and the Campaign to Ban Trophy Hunting, to name a few – passionately opposed to it and constantly campaigning to ban it or at least to get countries to ban imports of trophies from hunting, as happened unsuccessfully in Britain in 2022 and 2023. This section is limited to the role of trophy hunting in conservation, and I would recommend readers who want to go more deeply into this topic to read Nikolaj Bichel's and Adam Hart's comprehensive and masterly examination of the whole issue,[61] together with the IUCN's 2016 briefing paper on hunting and conservation.[62] The objections to trophy hunting are noted, but the primary purpose here is to examine how trophy hunting has been used in rhino conservation.

It is very clear from Chapters 7 and 9 that trophy hunting played a major role, however much its opponents may object on moral/ethical grounds, in the recovery of the white rhino in southern Africa and its translocation from the Hluhluwe-iMfolozi population. Since then, the growth of private ownership of rhinos has involved trophy hunting as a means of funding the cost of managing and protecting white and black rhinos in private ownership and, in Namibia's case in particular, to bring in funding for conservation, anti-poaching, community conservancies and rural livelihoods. In Zimbabwe, large conservancies created from former cattle ranches – Bubye Valley, Bubiana and Savé Valley Conservancy – have also played an important role in securing and breeding rhinos to restore depleted or exterminated populations with trophy hunting, though not now of rhinos, as a major income stream. As Milliken and Shaw conclude, 'trophy hunting is widely regarded as having been a positive force by contributing to biological management, range expansion, the generation of revenue for conservation authorities and incentives for wildlife conservation for a broad range of stakeholders'.[63] In 2012, the trophy hunting industry in South Africa alone involved 500 trophy hunting companies and 3,000 professional hunters supported by other wildlife professionals, veterinarians and taxidermists, with a total of 70,000 people directly employed or in businesses serving the hunting industry.[64] One argument made against trophy hunting is that it could be replaced by photographic safari tourism with no killing of animals involved. This, however, is not a realistic approach. As Leader-Williams wrote in 2009 (and it still applies today), many areas used for trophy hunting are not suitable for photographic safaris owing to the vegetation cover, terrain and lack of infrastructure, which do not enable satisfying game viewing or provide the scenery and the accommodation standard that safari tourists expect; moreover, hunting areas 'remain under conservation management because of the economic incentives that hunting provides ... Recreational hunting may also be an important use of private and communal lands that again remain under conservation management'. He notes that in Zimbabwe and Namibia, hunting on private or communal land has doubled wildlife habitat under conservation.[65]

The argument that trophy hunting reduces populations of endangered species such as black and white rhino and takes out prime breeding males is similarly problematic, as Milliken and Shaw found:

> Rather than hindering population growth, trophy hunting
> is regarded as having positively influenced White Rhino
> numbers and population performance. From a biological
> perspective, sport hunting results in the elimination of
> 'surplus' male animals that otherwise might engage in

fighting and/or kill other rhinos ... selective removals have
served to stimulate breeding performance so that South
Africa's sustained population growth rate for White Rhinos
has remained high. Sport hunting has importantly produced
incentives for the allocation or conversion of large land
areas in the private sector to wildlife-based land use for the
specific purpose of stocking and maintaining rhinos. This ...
has effectively resulted in another 20.5 million hectares of
land being made available for wildlife in South Africa.[66]

The most recent and perhaps strongest arguments for trophy hunting
as a rhino conservation tool have come from 't Sas Rolfes et al., with two
of the co-authors, Mike Knight and Richard Emslie, being veteran and
leading members of the IUCN AfRSG.[67] Their detailed and evidence-based
survey confirms that removal of old males through trophy hunting, far
from damaging the breeding prospects of rhino populations, 'can enhance
population demography and genetic diversity, encourage range expansion,
and generate meaningful socioeconomic benefits to help fund effective
conservation'.[68] They add that regulated trophy hunting of rhinos in southern
Africa has proved sustainable with small numbers, representing a very small
percentage of the populations hunted annually, and with a greater number of
rhinos now in countries that allow hunting than before controlled hunting
began. They contend strongly that 'Terminating this management option
and significant funding source could have negative consequences at a time
when rhinos are being increasingly viewed as liabilities ... conservation of
certain highly threatened species can be supported by cautiously selective
and limited legal hunting.' They believe that the moral objections to trophy
hunting ignore the very important consequential, one might say Benthamite
and utilitarian, advantages to conservation of quota-driven, regulated
hunting, stressing that once the stain of pseudo-hunts had been erased,
trophy hunting played a constructive role and off-take has been limited, with
2,358 white rhinos trophy hunted in South Africa and 61 in Namibia from
1972 to 2018, during which period numbers rose appreciably despite the
heavy poaching in the 2010s, with South Africa earning $154m from hunting
and Namibia earning $18.5m.[69] One of the arguments used against trophy
hunting as a source of income for conservation and for communal economic
development, as well as providing monetary incentives for people to support
rhino conservation, is that the income derived is small. The figure often in-
accurately cited is that only 3% of trophy benefits reach communities. This
figure is from a report by Economists at Large and repeated by a myriad of
anti-hunting NGOs and activists.[70] As Bichel and Hart (and 't Sas Rolfes et al.)

argue,[71] this figure was plucked out of context from a report on hunting for the International Council for Game, Wildlife Conservation and the FAO. The figure cited was a hypothetical one relating to one year in a particular hunting area in Tanzania, and did not include the whole array of benefits accruing to people from trophy hunting.[72]

Naidoo et al., in their study of the role of trophy hunting in conservation in Namibia, looked at income for 77 communal conservancies. They found that the conservancies received only 8–12% of revenue from tourist lodges but 30–75% of the trophy price from hunting, with tourism providing 20–50 jobs and hunting 8–10 jobs for conservancy members.[73] Jobs were the main benefit for communities from tourism, while 64.3% of benefits from trophy hunting were in the form of cash, most of which went to the communities; there was the added benefit of supplies of meat from hunting.[74] The loss of hunting revenue for conservancies would mean income would fall well below operating costs, damaging community benefits and conservation funding.[75]

Trophy hunting, given the distaste with which many view it and the regular campaigns to get Western countries to ban the imports of trophies, thereby reducing the appeal of it to hunters and risking its viability as a business that clearly does support conservation, is constantly under threat from actions and campaigns outside the rhino range states. However, I will give the last word on hunting to Save the Rhino, who take a balanced view and accept that trophy hunting can be a viable tool to support rhino conservation as long as it is strictly controlled with quotas based on biological principles. It can remove bulls past breeding age that may hinder breeding by younger males, and thereby promote genetic diversity within a rhino population, while the income can benefit anti-poaching and rhino management, and provide direct benefits for rhino owners and communal conservancies. Save the Rhino conclude, as I do, that:

> In an ideal world rhinos wouldn't be under such extreme threat and perhaps there would be no need for trophy hunting to generate funds. However, the reality is that rhino conservation is expensive and there are huge pressures to provide enough land and security measures. Parks and programmes that allow trophy hunting can use the funds generated to protect their rhino populations as a whole. Like it or not, trophy hunting has played a key role in the recovery of the Southern white rhino population in South Africa, helping the species recover from the brink of extinction … [and] has a valid role in overall rhino conservation strategies.[76]

Financial models to conserve rhinos

Beyond the campaign for a legalised international trade in rhino horn and the debates over the ethics and efficacy of trophy hunting, the search has continued for means of funding rhino conservation and providing income and empowerment to local communities whose buy-in to supporting rhino conservation is vital. One approach has been to look for investment and financial models that would bring in much-needed funds. One model that has achieved attention and support from bodies such as the ZSL, the World Bank and the Global Environment Facility (GEF, a multilateral environmental fund that provides grants and finance for projects related to biodiversity) is the proposed Wildlife Conservation Bond, or Rhino Bond. This aims to get investors to put money into conservation and then get a return on the investment, based on any increase in South African black rhino numbers. The scheme is backed by the International Bank for Reconstruction and Development, an arm of the World Bank. The target for investment is R152m, which would be used to increase black rhino populations in Addo Elephant NP and the Great Fish River Nature Reserve. The growth rates in the populations would be checked by the ZSL. If population growth is achieved, investors would get their stake back plus about 9% profit.[77]

It would be a payment-for-results mechanism, in which there is a financial incentive for achieving predetermined objectives, in this case increases in rhino numbers in the chosen reserves. Although it would have a single species focus, the bond scheme would be geared to conserving habitat and therefore using the rhino as an umbrella species in whose shade all wildlife in the chosen area would benefit.[78] It is hoped that the scheme will attract 'new, non-traditional donors to rhino conservation efforts, [and] direct them to some of the most important rhino populations'. The measure for a return on the investment would be

> a target net population growth rate that each site must
> aim to achieve ... over the course of the five years ... if the
> net population growth rate target is achieved across the
> sites, then the outcome-payers may pay back the investors
> their original investment plus the pre-agreed yield for
> achieving that target. If the rhino performance is just
> below the target, then the investors may receive back their
> original investment plus a smaller yield; and if the rhino
> performance is extremely poor, the investor may receive
> back less than their original investment. [79]

The project was formally launched in March 2022, with WC/Rhino Bonds available to investors as part of a Development Bond with a potential performance payment from the GEF.[80] So far there has been no further information on uptake or on the progress of rhino conservation in the chosen reserves, which have small rhino populations and are not really at the forefront of rhino conservation in South Africa.

Another move towards helping rhino owners came in November 2023, when the Wilderness Foundation Africa announced that the South African Department of Forestry, Fisheries and the Environment had agreed to provide tax incentives for the conservation of rhinos and lions under Biodiversity Management Agreements, with environment minister Barbara Creech initiating the deal with the Sustainable Finance Coalition under the terms of the Income Tax Act, allowing ordinary South Africans who are safeguarding threatened ecosystems or species to deduct all expenses related to their conservation efforts from their taxable income.[81] It was not clear quite how many private rhino owners would qualify for this, or the extent of the tax write-off when compared with the costs of security, management and so forth.

Community empowerment and ownership

The last section, and perhaps the most important when it comes to long-term conservation and recovery of Africa's rhino populations, concerns the role of communities that live alongside rhinos, whether along the boundaries of NPs and reserves or on private land or communal conservancies that have rhino populations. The approaches already outlined are important, and may work as part of toolkits for rhino conservation according to local circumstances. It is unlikely, though, that any one of them alone is going to work – and they will be best applied in varying combinations that meet the needs in each area. However, what is clear is that without the support and, preferably, the active engagement of communities, conservation success is likely to be unattainable in the short or long term. Where communities have been brought in willingly to take part in conservation, most notably in Namibia but also at the start of the Luangwa programme detailed in Chapter 7 and in the opening years of the CAMPFIRE programme in Zimbabwe, community support has been vital and has achieved success, with CAMPFIRE viewed in the 1990s as the model for community conservation involvement and empowerment.[82] Local communities could opt into CAMPFIRE schemes and negotiate versions of the sustainable-use strategy that would suit them, mixing hunting for meat and wildlife products for sale, sale of trophies and concessions for safari hunting. The plans for CAMPFIRE included the sale of ivory, but the CITES ivory ban in 1989 took away that income stream. However, a major drawback was that local district councils had a major role in the scheme's management in each area and could retain income. CAMPFIRE income was then used by the councils, with less going to communities and a reduced sense of ownership and empowerment on their part as their individual shares in income declined, so it was less successful than it could have been on a community basis and as an incentive to conserve habitat and wildlife rather than poach. The Luangwa scheme foundered as a result of the rent-seeking behaviour of traditional chiefs and local councils, which cut the benefit to communities and devalued the exercises. The success of the Namibian communal conservancies in conservation, particularly of black rhino, is clear, but it is necessary to repeat the SRT view that even in Namibia one of the

> biggest threats is poor conservancy governance. Without functioning conservancies, SRT cannot adequately monitor rhinos as their work heavily relies on conservancies reinvesting into conservation through the provision of rhino rangers for patrols and keeping rhino areas free from settlement and unsustainable tourism. Conservancies are also custodians of wildlife and have a responsibility to ensure transparency and good governance in their activities.[83]

Awareness of the importance of community needs, aspirations and grievances is key to getting people to support rhino conservation, to reject attempts to recruit them to poach and to work with conservation bodies to prevent poaching. In 1988, John Hanks of WWF commented at a conference in Zambia: 'We must have sufficient humility to approach local communities and find out their needs and aspirations before we start work. At all costs we must avoid [imposing] planning upon people, telling them what to do without any form of consultation whatsoever.'[84]

Where people do not benefit from conservation, notably around NPs and protected GRs, as has happened around KNP with bad outcomes, they 'generally feel marginalized and unheard; they do not benefit from the park so do not support the anti-poaching efforts. They dislike the Park's "fortress conservation" approach', and may aid or even become poachers.[85] However, where people feel a benefit, they develop more positive attitudes and 'can play a critical role in protecting animals from outsiders, acting as front-line guards on the edge of the Park'. Similarly, if parks staff are poorly paid or badly managed, they may turn a blind eye to poachers, help them to find rhinos and get horns out, or even turn to poaching themselves as has happened in many areas,[86] not least in KNP and recently Etosha. When rangers at KNP went on strike in 2012 over pay and working conditions, 26 rhinos were poached there in two months.[87]

If poor pay and conditions for staff or a lack of benefit for communities around NPs, GRs and private reserves alienates them, this may make individuals more inclined to side with the poachers. Over-militarised anti-poaching strategies may have similar effects, with communities feeling victimised and treated as criminals because of poaching by some local people. When that happens, all trust breaks down, and communities are likely to refrain from reporting incursions by poachers and to tip off poachers about anti-poaching patrols and the location of rhinos.[88] It is for this reason, as Duffy pointed out, that groups such as the Game Rangers Association of Africa have criticised militarised anti-poaching and the treatment of poaching as a war, emphasising that where army units, special forces and former military personnel are involved or APUs are almost converted into paramilitary units, they may be seen as occupying forces and not protectors of national wildlife resources.[89] Militarised anti-poaching may have short-term results, but almost always escalates the level of violence and the treatment of communities as hostile pools of possible poachers. Rather, as Muntifering et al. argue with regard to Namibia's experience, 'the military-style approach to governance typically does not enrich or motivate local people; illicit trade and organized crime often do', and so the answer is empowering and gaining the support of communities living with wildlife. This can create 'the social foundation that enforcement-based strategies require to be successful', and encourages the local communities to become a first line of defence and to feel a sense of ownership, gaining direct benefits.[90] The key is 'developing an economic and socio-political relationship between the rhinoceros and local communities that harnesses human values to deliver greater return on investment for rhinoceros conservation initiatives … Strategies that recognize individual and communal values, harness normative behaviour, and invest in social capital are likely to hold greater promise for changing and sustaining pro-rhinoceros behaviour.'[91] Sullivan is a critic of trophy hunting and of many aspects of community-based conservation projects, seeing them as 'opportunities arising from globalisation processes under neoliberalism' involving paternalism and colonialism, and as 'service-providers' for foreign tourists, but even she admits that 'initiatives have thus facilitated the means whereby rural people can access monetary, employment and other oppor-tunities … They have also generated important institutional frameworks for building and enhancing local infrastructural and governance structures related to livelihood activities connecting people and landscapes.'[92]

Community-based initiatives and systems such as those of the Namibian community conservancies may not be perfect and may have flaws in terms of governance, internal democracy and dependence on income from foreign tourists, hunters and support from foreign donors, but they have achieved considerable successes in Namibia in conservation of rhino and other wildlife, such as desert-adapted lions and elephants. They are not the finished article, but they provide a foundation that can be built on and models that can be replicated to develop community-based conservation where it is far from perfect, as in Kenya, and where rent-seeking institutional rivalries have devalued earlier brave efforts, as in Zambia and Zimbabwe. Combined with translocations, dehorning, regulated hunting and, perhaps in the distant future, a controlled international legal trade in non-mortality rhino horn that would reward conservation-supporting communities and locally supported anti-poaching, community involvement is vital. Communities, their well-being, sense of ownership of wildlife and their support for conservation are the key to the future of rhino conservation.

Notes

Introduction

1. Richard Emslie and Martin Brooks (1999) *Status Survey and Conservation Action Plan: African Rhino*. Gland, Switzerland: IUCN/SSC AfRSG, p. v.
2. Jasmine Sidney (1961) The past and present distribution of some African ungulates. *Transactions of the Zoological Society of London*, accepted 13 June 1961 and published 1965, p. 51. See also Save the Rhino (n.d.) The rise and fall of the Northern white rhino. https://www.savetherhino.org/rhino-species/white-rhino/the-last-northern-white-rhinos/#:~:text=As%20late%20as%201960%2C%20there,in%20Garamba%20National%20Park%2C%20DRC. Accessed 17 July 2023.
3. Emslie and Brooks (1999), p. 10.
4. Mike Knight, Keitumetse Mosweub and Sam M. Ferreira (2023) AfRSG chair report, *Pachyderm* 64: 13–30, pp. 13–16.
5. Forestry, Fisheries and Environment on Rhino Poaching (2024) https://www.gov.za/news/media-statements/forestry-fisheries-and-environment-rhino-poaching-27-feb-2024. Accessed 28 February 2024.

Chapter 1: Evolution, status and behaviour

1. Kristin Nowell (2012) Assessment of rhino horn as a traditional medicine. A report prepared for the CITES Secretariat. TRAFFIC. http://www.rhinoresourcecenter.com/pdf_files/138/1389957235.pdf, p. 1. Accessed 16 August 2023.
2. D.J. Pienaar, A.J. Hall-Martin and P.M. Hitchens (1991) Horn growth rates of free-ranging white and black rhinoceros. *Koedoe*, 34(2): 97–105, p. 98.
3. Ibid., p. 105.
4. D.H.M. Cumming, R.F. Du Toit and S.N. Stuart (1990) *African Elephants and Rhinos Status Survey and Conservation Action Plan*. Gland, Switzerland: IUCN/SSC African Elephant and Rhino Specialist Group, p. 3.
5. Esmond Bradley Martin and Chryssee Bradley Martin (1982) *Run Rhino Run*. London: Chatto and Windus, p. 11.
6. Ross Barnett (2019) *The Missing Lynx: The Past and Future of Britain's Lost Mammals*. London: Bloomsbury, p. 134.
7. Jonathan Kingdon (2013) Mammalian evolution in Africa. In Jonathan Kingdon et al. (eds) *Mammals of Africa, Volume 1*. London: A&C Black, 75–100, p. 78.
8. Ibid., p. 77.
9. Robin S. Reid (2012) *Savannas of Our Birth: People, Wildlife and Change in East Africa*. Berkeley: University of California Press, p. 83.
10. Ibid.
11. The Miocene epoch. https://ucmp.berkeley.edu/tertiary/miocene.php. Accessed 28 May 2023.

12. Shanlin Liu et al. (2021) Ancient and modern genomes unravel the evolutionary history of the rhinoceros family. *Cell* 184: 4874–85, p. 4874. See also Esperanza Cerdeñ (1998) Diversity and evolutionary trends of the Family Rhinocerotidae (Perissodactyla). *Palaeogeography, Palaeoclimatology, Palaeoecology* 141: 13–34, p. 13.

13. Cerdeñ (1998), p. 20.

14. Ibid.

15. Ibid., p. 13.

16. Claude Guérin (2003) Miocene Rhinocerotidae of the Orange River Valley Namibia. *Memoir Geological Survey Namibia*, 19: 257–81, p. 257.

17. Shanlin Liu et al. (2021), p. 4874.

18. Ibid.

19. Ibid.

20. Richard G. Klein and Katharine Scott (1986) Re-analysis of faunal assemblages from the Haua Fteah and other Late Quaternary archaeological sites in Cyrenaican Libya. *Journal of Archaeological Science* 13: 515–42, p. 523.

21. Norman Owen-Smith (1999) Ecological links between African savanna environments, climate change, and early hominid evolution. In Timothy G. Bromage and, Friedemann Schrenk (eds) *African Biogeography, Climate Change, and Human Evolution (Human Evolution Series)*. Oxford: Oxford University Press, pp. 138–149, p. 142.

22. Denis Geraads and Samir Zouhri (2021) A new late Miocene elasmotheriine rhinoceros from Morocco. *Acta Palaeontologica Polonica* 66(4): 753–65, p. 753.

23. Ibid.

24. Cumming et al. (1990), p. 3.

25. Kees Rookmaaker (2013) Rhinocerotidae. In Jonathan Kingdon and Michael Hoffmann (eds) *Mammals of Africa, Volume V*. London: Bloomsbury, pp. 444–5, p. 444.

26. Luca Pandolfia, Florent Rivals and Rivka Rabinovich (2020) A new species of rhinoceros from the site of Bethlehem: '*Dihoplus*' *bethlehemsis* sp. nov. (Mammalia, Rhinocerotidae). *Quaternary International* 537: 48–60, pp. 48–9.

27. Cumming et al. (1990), p. 3.

28. Claude Guerin (1987) A brief paleontological history and comparative anatomical study of the recent rhinos of Africa. *Pachyderm* 9: p. 5.

29. Fátima Sánchez-Barreiro et al. (2023) Historic sampling of a vanishing beast: population 3 structure and diversity in the black rhinoceros. *Molecular Biology and Evolution* 40(9): no page numbers. https://doi.org/10.1093/molbev/msad180. Accessed 1 August 2023.

30. Ibid.

31. Ibid.

32. Eric H. Harley et al. (2016) Comparison of whole mitochondrial genome sequences of northern and southern white rhinoceroses (*Ceratotherium simum*): the conservation consequences of species definitions. *Conservation Genetics* 17 (4): no page numbers. https://doi.org/10.1007/s10592-016-0861-2. Accessed 17 July 2023.

33. R. Emslie (2020) *Diceros bicornis* ssp. *longipes*, *The IUCN Red List of Threatened Species* 2020, https://dx.doi.org/10.2305/IUCN.UK.2020-1.RLTS.T39319A45814470.en. Accessed 23 August 2022.

34. Emslie and Brooks (1999), p. 2.

35. Emslie (2020).

36. C. Groves and C. Grubb (2011) *Ungulate Taxonomy*. Baltimore: The Johns Hopkins University Press, p. 317.

37. R. Emslie (2020) *Diceros bicornis. The IUCN Red List of Threatened Species* 2020. e.T6557A152728945. https://dx.doi.org/10.2305/IUCN.UK.2020-1.RLTS.T6557A152728945.en. Accessed 1 May 2023.

38. Norman Owen-Smith and Joel Berger (2006) Rhinoceroses. In David W. Macdonald (ed.) *The Princeton Encyclopedia of Mammals*, pp. 698–703. Princeton, NJ: Princeton University Press, p. 698.

39. Save the Rhino (n.d.) Black Rhino, https://www.savetherhino.org/rhino-info/rhino-species/black-rhinos/. Accessed 2 October 2023.

40. Simson !Uri#khob (2020) Attitudes and perceptions of local communities towards the re-introduction of black rhino (*Diceros bicornis bicornis*) into their historical range in northwest Kunene Region, Namibia. *Future Pasts Working Papers No. 8.* https://www.futurepasts.net/_files/ugd/5ba6bf_e142d4b9b70b4bfeb9c95f24a0c88bc1.pdf, p. 8. Accessed 6 February 2023.

41. Nick Carter (1965) *The Arm'd Rhinoceros.* London: Andre Deutsch, p. 13.

42. Emslie and Brooks (1999), p. 2.

43. L.C. Rookmaaker (2004) Historical distribution of the black rhinoceros (*Diceros bicornis*) in West Africa. *African Zoology* 39(1): 1–8, p. 1. See also T.J. Foose and R.W. Reece (eds) (1998) *AZA SSP Rhinoceros Masterplan.* AZA Rhino Advisory Group, p. 4.

44. R. Emslie (2020) *Diceros bicornis* ssp. *longipes. The IUCN Red List of Threatened Species* 2020, https://www.iucnredlist.org/species/39319/45814470. Accessed 2 October 2023.

45. Isabelle Lagrot, Jean-Francois Lagrot and Paul Bour (2007) Probable extinction of the western black rhino, *Diceros bicornis longipes*: 2006 survey in northern Cameroon. *Pachyderm* 43, July–December: 19–28, p. 19.

46. R. Emslie (2020) *Diceros bicornis* ssp. *minor. The IUCN Red List of Threatened Species* 2020: e.T39321A152729173, https://www.iucnredlist.org/species/39321/152729173. Accessed 2 October 2023.

47. Sánchez-Barreiro et al. (2023).

48. Sam Ferreira, Save the Rhino webinar, 27 September 2023. https://www.savetherhino.org/rhino-info/population-figures/. Accessed 2 October 2023.

49. Richard H. Emslie and Kerin Adcock (2013) Black rhinoceros. In Jonathan Kingdon and Michael Hoffmann (eds), *Mammals of Africa*, pp. 455–66, p. 456.

50. Kes Hillman Smith (1985) Black rhino behaviour, pp. 9–11. http://www.rhinoresource center.com/pdf_files/125/1255247232.pdf. Accessed 12 April 2023, p. 11.

51. M. 't Sas Rolfes et al. (2022) Legal hunting for conservation of highly threatened species: The case of African rhinos. *Conservation Letters* e12877, pp. 1–9. https://doi.org/10.1111/conl.12877. Accessed 31 July 2023, p. 3.

52. Halszka Hrabar and Johan T. du Toit (2005) Dynamics of a protected black rhino (*Diceros bicornis*) population: Pilanesberg National Park, South Africa. *Animal Conservation* 8: 259–267, p. 263.

53. Rudolf Schenkel and Lotte Schenkel-Hulliger (1969) *Ecology and Behaviour of the Black Rhinoceros (Diceros bicornis L.).* Hamburg: Verlag Paul Parey, pp. 14–15.

54. John Mukinya (1977) Feeding and drinking habits of the black rhinoceros in Masai Mara Game Reserve. *East African Wildlife Journal* 15: 125–38, p. 133.

55. Krisztián Gyöngyi and Morten Elmeros (2017) Forage choice of the reintroduced black rhino and the availability of selected browse species at Majete Wildlife Reserve, Malawi. *Pachyderm* 58: 40–50, p. 48.

56. Hillman Smith (1985), p. 11.

57. Ibid.

58. Emslie and Adcock (2013), p. 459.

59. Hillman Smith (1985), p. 10.

60. Eric Dinerstein (2011) Family Rhinocerotidae (Rhinoceroses). In Don E. Wilson and Russell A. Mittermeier (eds) *The Mammals of the World. 2. Hoofed Mammals*, pp. 144–81. Barcelona: Lynx Edicions, p. 165.

61. Nikki le Roex et al. (2019) Seasonal space-use and resource limitation in free-ranging black rhino. *Mammalian Biology* 99: 81–7, p. 82.

62. WWF (n.d.) 10 facts about rhino. https://www.wwf.org.uk/learn/fascinating-facts/rhinos#:~:text=The%20names%20of%20black%20and,have%20a%20pointy%20upper%20lip).Accessed 3 October 2023.

63. Kes Hillman-Smith, Mankoto ma Oyisenzoo and Fraser Smith (1986) A last chance to save the northern white rhino? *Oryx* 20(1): 20–6, p. 20.

64. E.L. Trouessart (1909) Le Rhinoceros blanc du Soudan. *Proceedings of the Zoological Society of London* 79 (1): 199.

65. Hillman-Smith, Oyisenzoo and Fraser Smith (1986), p. 20.

66. Ibid.

67. Emslie and Brooks (1999), p. 8.

68. T.J. Foose and R.W. Reece (eds) (1998) *AZA SSP Rhinoceros Masterplan*. AZA Rhino Advisory Group, p. 5.

69. International Rhino Foundation (2023) *State of the Rhino*. https://rhinos.org/about-rhinos/state-of-the-rhino/. Accessed 5 October 2023. See also Norman Owen-Smith (2013) White rhinoceros. In Kingdon and Hoffmann, *Mammals of Africa*, pp. 445–55, p. 447.

70. IUCN Red List. https://www.iucnredlist.org/species/4185/45813880. Accessed 5 October 2023.

71. Ian Player (1972) *The White Rhino Saga*. London: Collins, p. 139.

72. Norman Owen-Smith (1973) *The Behavioural Ecology of the White Rhinoceros*. PhD thesis, University of Wisconsin, p. xxi.

73. Player (1972), p. 140.

74. Owen-Smith (1973), p. xxi.

75. A.J. Hall-Martin et al. (1993) Determination of species and geographic origins of rhinoceros horn by isotopic analysis and its possible application to trade control. In Oliver A. Ryder (ed.) *Rhinoceros Biology and Conservation*, pp. 123–35. San Diego, CA: Zoological Society of San Diego, p. 128.

76. Sam M. Ferreira et al. (2019) Species-specific drought impacts on black and white rhinoceroses, *PLoS ONE* 14 (1): no page numbers.

77. M.A. Miller et al. (2018) Conservation of white rhinoceroses threatened by bovine tuberculosis, South Africa, 2016–2017. *Emerging Infectious Diseases* 12: 2373–5.

78. B.L. Penzhorn et al. (1994) *Proceedings of a symposium on rhinos as game ranch animals. Onderstepoort, Republic of South Africa*, 9–10 September, pp. 1–242, pp. i–iv.

79. Owen-Smith (1973), p. xxi.

80. Ibid.

81. Janet L. Richlow, John Kie and Joel Berger (1999) Territoriality and spatial patterns of white rhinoceros in Matobo National Park, Zimbabwe. *African Journal of Ecology* 37: 295–304, p. 296.

82. Ibid., p. 302.

83. D.J. Pienaar, J. du P. Bothma and G.K. Theron (1993) White rhinoceros range size in the south-western Kruger National Park, *Journal of Zoology* 229(4): 641–9, p. 641.

84. Ibid., p. 646.

85. Samuel G Penny et al. (2022): Changes in social dominance in a group of subadult white rhinoceroses (*Ceratotherium simum*) after dehorning. *African Zoology*: 1–11. https://doi.org/10.1080/15627020.2022.204615. Accessed 6 April 2022, pp. 1–2.

86. Ibid.

87. Ibid., p. 6.

88. Owen-Smith (2013), p. 450.

89. Dinerstein (2011), pp. 59–60.

90. Owen-Smith (2021), pp. 148–9.

91. M. Shrader, N. Owen-Smith and J.O. Ogutu (2006) How a mega-grazer copes with the dry season: food and nutrient intake rates by white rhinoceros in the wild. *Functional Ecology* 20: 376–84, p. 376.

92. C. Brain, O. Forge and P. Erb (2001) Lion predation on black rhinoceros (*Diceros bicornis*) in Etosha National Park. *African Journal of Ecology* 37(1): 107–9, p. 107.

93. Ibid.

94. Ibid., p. 453.

95. Bradley Martin and Bradley Martin (1982), p. 27.

96. Garth Owen-Smith (2010) *An Arid Eden*. Johannesburg: Jonathan Ball, p. 10.

97. Southern African Wildlife College (2022) An extraordinary hyena attack. https://wildlifecollege.org.za/an-extraordinary-hyena-attack/. Accessed 4 October 2022.

98. Ibid.

99. Schenkel and Schenkel-Hulliger (1969), p. 24.

100. Joel Berger and Carol Cunningham (1998) Behavioural ecology in managed reserves: gender-based asymmetries in interspecific dominance in African elephants and rhinos. *Animal Conservation* 1: 33–8, p. 35.

101. Player (1967), p. 147.

102. Liz Sly (1994) A murder mystery: Why were elephants slaughtering rhinos? – Lack of adult role models gets the blame. *Seattle Times*, 23 October. https://archive.seattletimes.com/archive/?date=19941023&slug=1937416. Accessed 6 October 2023.

103. Berger and Cunningham (1998), p. 36.

104. Niki Moore (2002) Operation father figure. *Mail and Guardian*, 24 May. https://mg.co.za/article/2002-05-24-operation-father-figure/#:~:text=classic%20murder%20mystery.-,In%201994%20rhinos%20were%20being%20found%20dead%20in%20the%20Pilanesberg,revenge%20on%20these%20smaller%20animals. Accessed 6 October 2023.

105. Ibid., p. 26.

106. Daphne Sheldrick (2012) *An African Love Story*. London: Penguin, pp. 116–18.

107. M. Landman, D.S. Schoeman and D.I.S. Kerley (2013) Shift in black rhinoceros diet in the presence of elephant: evidence for competition? *PLoS ONE* 8(7): 1–8, p. 1.

108. Ibid., p. 7.

109. Owen-Smith (1999), p. 148.

110. L.S.B. Leakey (1961) *The Progress and Evolution of Man in Africa*. Oxford: Oxford University Press, pp. 9 and 40.

Chapter 2: From 20,000 BCE to European penetration

1. Owen-Smith (2010), p. 389.

2. Norman Owen-Smith (2021) *Only in Africa: The Ecology of Human Evolution*. Cambridge: Cambridge University Press, p. 315.

3. L.C. Rookmaaker (2004) Historical distribution of the black rhinoceros (*Diceros bicornis*) in West Africa. *African Zoology* 39 (1): 1–8, p. 1.

4. Mary Leakey (1983) *Africa's Vanishing Art*. New York: Doubleday, pp. 20–1.

5. Ibid., pp. 29 and 31.

6. Reid (2012), p. 95.

7. Christopher Ehret (2016) *The Civilizations of Africa: A History to 1800*. Charlottesville, Virginia: University of Virginia Press, p. 37.

8. Ibid., pp. 41–2.

9. John Iliffe (2007) *Africans: The History of a Continent*. Cambridge: Cambridge University Press, 2nd edition, pp. 12–3.

10. Ehret (2016), pp. 72–5.

11. Christopher Ehret (1998) *An African Classical Age: Eastern and Southern Africa in World History 1000 B.C. to A.D. 400*. Oxford: James Currey, p. 6

12. Lawrence Barham and Peter Mitchell, *The First Africans: African archaeology from the earliest toolmakers to most recent foragers*. Cambridge: Cambridge University Press, Kindle edition, p. 344.

13. Ehret (2016), pp. 79 and 86.

14. Iliffe (2007), p. 17.

15. Reid (2012), p. 96.

16. Juliet Clutton-Brock (1994) The legacy of Iron Age dogs and livestock in Southern Africa. *Azania: Archaeological Research in Africa* 29–30 (1): 161–7, p. 164; Frank W. Marlowe (2010) *The Hadza Hunter-Gatherers of Tanzania*. Berkeley: University of California Press, p. 18.

17. Clutton-Brock (1994), p. 27; R. Woodroffe and L.G. Frank (2005) Lethal control of African lions (*Panthera leo*): local and regional population impacts. *Animal Conservation* 8: 91–8.

18. Edwin N. Wilmsen (2017) Baubles, bangles and beads: commodity exchange between the Indian Ocean region and Interior Southern Africa during 8th–15th centuries. *Journal of Southern African Studies* 43(5), 913–26, p. 920.

19. Clive A. Spinage (2012) *African Ecology: Benchmarks and Historical Perspectives.* Heidelberg: Springer, p.72

20. Dale J. Osborn with Jana Osbornová (1998) *The Mammals of Ancient Egypt. Natural History of Egypt, Volume IV.* Warminster: Aris and Phillips, p. 3.

21. Richard Carrington (1972) Animals in Egypt. In A. Houghton Brodrick (ed.), *Animals in Archaeology*, pp. 69–89. London: Barrie and Jenkins, pp. 74–5.

22. Osborn (1998), p. 3.

23. Martin (1982), pp. 12–13.

24. T.H. White (trans. and ed.) (1984) *The Book of Beasts.* New York: Dover Publications, pp. 20–1.

25. Peter I. Bogucki (2008) *Encyclopedia of Society and Culture in the Ancient World.* New York: Facts On File, p. 389.

26. Adu Boahen (1962) The caravan trade in the nineteenth century. *Journal of African History* 2: 349–59, p. 349.

27. Ibid.

28. Martin, 1982, p. 13.

29. Kees Rookmaaker (2013) Rhinocerotidae. In Kingdon and Hoffmann, *Mammals of Africa*, pp. 444–45.

30. P. Salama (1990) The Sahara in classical antiquity. In G. Mokhtar (ed.) *General History of Africa, II: Ancient Civilizations of Africa*, pp. 286–95. London: James Currey/UNESCO, p. 293.

31. Roland Auguet (1972) *Cruelty and Civilization: The Roman Games.* London: Routledge, p. 84.

32. Ibid., p. 85.

33. H. Rackham (ed. and trans.) (1958) *Pliny's Natural History.* Cambridge, MA: Harvard University Press.

34. John Bostock and H.T. Riley (eds) (1855) *The Natural History. Pliny the Elder.* London. Taylor and Francis, http://www.perseus.tufts.edu/hopper/text?doc=Perseus%3Atext%3A1999.02.0137%3Abook%3D8%3Achapter%3D29#note1. Accessed 27 June 2022.

35. T.V. Buttrey (2007) Domitian, the rhinoceros, and the date of Martial's *Liber De Spectaculis. Journal of Roman Studies* 97: 101–12, p. 101.

36. Ibid., p. 106.

37. Ibid.

38. Oppian (1928) *Cynegetica*, published in the Loeb Classical Library, no page numbers. https://penelope.uchicago.edu/Thayer/E/Roman/Texts/Oppian/Cynegetica/4*.html. Accessed 15 August 2022.

39. Aelian (n.d.) On the nature of animals. http://www.attalus.org/info/aelian.html. Accessed 27 June 2022, Book 17, p. 44.

40. Wilfred Harvey Schoff (1912) *The Periplus of the Erythraean Sea: Travel and Trade in the Indian Ocean.* New York: Longmans Green, p. 9.

41. Ibid.

42. Basil Davidson (1998) *West Africa Before the Colonial Era. A History to 1850.* London: Pearson, p. 19.

43. Martin (1982), p. 14.

44. Ibid., p. 53.

45. Ibid.

46. Paul Lane (2004) The 'moving frontier' and the transition to food production in Kenya, Azania. *Archaeological Research in Africa* 39(1): 243–64, p. 248.

47. Ibid., p. 252.

48. Philip Curtin et al. (1995) *African History From Earliest Times to Independence.* London: Longman, p. 104.

49. Diane P. Gifford et al. (1980) Evidence for predation and pastoralism at prolonged drift: a pastoral Neolithic site in Kenya. *Azania: Archaeological Research in Africa* 15(1): 57–108, pp. 64–5.

50. M. Tefera (2004) Recent evidence of animal exploitation in the Axumite epoch, 1st–5th centuries AD. *Tropical Animal Health and Production* 36: 105.

51. F. Anfrey (1990) The civilization of Axum from the first to the seventh century. In Mokhtar, *General History of Africa*, pp. 203–13, p. 213.

52. David Wright (2003) Archaeological investigations of three pastoral Neolithic sites in Tsavo National Park, Kenya. *Azania: Archaeological Research in Africa* 38 (1): 183–8, p. 187.

53. Ibid., pp. 107–9.

54. J.E.G. Sutton (1973) *Early Trade in Eastern Africa*. Nairobi: East African Publishing House, summary page.

55. Ibid., p. 17.

56. A.M.H. Sheriff (1990) The East African coast and its role in maritime trade. In Mokhtar, *General History of Africa*, pp. 306–12, p. 311.

57. Ehret (1998), p. 276.

58. Ehret (2016), p. 202.

59. Daniel Martin Varisco (1989) Beyond rhino horn – wildlife conservation for North Yemen. *Oryx* 23(4): 215–9, p. 216.

60. Iliffe (2007), p. 54.

61. Varisco (1989), p. 216.

62. I. Hrbek (1992) *Africa from the Seventh to the Eleventh Century*. London: James Currey/UNESCO, p. 293.

63. V Matveiev (1997) The development of Swahili civilization. In J. Ki-Zerbo and D.T. Niane (eds) *General History of Africa, Volume IV: Africa from the Twelfth to the Sixteenth Century*, pp. 181–90. Oxford: James Currey/UNESCO, p. 183.

64. Keith Somerville (2019) *Ivory: Power and Poaching in Africa*. London: Hurst and Company (updated edition), p. 22.

65. Philip Snow (1988) *The Star Raft: China's Encounter with Africa*. London: Weidenfeld & Nicolson, p. 5.

66. Ibid., p. 25.

67. Teeputai Saengas (2022) Rhino horn art and ancient Chinese cultural society. *Journal of Positive School Psychology* 6(9): 574–83, p. 583. Available at http://journalppw.com.

68. Bradley Martin and Bradley Martin (1982), p. 53.

69. Ibid.

70. Ibid., 578. See also N.F. Singer (1991) Rhino horn and elephant ivory. *Arts of Asia* 21 (5): 98–105.

71. Saengas (2022), p. 577.

72. Emslie and Brooks (1999), p. 28.

73. Alfredo González-Ruibal et al. (2014) Late hunters of western Ethiopia: the sites of Ajilak (Gambela), c. AD 1000–1200. *Azania: Archaeological Research in Africa* 49(1): 64–101, p. 95.

74. Leo Africanus (aka Al-Hassan ibn-Mohammed al-Wezaz al-Fasi) (1896) *The History and Description of Africa of Leo Africanus, Volume 1*. London: Hakluyt Society. Reprinted by Forgotten Books, p. 34.

75. Ibid., p. 36.

76. Isabel Boavida et al. (eds) (2011) *Pedro Páez's History of Ethiopia, Volume I, 1622*. Farnham: Hakluyt Society, p. 228.

77. Ibid.

78. L.C. Rookmaaker and R. Kraft (2011) The history of the unique type of *Rhinoceros cucullatus*, with remarks on observations in Ethiopia by James Bruce and William Cornwallis Harris (Mammalia, Rhinocerotidae). *Spixiana* 34(1): 133–44, p. 133.

79. Ibid.

80. Ibid., p. 134.

81. L.C. Rookmaaker (2004) Historical distribution of the black rhinoceros (*Diceros bicornis*) in West Africa. *African Zoology* 39(1): 1–8, p. 1.

82. Basil Davidson (1998) *West Africa Before the Colonial Era. A History to 1850*. London: Longman, p. 10.

83. Keith Somerville (2017) *Africa's Long Road Since Independence: The Many Histories of a Continent*. London: Penguin, pp. 167–70.

84. Curtin et al. (1995), pp. 84–5.

85. Ibid., p. 69.

86. Davidson (1998), p. 193.

87. Alan Barnard (1992) *Hunters and Herders of Southern Africa. A Comparative Ethnography of the Khoisan Peoples*. Cambridge: Cambridge University Press, pp. 30–2.

88. UNESCO (n.d.) Tsodilo, https://whc.unesco.org/en/list/1021/. Accessed 28 July 2022.

89. Ibid.

90. Laurens van der Post and Jane Taylor (1984) *Testament to the Bushmen*. London: Viking, p. 37.

91. H.H. Roth (1967) White and black rhinoceros in Rhodesia. *Oryx* 9(3): 217–31, pp. 217–18.

92. Sian Sullivan and Jeff Muntifering (2020) A foreword: historicising black rhino in west Namibia. In !Uri︤khob, *Attitudes and perceptions of local communities*, p. 2.

93. Eugene Joubert (1971) The past and present: distribution and status of the black rhinoceros (*Diceros bicornis* Linn. 1758) in South West Africa. *Madoqua* 1(4): 33–43, pp. 34–6.

94. Ibid., p. 36.

95. Personal communication from San trackers in Botswana, 2015 and 2018.

96. Ehret (2016), 165.

97. Ibid., p. 271.

98. Hrbek (1992), p. 318.

99. David Birmingham (1983) Society and economy before AD 1400. In David Birmingham and Phyllis Martin (eds) *History of Central Africa, Volume One*. London: Longman, pp. 1–29, p. 24.

100. John Giblin (2016) Meet the 800-year-old golden rhinoceros that challenged apartheid South Africa. *The Conversation*, 16 September. https://theconversation.com/meet-the-800-year-old-golden-rhinoceros-that-challenged-apartheid-south-africa-64093. Accessed 22 August 2022.

101. Roland Oliver and Anthony Atmore (1975) *Medieval Africa 1250–1800*. Cambridge: Cambridge University Press, p. 203.

102. Graeme Barker (1978) Economic models for the Manekweni Zimbabwe, Mozambique. *Azania: Archaeological Research in Africa* 13(1): 71–100, p. 71.

103. Somerville (2019), p. 20.

104. Emslie and Brooks (1999), p. 26.

105. Edward A. Alpers (1975) *Ivory and Slaves: Changing patterns of international trade in East Central Africa to the later nineteenth century*. Berkeley, CA: University of California Press, pp. 58 and 63.

106. Peter Delius, Tim Maggs and Maria Schoeman (2012) Bokoni: old structures, new paradigms? Rethinking pre-colonial society from the perspective of the stone-walled sites in Mpumalanga. *Journal of Southern African Studies* 38(2): 399–414, p. 399.

107. Ibid., p. 411.

108. Ibid., p. 413.

109. Somerville (2019), pp. 12–24. See also Philip Curtin et al. (1995) *African History. From Earliest Times to Independence*. London: Longman, pp. 206 and 360; John Iliffe (2007) *Africans: The History of a Continent*. Cambridge: Cambridge University Press, p. 74 for West Africa, p. 187 for East Africa, p. 171 for Ethiopia, p. 127 for southern Africa.

110. Jan van Riebeeck (ed. H.C.V. Leibrandt) (1897) *Precis of the Archives of the Cape of Good Hope, December 1651–December 1653, Riebeeck's Journal, Part I*. Cape Town: W.A. Richards & Sons, Government Printers, p. 78.

111. J.L. Cloudsley-Thompson (1967) *Animal Twilight: Man and game in eastern Africa*. London: G.T. Foulis and Co., p. 52.

112. Ibid., p. 89.

113. Jan van Riebeeck (ed. H.C.V. Leibrandt) (1897) *Precis of the Archives of the Cape of Good Hope, December 1651–December 1653, Riebeeck's Journal, Part II*. Cape Town: W.A. Richards & Sons, Government Printers, p. 158.

114. Ibid., p. 186.

115. Jan van Riebeeck (ed. H.C.V. Leibrandt) (1897) *Precis of the Archives of the Cape of Good Hope, December 1651–December1653, Riebeeck's Journal, Part III.* Cape Town: W.A. Richards & Sons, Government Printers, p. 33.

116. Ibid., p. 230.

117. L.C. Rookmaaker (2005) Review of the European perception of the African rhinoceros. *Journal of Zoology* 265: 365–376, p. 367.

118. Ibid., p. 367.

119. Ibid.

120. Sullivan et al. (2021), pp. 6–7.

121. Joubert (1971), p 34.

122. Heinrich Vedder (1966) *South West Africa in Early Times.* London: Frank Cass, digital bookshelf edition, p. 36.

123. Martyn Woodward (2011) *A Monstrous Rhinoceros (as from life): Toward (and beyond) the epistemological nature of the enacted pictorial image.* Plymouth: Transtechnology Research, Plymouth University. http://www.rhinoresourcecenter.com/pdf_files/157/1572634434.pdf. Accessed 27 June 2022, pp. 10–11.

124. E. Gombrich (1960) *Art and Illusion: A Study in the Psychology of Pictorial Representation.* London: Phaidon, pp. 66–7.

Chapter 3: European penetration and occupation

1. Reid (2012), p. 105.

2. Bradley Martin and Bradley Martin (1982), p. 32.

3. Ibid.

4. Nigel Leader-Williams (1992) *The World Trade in Rhino Horn: A Review.* Cambridge: TRAFFIC. https://portals.iucn.org/library/sites/library/files/documents/Traf-004.pdf, accessed 12 September 2023, p. 6 See also Ronald Orenstein (2013) *Ivory, Horn and Blood: Behind the Elephant and Rhinoceros Poaching Crisis.* Buffalo, NY: Firefly, p. 67.

5. Convention for the Preservation of Wild Animals, Birds, and Fish in Africa, 19 May 1900 (1904) *Journal of the Society for the Preservation of the Fauna of the Empire* (henceforth *Journal*) 2: 29.

6. Ibid., p. 32.

7. L. C. Rookmaaker (2005) Review of the European perception of the African rhinoceros. *Journal of Zoology* 265: 365–76, p. 372.

8. Rookmaaker (2005), p. 372.

9. W.H. Drummond (1876) On the African rhinoceroses, *Proceedings of the Zoological Society of London* (henceforth *Proceedings*) 8: 109–14, p. 109.

10. F.C. Selous (1881) On the South African rhinoceroses. *Proceedings* 49(3): 725.

11. R. Emslie (2020) *Diceros bicornis* ssp. *longipes, The IUCN Red List of Threatened Species 2020.* https://dx.doi.org/10.2305/IUCN.UK.2020-1.RLTS.T39319A45814470.en. Accessed 23 August 2022.

12. Emslie and Brooks (1999), p. 2.

13. Player (1972), p. 32.

14. Emslie and Brooks (1999), pp. 3–4.

15. Ibid.

16. Abdul Sherrif (1987) *Slaves, Spices & Ivory in Zanzibar.* Oxford: James Currey, p. 12.

17. Ibid.

18. Somerville (2017), pp. 6–7.

19. Reid (2012), p. 107.

20. Thomas P. Ofcansky (2002), *Paradise Lost: A history of game preservation in East Africa,* Morganstown, West Virginia: University of West Virginia Press, p. 3.

21. Extract from Sir Charles Eliot's Reports for the British East Africa Protectorate for the years 1902 and 1903. *Journal,* 1904, Appendix 7, p. 50.

22. Huxley to Ofcansky, in Ofcansky (2002), p. 11.
23. Ibid., p. 50.
24. *Journal* (1905), p. 56.
25. Ibid., p. 85.
26. *Journal* (1906), p. 34.
27. Ian Parker (2001) Jackson and Percival. In Ian Parker and Stan Bleazard (eds) *An Impossible Dream*, pp. 13–31. Kinloss: Librario, p. 13.
28. J.H. Patterson (1907) *The Man-Eaters of Tsavo*. London: Macmillan.
29. Parker (2001), p. 22.
30. Abel Chapman (1908) *On Safari: Big game hunting in British East Africa*. London: Longmans, Green. Kindle edition, loc. 841.
31. Ibid., loc. 1264.
32. Ibid., loc. 1316.
33. Kálmán Kittenberger (1989) *Big Game Hunting and Collecting in East Africa*. New York: St Martin's Press, pp. 45–6.
34. Major Robert W. Foran (1933, reprinted in 2017) *Kill or be Killed: The rambling reminiscences of an amateur hunter*. London: Hutchinson & Co., p. 187.
35. Ibid., p. 188.
36. Ibid., p. 193.
37. John M. Mackenzie (1987) Chivalry, social Darwinism and ritualised killing: the hunting ethos in Central Africa up to 1914. In David Anderson and Richard Grove (eds), *Conservation in Africa: People, policies and practice*, pp. 41–61. Cambridge: Cambridge University Press, p. 41.
38. Ibid., p. 3.
39. Somerville (2019), p. 67.
40. Theodore Roosevelt (1910) *African Game Trails: An account of the African wanderings of an American hunter-naturalist*. New York: Charles Scribner's Sons, Kindle edition.
41. Frederick Jackson (1930, reprinted 1969) *Early Days in East Africa*. London: Dawsons, p. 381.
42. Brian Herne (1999) *White Hunters: The golden age of African safaris*. New York: Henry Holt and Co, p. 69.
43. Roosevelt, 1910, Ibid., loc. 1076.
44. Ibid., loc. 2111.
45. Ibid., loc. 2825.
46. Chauncy Hugh Stigand (1913) *The Game of British East Africa*. London: Horace Cox, Kindle edition, loc. 471.
47. Ibid., loc. 5650.
48. Ibid., loc. 836.
49. Herne (1999), p. 131.
50. Stigand (1913), loc. 7260.
51. *Journal* (1913), p. 22.
52. Colony and Protectorate of Kenya (1922) An ordinance No 58 of 1921, Game Ordinance. *Journal*: 52–5.
53. Captain Keith Caldwell (1924) Game preservation, its aims and objects. *Journal*: 45–56, p. 46.
54. C.W. Hobley (1922) The fauna of East Africa and its future. *Proceedings* 92(1): 6.
55. Ibid.
56. Ibid.
57. *Journal* (1923): 68 and 70.
58. Caldwell (1924), p. 51.
59. Ibid., p. 53.
60. *Journal* (1928): 80–1.
61. Parker (2001), pp. 28–9.
62. Ian Parker (2004) *What I Tell You Three Times Is True*. Kinross: Librario, p. 64.

63. Ibid.
64. Keith Caldwell (1937) African Game Preservation. *Journal* 32 (1937): 34.
65. Peter T. Dalleo (1979) The Somali role in organized poaching in Northeastern Kenya. *International Journal of African Historical Studies* 12(3): 472–82, p. 472.
66. Ibid.
67. *Journal* 42 (1941): 32–3.
68. *Journal* (1928): 68.
69. *Journal* (1930): 51.
70. Ibid., p. 52.
71. Ibid.
72. Ibid., p. 53.
73. Kenya Colony, Game Department Annual Report for 1932, 1933 and 1934, *Journal* 27 (1936): 36.
74. *Journal* 31 (1937): 32.
75. Ibid., pp. 34–5.
76. R.W.G. Hingston (1930) Report on a mission to East Africa for the purpose of investigating the most suitable methods of ensuring the preservation of its indigenous fauna, *Journal* 12: 41.
77. Keith Caldwell (1938) Troubles of a Game Warden, *Journal* 33: 81.
78. Mervyn Cowie (1961) *Fly, Vulture*. London: George G. Harrap & Co., p. 32.
79. Ibid., p. 215.
80. Ibid., p. 216.
81. Anthony Cullen and Sydney Downey (1960) *Saving the Game*. London: Jarrolds, p. 169.
82. Ibid., p. 52.
83. Peter Jenkins (2001) Three stalwarts. In Parker and Bleazard, *An Impossible Dream*, pp. 33–8, p. 34.
84. Rodney Elliott (2001) Game ranger, Maralal. In Parker and Bleazard, *An Impossible Dream*, pp. 47–57, pp. 56–7.
85. Cullen and Downey (1960), pp. 85, 89.
86. *Journal* 51 (1945): 10.
87. Parker (2004), p. 93.
88. Daphne Sheldrick (1973) *The Tsavo Story*. London: Collins and Harvill, p. 17.
89. Ibid., pp. 22–5.
90. Keith Caldwell (1951) Report on a visit to East Africa, *Oryx*, 4 December: 178.
91. Ian Parker and Mohamed Amin, *Ivory Crisis*. London: The Hogarth Press, 1983, pp. 48–50.
92. Cullen and Downey (1960), p. 58.
93. Edward I. Steinhart (2006) *Black Poachers, White Hunters: A Social History of Hunting in Colonial Kenya*. Oxford: James Currey, p. 196.
94. Parker (2004), p. 95.
95. An account by the secretary of his visit with his wife to east and central Africa between June and October, 1957 (1957) *Oryx* 4: 265.
96. Cullen and Downey (1960), p. 55.
97. Parker (2004), p. 78.
98. Sheldrick (1973), p. 113.
99. Ibid., p. 115.
100. Parker (2004), p. 129.
101. Sir Richard Burton (1860) *The Lake Regions of Central Africa: From Zanzibar to Lake Tanganyika, Volume I*. London and New York: Harper, Kindle edition; Sir Richard Burton (1860) *The Lake Regions of Central Africa: From Zanzibar to Lake Tanganyika, Volume II*, London: Longman Green, Kindle edition.
102. Burton (1860) *Volume II*, loc. 249.
103. Ibid., loc. 4614.
104. Ibid., loc. 6056.

105. Henry M. Stanley (1879) *Through the Dark Continent, Volume I*, Reprinted 1988 New York: Dover Publications, p. 89.

106. Bradley Martin (1982), p. 76.

107. Stanley (1879), p. 365.

108. Ibid., p. 367.

109. John Hanning Speke (1863) *Journal of the Discovery of the Source of the Nile*, Kindle edition, loc. 10. https://read.amazon.co.uk/?asin=B0082RVBP6&ref_=kwl_kr_sea_2&language=en-GB. Accessed 21 July 2022.

110. Ibid., loc. 966.

111. Ibid., loc. 2652.

112. Lawrence E.Y. Mbogoni (2013) *Aspects of Colonial Tanzania History*. Dar es Salaam: Mkukina Nyota Publishers, p. 35.

113. Extract from report on Tanganyika Territory, *Journal* (1922): 47.

114. Jan Bender Shetler (2007) *Imagining Serengeti. A history of landscape memory in Tanzania from the earliest times to the present*. Athens: Ohio University Press, p. 180.

115. K.M. Homewood and W.A. Rodgers (1991) *Maasailand Ecology*. Cambridge: Cambridge University Press, p. 70.

116. Ibid.

117. C.F. Swynnerton (1926) The working of the Game Ordinance, Tanganyika Territory. *Journal*: 33.

118. *Journal* 21 (1934): 66.

119. Extracts from Game Department Report, 1935. *Journal* 27 (1936): 21.

120. Tanganyika Territory Game Report for 1935. *Journal* 30 (1937): 75.

121. Lawrence E.Y. Mbogoni (2013) *Aspects of Colonial Tanzania History*. Dar es Salaam: Mkuki na Nyota, p. 41.

122. *Journal* 43 (1941): 32.

123. *Oryx* 4 (1957): 256.

124. Homewood and Rodgers (1991), p. 138.

125. Myles Turner (1987) *My Serengeti Years: The memoirs of an African game warden*. London: Elm Tree Books, pp. 49–50.

126. John M. Mackenzie (1988) *The Empire of Nature: Hunting conservation and British imperialism*. Manchester: Manchester University Press, p. 151.

127. John Tosh, The Northern Interlacustrine Region. In Richard Gray and David Birmingham (eds) *Pre-Colonial Trade Essays on Trade in Central and Eastern Africa before 1900*, 103–18, p. 113. London, 1970.

128. Ofcansky (2002), p. 28.

129. *Journal* (1905): 60.

130. Kittenberger (1989), p. 118.

131. *Journal* (1921): 28.

132. *Journal* (1926): 78.

133. *Journal* (1930): 19.

134. Ibid.

135. Ibid., p. 43.

136. *Journal* 12 (1930): 47.

137. *Journal* 22 (1934): 43.

138. *Journal* 24 (1935): 38.

139. Ibid., p. 51.

140. *Journal* 32 (1937): 61.

141. *Journal* 33 (1938): 61.

142. *Journal* 41 (1940): 21–2.

143. *Oryx* 2 (1950): 97.

144. Ibid., p. 97.

145. Eric L. Edroma (1982) White rhino extinct in Uganda. *Oryx* 16(4): 352–5, p. 352.

146. Ibid., p. 353.

147. Harald George Carlos Swayne (1894) *Seventeen Trips Through Somaliland: A record of exploration & big game shooting, 1885 to 1893: Being the narrative of several journeys in the hinterland of the Somali coast protectorate*. London: Rowland Ward. Kindle edition, loc. 56.

148. Ibid., loc. 167.

149. Ibid., loc. 2052–75.

150. Ibid., loc. 2297.

151. H.G.C. Swayne (1894) Further field-notes on the game-animals of Somaliland. *Proceedings* 62(1): 321–2.

152. *Journal* (1909): 113.

153. Ralph Evelyn Drake-Brockman (1910) *The Mammals of Somaliland*. London: Hurst and Blackett, p. 106.

154. Bradley Martins (1982), p. 41.

155. Samuel White Baker (n.d.) *The Albert N'Yanza, Great Basin of the Nile*. Kindle edition, loc. 1853.

156. Samuel White Baker (1867) *The Nile Tributaries of Abyssinia, and the Sword Hunters of the Hamran Arabs*. Delhi: Open Books. Kindle edition, loc. 1845.

157. Ibid., loc. 3706.

158. J.L. Cloudsley-Thompson (1967) *Animal Twilight: Man and game in eastern Africa*. London: G.T. Foulis and Co., p. 27.

159. Ibid., pp. 39–41.

160. *Proceedings* (8 March 1864): 106.

161. Edroma (1982), p. 352.

162. Lord Cromer (1902) Extract from Lord Cromer's Report for Egypt and the Sudan for the year 1902. *Journal* (1904): 61.

163. *Journal* (1907): 84.

164. *Journal* (1908): 133–4.

165. Roosevelt (1910), loc. 4626.

166. Ibid., loc. 4650.

167. *Journal* (1923): 63.

168. Emslie and Brook (1999), p. 9.

169. *Special Sudan Gazette, Supplement to No. 54, December, 1903. Published by authority of the Sudan Government*. An Ordinance for the Preservation of Wild Animals and Birds, published in the *Journal* 1 (1904): 11–27.

170. The Sudan, *Journal* 53 (1946): 36.

171. Edroma, 1982, p. 352.

172. P.Z. Mackenzie (1953) Rhino traps and rhino horns. *Sudan Wildlife and Sport* 3(1). http://www.rhinoresourcecenter.com/pdf_files/118/1184848667.pdf, pp. 5–6. Accessed 3 November 2022.

173. Cloudsley-Thompson (1967), p. 19.

174. Mackenzie (1953), pp. 5–6.

175. Spinage (2012), p. 432.

176. E.L. Trouessart (1909) Le Rhinoceros blanc du Soudan. *Proceedings* 79(1): 199.

177. Ibid.

178. Marcus Daly (1937) *Big Game Hunting and Adventure: 1897–1936*. London: Macmillan. Cited by Bradley Martins (1982), p. 41.

179. L. Blancou (1960) Destruction and protection of the fauna of French Equatorial and of French West Africa. II. The larger animals. *African Wildlife* 14(2). http://www.rhinoresourcecenter.com/pdf_files/129/1298077363.pdf, pp. 293–4. Accessed 3 November 2022.

180. Mbayma Atalia (1993) Strategies for the conservation of rhino in Zaire. In Ryder (ed.) *Rhinoceros Biology and Conservation*, pp. 178–82, p. 178.

181. Foran (1933), p. 209.

182. Kai Curry-Lindahl (1972) War and the white rhinos. *Oryx* 11(4): 263–7, p. 263.

183. L.C. Rookmaaker (2004) Historical distribution of the black rhinoceros (*Diceros bicornis*) in West Africa. *African Zoology* 39(1): 1–8, p. 1.

184. A.H. Haywood (1932) Nigeria: preservation of wildlife. *Journal* 17: 32.

185. Ibid., p. 39.

186. D.R. Rosevear (1953) *Checklist and Atlas of Nigerian Mammals.* Lagos: Government printer, p. 120.

Chapter 4: Rhino populations decimated by the arrival of Europeans and firearms in southern Africa

1. Owen-Smith (2010), p. 390.

2. Ian Player (1967) Translocation of white rhinoceros in South Africa. *Oryx* 9(2): 137–50, p. 137.

3. Kees Rookmaaker (2013) Rhinocerotidae. In Kingdon and Hoffmann, *Mammals of Africa*, pp. 444–54, p. 453.

4. Cited by Emslie and Brook (1999), p. 10.

5. Selous (1881), p. 725.

6. Ibid., p. 242.

7. Norman Etherington (2016) *Big Game Hunter: A biography of Frederick Courteney Selous.* London: Robert Hale, p. 99.

8. Chapman (1908), p. 129.

9. Abel Chapman (1922) The white rhinoceros of the Sudan. *Journal*, p. 44.

10. One of the few from the nineteenth century is Major Serpa Pinto (1881) *How I Crossed Africa: From the Atlantic to the Indian ocean, through unknown countries; discovery of the Great Zambesi Affluents* (translated by Alfred Elwes). Philadelphia: J.B. Lippincott and Co.

11. Joachim John Monteiro (1875) *Angola and the River Congo. Volume II.* London: Macmillan, p. 79.

12. Brian John Huntley (2023) *Ecology of Angola: Terrestrial biomes and ecoregions.* Springer eBook. https://link.springer.com/book/10.1007/978-3-031-18923-4. Accessed 12 March 2023, p. 3.

13. Brian J. Huntley (2017) *Wildlife at War in Angola.* Pretoria: Protea Book House, p. 136.

14. G.C. Shortridge (1934) *The Mammals of South West Africa, Volume I.* London: William Heinemann, p. 414.

15. S. Newton da Silva (1952) Wildlife and its protection in Angola. *Oryx* 7: 346.

16. Ibid.

17. Fred Morton and Robert Hitchcock (2014) Tswana hunting: continuities and changes in the Transvaal and Kalahari after 1600. *South African Historical Journal* 66(3): 418–439, p. 418.

18. Ibid., p. 420.

19. William J. Burchell (1824) *Travels in the Interior of Southern Africa, Vol. II.* London: Longman, Hurst, Rees, Orme and Brown, p. 320.

20. Ibid., p. 431.

21. Barry Morton (1997) The hunting trade and the reconstruction of northern Tswana societies after the Difaqane, 1838–1880. *South African Historical Journal* 36: 220–39, p. 226.

22. Ibid., p. 228.

23. L.C. Rookmaaker (2005) Review of the European perception of the African rhinoceros. *Journal of Zoology* 265: 365–76, p. 372.

24. W. Edward Oswell (1900) *William Cotton Oswell, Hunter and Explorer: Volume I.* New York: Doubleday (Reprinted by Lightning Source, n.d.), p. 138.

25. Charles John Andersson (1857) *Lake Ngami: Or Explorations and Discoveries During Four Years' Wanderings in the Wilds of South Western Africa.* New York: Dix, Edwards & Company. Kindle edition, loc. 6405.

26. Player (1972), p. 31.

27. William Charles Baldwin (1863) *African Hunting from Natal to the Zambesi*. New York: Harper and Brothers, p. 173.

28. Ibid., p. 282.

29. Selous (1881), p. 725.

30. Player (1972), p. 32.

31. *Journal* (1905): 76.

32. Morton and Hitchcock (2014), p. 436.

33. Robert Kay and June Kay (1962) Preservation in N'Gamiland. *Oryx* 5 (October): 285–6.

34. J.J. Mallinson (1962) Dangers involved in the exploitation of game in N'Gamiland. *Oryx* 5 (October): 288.

35. Somerville (2016), p. 151; June Kay (1963), Moremi Wildlife Reserve, Okavango. *Oryx* (August): 2.

36. Personal communication, Botswana Department of Wildlife and National Parks officials, with the Kalahari Conservation Society and the Khama Rhino Sanctuary in 1993 and 2015.

37. Alfred Sharpe (1904) Game census of the British central Africa Protectorate. *Journal* 1: 74.

38. *Journal* (1908): 76.

39. *Journal* (1927): 111–12.

40. F. Vaughan Kirby (1899) *Sport in East Central Africa: Being an account of hunting trips in Portuguese and other Districts of East Central Africa*. London: Rowland Ward, p. 29.

41. Ibid., pp. 40–3.

42. Hingston (1930), p. 28.

43. Joubert (1971), p. 36.

44. Ibid.

45. Sian Sullivan et al. (2021) Historicising black rhino in Namibia: colonial-era hunting, conservation custodianship, and plural values, *Future Pasts Working Paper no 13*, https://www.futurepasts.net/fpwp13-sullivan-urikhob-kotting-muntifering-brett-2021, accessed 1 August 2024, p. 5.

46. Ibid.

47. Sullivan and Muntifering (2020), p. 12.

48. Sullivan et al. (2021), pp. 8–9.

49. Cited by Sullivan et al. (2021). See also *Journey from Cape Town – Walvis Bay (overland) – Niais (central Nambia) by James Edward Alexander 1836–37*. Compiled by Sian Sullivan for Future Pasts and Etosha-Kunene Histories, James Edward Alexander 080520.docx (google.com). Accessed 6 February 2023, no page numbers.

50. *Journey from Cape Town.*

51. Ibid.

52. Sullivan and Muntifering (2020), p. 4.

53. Heinrich Vedder (1966) *South West Africa in Early Times*. London: Frank Cass, digital bookshelf edition, p. 285.

54. Cited by Sullivan et al. (2021) pp. 11–2.

55. Francis Galton (1853) *Narrative of an Explorer in Tropical South Africa* (second edition, 1889). https://galton.org/books/south-west-africa/, no page numbers. Accessed 10 May 2023.

56. Francis Galton (2012, first published 1853) *Francis Galton's Narrative of His Exploration of Namibia in 1851 (Annotated)*. First digital version publication 2012: Keith Irwin, PO Box 2480, Swakopmund, Namibia, loc. 4338.

57. Andersson (1857), loc. 919.

58. Ibid., loc. 3230.

59. Ibid.

60. Ibid., loc. 1183.

61. Vedder (1966), p. 417.

62. Joubert (1971), pp. 36–7 and 40.

63. Christo Botha (2005) People and the environment in colonial Namibia. *South African Historical Journal* 52: 170–190, p. 179.

64. Ibid.

65. Ibid., p. 180.

66. The emergence of Etosha National Park in Namibia and its impact on the Hai//Om people. https://cases.open.ubc.ca/the-emergence-of-etosha-national-park-in-namibia-and-its-impacts-on-the-hai-om-people/#:~:text=In%201920%2C%20South%20Africa%20agreed,1967%2C%20becomes%20Etosha%20National%20Park. Accessed 11 May 2023.

67. Owen-Smith (2010), p. 7.

68. Sullivan and Muntifering (2020), p. 10.

69. G.C. Shortridge (1934) *The Mammals of South West Africa, Volume I*. London: William Heinemann, p. 413.

70. Sir John Barrow (1806) Travels into the interior of southern Africa: in which are described the character and the condition of the Dutch colonists of the Cape of Good Hope, and ... in the animal, mineral and vegetable... London: T. Cadell and W. Davies. Reprinted in 2017 by Hard Press, Kindle edition, loc. 724.

71. William J. Burchell (1822) *Travels in the Interior of Southern Africa, Volume I*. London: Longman, Hurst, Rees, Orme, and Brown, p. 243.

72. Burchell (1824), p. 73.

73. Cited by Rookmaaker (2005), p. 370.

74. Major Robert W. Foran (1933, reprinted in 2017) *Kill or be Killed: The Rambling Reminiscences of an Amateur Hunter*. London: Hutchinson & Co., pp. 191 and 208.

75. Martin Hall (1977) Shakan pitfall traps: hunting technique in the Zulu kingdom. *Annals of the Natal Museum* 23 (1): 8.

76. E.A. Ritter (1978) *Shaka Zulu*. Harmondsworth: Penguin, p. 229.

77. Cited by Player (1972), p. 23.

78. Baldwin (1863), p. 19.

79. W.H. Drummond (1875) *Rough Notes on the Large Game and Natural History of South and South-East Africa*. New Delhi: Isha Books, p. 29.

80. Ibid., pp. 84–5.

81. Player (1972), p. 33.

82. Rookmaaker (2005), p. 372.

83. Ibid.

84. An extract from a correspondent at Barberton in South Africa. *Journal* (1904): 40–1.

85. Major J. Stevenson-Hamilton (1905) Game Preservation in the Transvaal. *Journal*: 21.

86. Ibid., p. 32.

87. *Journal* (1909): 64.

88. Siyabona Africa (n.d.) History and geography of Kruger National Park. https://www.krugerpark.co.za/Kruger_National_Park_Regions-travel/history-geography-kruger-park.html. Aacessed 3 October 2022.

89. Cited by Ian Player (1967) Translocation of white rhinoceros in South Africa. *Oryx* 9(2): 137–50, p. 137.

90. Sabi (1922) Empire fauna 1922. *Journal*: 41.

91. Herbert Lang (1924) Threatened extinction of the white rhinoceros (*Ceratotherium simum*). *Journal of Mammalogy* 5(30): 173–80, p. 173.

92. C.W. Hobley (1926) The Zululand Game Reserves. A summary of the position and some suggestions. *Journal*: 43.

93. Ibid., p. 44.

94. *Journal* 24 (1934): 64–5.

95. *Journal* 30 (1937): 89–90.

96. Ibid., p. 80.

97. Ibid., p. 91.

98. Report of the Zululand Game Reserves and Parks Board for the year ending 31st March 1941. *Journal* 43 (1941): 37–8.

99. Ibid., p. 38.

100. Report of the Zululand Game Reserves and Parks Board for the year ending 31st March 1943. *Journal* 48 (1943): 21.

101. *Journal* 56 (1948): 17.
102. *Oryx* 5 (1952): 239.
103. Player, 1972 p. 20.
104. Ibid., p. 37.
105. Ibid., p. 107.
106. Gibson (1999), p. 23.
107. Richard Gray and David Birmingham (1970) Some economic and political consequences of trade in Central and Eastern Africa in the pre-colonial period. In Richard Gray and David Birmingham (eds) *Pre-Colonial African Trade: Essays on Trade in Central and Eastern Africa Before 1900*, pp. 1–23. London: Oxford University Press, p. 5.
108. See, for example, Thayer Scudder (1971) *Gathering Among African Woodland Savannah Cultivators*. Lusaka: Institute for African Studies, Zambia papers no. 5.
109. Simon Kumwenda (2021) Whose heritage? The state, local communities and game in South Luangwa National Park (SLNP) of Eastern Zambia, 1890–2001. University of Zambia, Master of Arts in History dissertation. http://dspace.unza.zm/bitstream/handle/123456789/7292/Main%20Document.PDF?sequence=1. Accessed 23 January 2023, p. 3.
110. Somerville (2017), p. 32.
111. *Journal* (1905): 74.
112. *Journal* (1927): 68–9.
113. Northern Rhodesia Game Ordinance no. 19, 1925. *Journal* (1925): 62.
114. Fraser F. Darling (1960) *Wildlife in an African Territory: A study made for the Game and Tsetse Control Department of Northern Rhodesia*. Oxford: Oxford University Press, pp. 120–1.
115. Kumwenda (2021), pp. 47–8.
116. Captain C.R.S. Pitman (1936) A report on the Faunal Survey of Northern Rhodesia. *Journal* 28: 26.
117. Emslie and Brooks (1999), p. 57.
118. Kumwenda (2021), p. 56.
119. *Oryx* 1 (1950): 15.
120. Gibson (1999), p. 27. For greater detail, see Mwelwa C. Musambachime (1992) Colonialism and the environment in Zambia, 1890–1964. In Samuel N. Chipungu (ed.) *Guardians in Their Time: Experiences of Zambians Under Colonial Rule, 1890–1964*. London: Macmillan.
121. Northern Rhodesia (1964) *Fauna Conservation, Chapter 241 of the LWS*. Lusaka: Government printer.
122. Gibson (1999), p. 29.
123. Baldwin (1863), pp. 179, 393.
124. Frederick Courteney Selous (2001, originally published in 1881) *A Hunter's Wanderings in Africa*. New York: Alexander Books/Resnick Library of African Adventure (covers the period from early 1873); Frederick Courteneya Selous (1984, originally published 1893) *Travel and Adventure in South-East Africa*. London: Century Publishing, p. xiii (covers the period from 1882 to 1887); F.C. Selous (1908) *African Nature Notes and Reminiscences*. London: Macmillan.
125. Selous (2001).
126. Selous (1984), p. 159.
127. John M. Mackenzie (1988) *The Empire of Nature. Hunting Conservation and British Imperialism*. Manchester: Manchester University Press, p. 128.
128. Sabelo J. Ndlovu-Gatsheni (2009) Mapping cultural and colonial encounters, 1880s-1930s. In Brian Raftopoulos and Alois Mlambo (eds) *Becoming Zimbabwe*. Harare: Weaver Press/Jacana, 39–74, pp. 44–6; Neil Parsons (1993) *A New History of Southern Africa* (2nd edn). London: Macmillan, pp. 178–81.
129. Graham Child (1996) *Wildlife and People: The Zimbabwean success*. Harare: Wisdom, p. 50.
130. R.T. Coryndon (1894) On the occurrence of the White or Burchell's Rhinoceros in Mashonaland. *Proceedings* 62(1): 329.
131. Ibid., p. 330.

132. G.C. Shortridge (1934) *The Mammals of South West Africa, Volume I*. London: William Heinemann, p. 429.

133. E.C. Chubb (1909) The mammals of Matabeleland. *Proceedings* 79(1): 124.

134. Roben Mutwira (1989) Southern Rhodesian Wildlife Policy (1890–1953): a question of condoning game slaughter? *Journal of Southern African Studies* 15(2), Special Issue on the Politics of Conservation in Southern Africa, 250–62, 250.

135. Ibid.

136. Mutwira (1989), p. 258.

137. Somerville (2016), p.87.

138. Mutwira (1989), p. 260.

139. I.D.M. (1952) The wildlife situation in Southern Rhodesia. *Oryx* 7: 352,

Chapter 5: The modern trade in rhino horn from 1960 to the present, legal and illegal

1. Personal communication with Lucy Vigne, 9 November 2023.

2. Tom Inskipp and Sue Wells (1979) *International Trade in Wildlife*. London: International Institute for Environment and Development/Routledge, p. 5.

3. CITES (1976) First meeting of the Conference of the Parties. https://cites.org/sites/default/files/eng/cop/01/E01-Cred-01.pdf accessed 16 May 2023, pp. 80–2.

4. CITES (n.d.) *Appendix*. https://cites.org/eng/node/130903 accessed 16 May 2023.

5. Ibid.

6. Julian Rademeyer (2012) *Killing for Profit: Exposing the illegal rhino horn trade*. Cape Town: Penguin Random House/Struik. Kindle edition, loc. 1863.

7. Ibid.

8. Mike Knight (2016) AfRSG report. *Pachyderm* 57: 21–2.

9. Lucy Vigne and Esmond Bradley Martin (2018) *Illegal Rhino Horn Trade in Eastern Asia Still Threatens Kruger's Rhinos*. Port Lympne: The Aspinall Foundation, p. 69.

10. Michael 't Sas Rolfes (2012) The rhino poaching crisis: a market analysis. http://www.rhino-economics.com/wp-content/uploads/2012/04/The-Rhino-Poaching-Crisis-by-Michael-t-Sas-Rolfes-Final.pdf. Accessed 20 June 2023, p. 8.

11. Nigel Leader-Williams (2005) Regulation and protection: successes and failures in rhino conservation. In Frank Oldfield (ed.) *Environmental Change: Key Issues and Alternative Approaches*. Cambridge: Cambridge University Press, pp. 89–99, pp. 93–6.

12. Michael 't Sas Rolfes (1995) *Rhinos: Conservation, Economics and Trade-offs*. London: IEA Environment Unit, pp. 5, 50.

13. Save the Rhino (n.d.) *Rhino Species*. https://www.savetherhino.org/rhino-info/rhino-species/. Accessed 9 October 2023.

14. E.B. Martin and T.C.I. Ryan (1990) How much rhino horn has come onto the international markets since 1970? *Pachyderm* 13: 20–6, p. 21.

15. D. Western (1989) The undetected trade in rhino horn. *Pachyderm* 11: 24–6, p. 24.

16. 't Sas Rolfes (2012), p. 8.

17. Ibid.

18. Ibid., p. 11.

19. Ibid., p. 14.

20. Leader-Williams (2005), p. 95.

21. Ibid., p. 80.

22. Save the Rhino (2023) World Rhino Day 2023: new numbers and challenges, 22 September 2023. https://www.savetherhino.org/africa/world-rhino-day-2023-new-numbers-and-challenges/. Accessed 9 October 2023.

23. Nigel Leader-Williams (1992) *The World Trade in Rhino Horn: A Review*. Cambridge: TRAFFIC, https://portals.iucn.org/library/sites/library/files/documents/Traf-004.pdf. Accessed 12 September 2023, p. 4.

24. Ibid.

25. Vigne and Martin (2018), pp. 55–6.
26. T. Milliken, K. Nowell and J.B. Thomsen (1993) *The Decline of the Black Rhino in Zimbabwe: Implications for Rhino Conservation*. Cambridge: TRAFFIC, p. 8.
27. Milliken, Nowell and Thomsen (1993), p. 8.
28. Ed Stoddard (2022) Rhino horn illicit trade driven by demand for luxury carvings, not medicine – new report, *Daily Maverick*, 27 October 2022, https://www.dailymaverick.co.za/article/2022-10-27-rhino-horn-illicit-trade-driven-by-demand-for-luxury-carvings-not-medicine-new-report/ accessed 18 September 2023,
29. Rademeyer (2012), loc. 4419–56.
30. Rosaleen Duffy (2010) *Nature Crime: How We're Getting Conservation Wrong*. New Haven: Yale University Press, p. 117.
31. Cumming, Du Toit and Stuart (1990), p. 51.
32. Ibid., loc. 4431.
33. Stoddard (2023).
34. Ibid.
35. Leader-Williams (1992), p. 6.
36. I.C.S. Parker and E.B. Martin (1979) Trade in African rhino horn. *Oryx* 15: 153–8, pp. 154–5.
37. Milliken, Nowell and Thomsen (1993), p. 9.
38. Ibid., pp. 14–6.
39. Personal communication with investigative journalist Stephen Ellis, December 2014, and with Brigadier Jan Breytenbach, commander of 32 Battalion of the SADF, November 1990.
40. Esmond Bradley Martin and Chryssee Bradley Martin (1987) Combating the illegal trade in rhinoceros products. *Oryx* 21(3): 143–8, pp. 143–4.
41. Milliken, Nowell and Thomsen (1993), p. 9.
42. Ibid., p. 10.
43. Ibid., p. 12.
44. Ibid.
45. Save the Rhino (2018) *A Legal Trade in Rhino Horn*. https://www.savetherhino.org/thorny-issues/legal-trade-in-rhino-horn/#:~:text=The%20international%20trade%20in%20rhino,has%20been%20banned%20since%201977. Accessed 7 November 2022.
46. Ed Stoddard (2017) South Africa's top court lifts ban on domestic sales in rhino horn. Reuters, 5 April. https://www.reuters.com/article/us-wildlife-rhinos-safrica-idUSKBN1771VP. Accessed 7 November 2022.
47. Bradley Martin and Bradley Martin (1987), pp. 143–8, p. 143.
48. Personal communication with Ian Parker, Richard Bell and Esmond Bradley Martin.
49. Cumming, Du Toit and Stuart (1990), p. 52.
50. L. Vigne and E.B. Martin (1989) Taiwan: the greatest threat to the survival of Africa's rhinos. *Pachyderm* 11: 23–5. p. 24.
51. Ibid.
52. E.B. Martin (1989) Report on the trade in rhino products in eastern Asia and India. *Pachyderm* 11: 13–20, p. 13.
53. Ibid.
54. Ibid., pp. 13–4.
55. Ibid.
56. Ibid.
57. Emslie and Brooks (1999), pp. 29–30.
58. Ibid., p. 30.
59. Ibid.
60. Personal communication with Lucy Vigne, 9 October 2023.
61. Emslie and Brooks (1999), p. 28.
62. Parker and Martin (1979), p. 153.
63. Emslie and Brooks (1990), p. 28.
64. Parker and Bradley Martin (1979), p. 153.

65. Esmond Bradley Martin (1987) The Yemeni rhino horn trade. *Pachyderm* 8: 13–16, p. 14.

66. Esmond Martin and Lucy Vigne (2007) Rising price for rhino horn in Yemen puts pressure on East Africa's rhinos. *Oryx* 41 (4): 431.

67. Lucy Vigne and Esmond Martin (2000) Price for rhino horn increases in Yemen. *Pachyderm* 28: 91–100, pp. 91–2.

68. Ibid., p. 92.

69. Ibid., pp. 92–3.

70. Lucy Vigne, Esmond Martin and Benson Okita-Ouma (2007) Increased demand for rhino horn in Yemen threatens eastern Africa's rhinos. *Pachyderm* 43: 73–86, p. 76.

71. Ibid., pp. 77–9.

72. Lucy Vigne and Esmond Martin (2013) Increasing rhino awareness in Yemen and a decline in the rhino horn trade. *Pachyderm* 53: 51–8, p. 51.

73. Personal communication with Lucy Vigne, 9 October 2023.

74. Cumming, Du Toit and Stuart (1990), p. 51.

75. Ibid., p. 52.

76. Cecilia Song and Tom Milliken (1990) The rhino horn trade in South Korea: still cause for concern. *Pachyderm* 13: 5–11, p. 5.

77. Ibid., p. 6.

78. Bradley Martins (1987), p. 145.

79. Martin, 1989, p. 16.

80. Ibid.

81. Esmond Bradley Martin and Chrissee Bradley Martin (1989) The Taiwanese connection—a new peril for rhinos. *Oryx* 23 (2): 76–81.

82. Ibid., p. 76.

83. Ibid.

84. US Fish and Wildlife Service (n.d.) Pelly Amendment to the Fishermen's Protective Act. https://www.fws.gov/law/pelly-amendment-fishermens-protective-act#:~:text=The%20Pelly%20Amendment%20to%20the,undermines%20the%20effectiveness%20of%20any. Accessed 16 October 2023.

85. Martin (1989), pp. 78–9.

86. Ibid., p. 16.

87. Chris R. Shepherd, Thomas N.E. Gray and Vincent Nijman (2018) Rhinoceros horns in trade on the Myanmar–China border. *Oryx* 52 (2): 393–5, p. 393.

88. Personal communication with Lucy Vigne.

89. Martin (1989), p. 18.

90. Esmond Bradley Martin (1993) The present-day trade routes and markets for rhinoceros products. In Ryder (ed.) *Rhinoceros Biology and Conservation*, pp. 1–9, p. 4.

91. Lucy Vigne (2013) Recent findings on the ivory and rhino-horn trade in Lao People's Democratic Republic. *Pachyderm* 54: 36–44, p. 36.

92. Ibid.

93. Rademeyer (2012), loc. 2559–2715.

94. Ibid.

95. Staff reporter (2018) Fury at release of rhino 'pseudo-hunt' kingpin. *Mail & Guardian*, 18 September. https://mg.co.za/article/2018-09-13-fury-at-release-of-rhino-pseudo-hunt-kingpin/#:~:text=Lemtongthai%2C%20a%20Thai%20national%2C%20was,years%20of%20his%20jail%20sentence. Accessed 17 October 2023.

96. South Africa gives rhino poacher 40-year jail term. BBC (9 November 2012). https://www.bbc.co.uk/news/world-africa-20267967. Accessed 4 April 2023.

97. Vigne and Bradley Martin (2018), p. 12.

98. Emslie and Brooks (1999), p. 27.

99. Ibid.

100. Martin (1993), p. 6.

101. Ibid.

102. Ibid., pp. 77–78.

103. Vigne and Martin (1989), p. 24.

104. Ibid.

105. Louise Shelley and Kasey Kincaid (2018) The convergence of trade in illicit rhino horn and elephant ivory with other forms of criminality. In William D. Moreto and Stephen F. Pires (eds) *Wildlife Crime: An Environmental Criminology and Crime Science Perspective*. Durham, NC: Carolina Academic Press, pp. 109–134, pp. 117–18.

106. Ibid.

107. Ibid., p. 19.

108. Richard H. Emslie et al. (2019) *African and Asian Rhinoceroses – Status, Conservation and Trade. A report from the IUCN Species Survival Commission (IUCN SSC) African and Asian Rhino Specialist Groups and TRAFFIC to the CITES Secretariat pursuant to Resolution Conf. 9.14 (Rev. CoP17)*, 1 IUCN SSC AfRSG, 2 IUCN SSC Asian Rhino Specialist Group (AsRSG), 3 TRAFFIC, 4 Nelson Mandela University, South Africa. http://www.rhinore-sourcecenter.com/pdf_files/156/1560170144.pdf. Accessed 7 June 2022, p. 12.

109. Ibid.

110. Xinhuanet (2018) China to continue strict bans on trading of rhino, tiger products: official. *Xinhuanet*, 13 December. https://www.xinhuanet.com/english/2018-12/13/c_137671675.htm. Accessed 17 October 2023.

111. Griffiths University (2018) Conservation conundrum: why China's bid to legalise rhino horn trade is no black and white issue. https://news.griffith.edu.au/2018/11/26/conservation-conundrum-why-chinas-bid-to-legalise-rhino-horn-trade-is-no-black-and-white-issue/. Accessed 21 August 2023.

112. Ibid.

113. Emslie et al. (2019), p. 7. See also Mike Knight (2019) AfRSG report, *Pachyderm* 60: 19.

114. Knight (2019), p. 19.

115. Emslie et al. (2019), p. 8.

116. Knight (2019), p. 21.

117. Emslie et al. (2019), p. 8.

118. Tom Milliken and Jo Shaw (2012) *The South Africa–Viet Nam Rhino Horn Trade Nexus: A deadly combination of institutional lapses, corrupt wildlife industry professionals and Asian crime syndicates*. Johannesburg: TRAFFIC. https://www.traffic.org/site/assets/files/2662/south_africa_vietnam_rhino_horn_nexus.pdf, p. 15. Accessed 15 April 2023.

119. Ibid.

120. Ibid., pp. 11–12.

121. Rademeyer (2012), loc. 65.

122. Milliken and Shaw (2012), p. 15.

123. Ibid., pp. 76–7; Moses Montesh (2012) *Rhino Poaching: A New Form of Organised Crime*. School of Criminal Justice, Department of Police Practice, College of Law Research and Innovation Committee of the University of South Africa, Pretoria, p. 7.

124. Personal communication from Lucy Vigne, 9 November 2023.

125. Rademeyer (2012), loc. 1745–71.

126. Ibid., loc. 1777.

127. Cited by Milliken and Shaw (2012), pp. 76–7.

128. Cited by Rademeyer (2012), loc. 2325.

129. Milliken and Shaw (2012), pp. 16–17, 57.

130. Ibid., pp. 16–17.

131. Rademeyer (2012), loc. 4191.

132. Milliken and Shaw (2012), pp. 16–17.

133. Ibid., p. 61.

134. Ibid., p. 81.

135. Dung Nguyen (2023) Wildlife internet trade in Viet Nam. *The Horn*, Save the Rhino, December. SR3498_TheHorn23-v6.1-ISSU-Eds.pdf (savetherhino.org). Accessed 6 December 2023.

136. Ibid.

137. Ibid.
138. Moses Montesh (2012) *Rhino Poaching: A new form of organised crime.* School of Criminal Justice, Department of Police Practice, College of Law Research and Innovation Committee of the University of South Africa (2012), p. 2.
139. Keith Somerville (2012) Rhino poaching in South Africa: organised crime and economic opportunity driving trade. *African Arguments* 29 October. https://africanarguments.org/2012/10/rhino-poaching-in-south-africa-organised-crime-and-economic-opportunity-driving-trade-by-keith-somerville/. Accessed 18 October 2023; Fiona MacLeod (2012) Vets charged for illegal use of tranquillisers. *Mail and Guardian*, 2 March. https://mg.co.za/article/2012-03-02-vets-charged-for-illegal-use-of-tranquillisers/. Accessed 19 June 2023.
140. Montesh (2012), p. 2.
141. Personal communication, Xolani Nicholus Funda, 30 August 2016.
142. Ashwell Glasson (2020) The rhino rifle syndicates. *Africa in Fact*, 17 February. http://the-eis.com/elibrary/sites/default/files/downloads/literature/SA_2020-02%20South%20Africa%20_GGA.pdf, pp. 2–4. Accessed 5 September 2023.
143. Ibid.
144. Ibid., p. 7.
145. Julian Rademeyer (2023) Landscape of fear: crime, corruption and murder in greater Kruger. *Enact* 36, January: 1; Montesh (2012), p. 7.
146. Alastair Nelson (2023) *Convergence of Wildlife Crime and Other Forms of Transnational Organized Crime in Eastern and Southern Africa.* Geneva: Global Initiative against Organized Crime, p. 2.
147. Ibid., p. 10.
148. 't Sas Rolfes (2012).
149. Ibid.
150. Rosaleen Duffy et al. (2016) Toward a new understanding of the links between poverty and illegal wildlife hunting. *Conservation Biology* 30(1): 14–22; Rosaleen Duffy and Freya A.V. St John (2013) *Poverty, Poaching and Trafficking: What are the links?* Evidence on Demand. https://dx.doi.org/10.12774/eod_hd059.jun2013.duffy. Accessed 12 June 2023.
151. Duffy and St John (2013), p. 2.
152. Duffy et al. (2016), pp. 15–16.
153. Ibid.
154. Rosaleen Duffy (2022) *Security and Conservation: The politics of the illegal wildlife trade.* New Haven: Yale University Press, p. 8.
155. See A.M. Hübschle (2017) The social economy of rhino poaching: of economic freedom fighters, professional hunters and marginalized local people. *Current Sociology* 65(3): 427–47.
156. Duffy and St John (2013), p. 3.
157. Elizabeth M. Naro et al. (2020) Syndicate recruitment, perceptions, and problem solving in Namibian rhinoceros protection. *Biological Conservation* 243: 1–9, p. 4.
158. Ibid., p. 6.
159. Ibid.

Chapter 6: From independence to 2005

1. Sidney (1961), p. 51; see also Save the Rhino (n.d.).
2. Rookmaaker (2013), p. 453.
3. Kes Hillman (1980) Rhinos in Africa now. *SWARA* 3(1): 22–4, p. 23.
4. Emslie and Brooks (1999), p. 10.
5. Cumming, Du Toit and Stuart (1990), p. 4.
6. Emslie and Brooks (1999), p. 31.
7. See Somerville (2017), pp. 139–44, 375–88; Somerville (2019), pp. 111–20, 239–43.
8. Emslie and Brooks (1999), p. 31.
9. Bradley Martin and Bradley Martin (1982), p. 7.

10. D. Western and L. Vigne (1985) The deteriorating status of African rhinos. *Oryx* 19: 215–20, p. 216.

11. Ibid.

12. Ibid., p. 218.

13. Milliken, Nowell and Thomsen (1993), p. 1.

14. Hillman (1980), p. 243.

15. Orenstein (2013), p. 72.

16. Milliken, Nowell and Thomsen (1993), p. 50.

17. Ibid., p. 6.

18. Milliken and Shaw (2012), p. 44.

19. Ibid.

20. Ibid.

21. Raymond Bonner (1993) *At the Hand of Man: Peril and Hope for Africa's Wildlife*. London: Simon & Schuster, p. 195.

22. Personal communication with Tanzanian politicians and wildlife personnel in 1986 and 1991, who wished to remain anonymous.

23. Emslie and Brooks (1999), p. 33.

24. Ibid., p. 31.

25. Ibid., p. 29.

26. Personal communication from John Hanks (former Director of WWF Africa Programme) and Pelham Jones of PROA, Johannesburg, August 2019.

27. Emslie and Brooks (1999), p. 18.

28. Milliken and Shaw (2012), p. 18.

29. Simon A.H. Milledge (2007) Illegal killing of African rhinos and horn trade, 2000–2005: the era of resurgent markets and emerging organized crime. *Pachyderm* 43: 96–107, p. 96.

30. Ibid., p. 104.

31. Bradley Martin (1993), p. 1.

32. Ibid.

33. C.G. Gakahu (1993) African rhinos: current numbers and distribution. In Ryder (ed.), *Rhinoceros Biology and Conservation*, pp. 160–3, p. 161.

34. Clark C. Gibson (1999) *Politicians and Poachers*. Cambridge: Cambridge University Press, p. 42.

35. Ian S.C. Parker and A.D. Graham (2019) Part II: auctions and export from Mombasa 1960–1978: elephant ivory, rhino horn and hippo teeth. *Pachyderm* 61: 163.

36. Ibid.

37. Ian S. C. Parker and Esmond Bradley Martin (1979) Trade in African rhino horn. *Oryx* 15(20): 153–8, p. 153.

38. Ibid., p. 157.

39. See Michaela Wrong (2009) *It's Our Turn to Eat*. London: Fourth Estate, for a detailed and insider account of corruption in Kenya; Somerville (2017), pp. 221–32, 334–5.

40. Personal communication with Ian Parker.

41. Ibid., p. 73. See also Rodger Yeager and Norman N. Miller (1986) *Wildlife Wild Death: Land Use and Survival in Eastern Africa*. Albany: State University of New York Press, pp. 88–9.

42. Personal communication with Ian Parker, and with a British hunter and taxidermist who worked in Kenya at the time but wanted to remain anonymous.

43. Gibson (1999), p. 73.

44. Ian Parker 1979) *The Ivory Trade*. Washington: US Fish and Wildlife Service, pp. 117–18 (Ian Parker Collection of East African Wildlife Conservation: the Ivory Trade. http://ufdc.ufl.edu/AA00020117/00001/3j University of Florida. Accessed 5 March 2015).

45. Personal correspondence with Richard Leakey, the veteran conservationist and former head of the KWS, 2014 and 2015.

46. Ian Parker (1974) *EBUR*, Nairobi, October 1974, typescript from Ian Parker Collection of East African Wildlife Conservation: The Ivory Trade, University of Florida, http://ufdc.edu/AA00020117/00011. Accessed 19 March 2015.

47. Somerville (2019), pp. 111, 114.
48. Martin (1982), p. 45.
49. Ibid.
50. Ibid., p. 49.
51. Parker (2004), pp. 190–1.
52. Ibid., p. 192.
53. Ibid., p. 195.
54. Sheldrick (2012), p. 239.
55. Jan Fox (2013) Ivory and corruption: an interview with Dr Richard Leakey. Going Places Africa, 1 August. https://goingplacesafrica.com/2013/08/01/ivory-and-corruption-an-interview-with-dr-richard-leakey/. Accessed 17 October 2022.
56. Martin Meredith (2001) *Africa's Elephant: A biography*. London: Hodder and Stoughton, p. 203.
57. Richard Leakey and Virginia Morell (2001) *Wildlife Wars: My fight to save Africa's natural treasures*. London: Macmillan, p. 1.
58. Ibid., p. 65.
59. Douglas-Hamilton (1988), p. 329.
60. Bradley Martin (1982), p. 49.
61. Emslie and Brooks (1999), p. 47.
62. Ibid.
63. *Traffic Bulletin* (1990), 9 March, p. 19.
64. Fox (2013).
65. Reid (2012), p. 196.
66. David Western (1982) Patterns of depletion in a Kenya rhino population and the conservation implications. *Biological Conservation* 24: 147–156, p. 147.
67. D. Western and D.M. Sindiyo (1970) The status of the Amboseli rhino population. *East African Wildlife Journal* 10: 43–57.
68. David Collett (1987) Pastoralists and wildlife: image and reality in Kenya Maasailand. In David Anderson and Richard Grove (eds) *Conservation in Africa: People, policies and practice*, pp. 129–48. Cambridge: Cambridge University Press, p. 144.
69. Western (1982), p. 148.
70. Ibid., p. 152.
71. Ibid.
72. Ibid., p. 155.
73. W.K. Lindsay (1987) Integrating parks and pastoralists: some lessons from Amboseli. In Anderson and Grove (eds) *Conservation in Africa*, pp. 149–167, p. 160.
74. Álvaro Fernández-Llamazares et al. (2020) Historical shifts in local attitudes towards wildlife by Maasai pastoralists of the Amboseli Ecosystem (Kenya): insights from three conservation psychology theories. *Journal for Nature Conservation* 53: 1–11, p. 6.
75. Parker (2004), pp. 132–3.
76. Martin (1982), p. 43.
77. Ibid., p. 133.
78. Meredith (2001), p. 203.
79. Martin (1982), p. 45.
80. Keith Somerville (2017) *Africa's Long Road to Independence*. London: Penguin, pp. 85–6.
81. Cited by Parker (2004), p. 172.
82. Ibid., p. 173.
83. Ibid., p. 134.
84. Ibid., p. 140.
85. Ian Meredith Hughes (1988) *Black Moon, Jade Sea*. London: Clifford Frost, pp. 43–4.
86. Ibid., p. 65.
87. Ibid., pp. 200–1.
88. P.H. Hamilton and J.M. King (1969) The fate of black rhinoceroses released in Nairobi National Park. *African Journal of Ecology* 7(1): no page numbers.

89. Gibson (1999), p. 74.

90. Personal communication with Ian Parker.

91. Personal communications with Richard Leakey and Ian Parker. See also Fox (2013).

92. Cedric Khayale et al. (2021) Kenya's first White Rhino Conservation and Management Action Plan. *Pachyderm* 62: 112–18, p. 112.

93. Emslie and Brooks (1999), p. 46.

94. C.G. Gakahu (1989) Sanctuaries offer a future for black rhinos in Kenya. *Pachyderm* 11: 32.

95. Benson Okita-Ouma et al. (2008) Minimizing competition by removing elephants from a degraded Ngulia rhino sanctuary, Kenya. *Pachyderm* 44: 80–7, p. 80.

96. Ibid.

97. Ibid., p. 81.

98. Cumming, Du Toit and Stuart (1990), p. 57.

99. Iain and Oria Douglas-Hamilton (1992) *Battle for the Elephants*. London: Doubleday, 1992, p. 253.

100. Ibid.

101. Ibid.

102. R.A. Brett (1990) The black rhino sanctuaries of Kenya. *Pachyderm* 13: 31–34, p. 31.

103. John Mukinya (1977) Feeding and drinking habits of the black rhinoceros in Masai Mara Game Reserve. *East African Wildlife Journal* 15: 125–38, p. 125.

104. Brett (1990), p. 32.

105. Milliken, Nowell and Thomsen (1993), p. 43.

106. Benson Okita-Ouma et al. (2009) Density dependence and population dynamics of black rhinos (*Diceros bicornis michaeli*) in Kenya's rhino sanctuaries. *African Journal of Ecology* 48: 791–9, p. 792.

107. M. Morgan-Davies (1996) Status of the black rhinoceros in the Masai Mara National Reserve, Kenya. *Pachyderm* 21: 38–45, p. 38.

108. Alan Birkett (2002) The impact of giraffe, rhino and elephant on the habitat of a black rhino sanctuary in Kenya. *African Journal of Ecology* 40: 276–82, p. 281.

109. F.J. Patton et al. (2010) The colonization of a new area in the first six months following 'same-day' free release translocation of black rhinos in Kenya. *Pachyderm* 47: 66–79, p. 66.

110. Personal communication with wildlife trade specialist Daniel Stiles, who worked with the chimpanzees at Sweetwater and lived at Ol Pejeta, 1 November 2023.

111. Rodney Elliott (1992) Lake Nakuru Black Rhinoceros Sanctuary. *Oryx* 24(3): 174.

112. Felix Patton and Martin Jones (2007) Determining minimum population size and demographics of black rhinos in the Salient of Aberdare National Park, Kenya. *Pachyderm* 45: 63–72, p. 63.

113. Ibid., pp. 66, 71.

114. Vigne, Martin and Okita-Ouma (2007), pp. 73–4.

115. Ibid., p. 75.

116. Hans Klingel and Ute Klingel (1966) The rhinoceroses of Ngorongoro Crater. *Oryx* 8(5): 302–6, p. 302.

117. H.A. Fosbrooke (1965) Success story at Ngorongoro. *Oryx* 8(3): 164–8, p. 167.

118. Clive A. Spinage (2012) *African Ecology: Benchmarks and historical perspectives*. Heidelberg: Springer, p. 613.

119. Ibid., p. 735.

120. M. Borner (1981) Black rhino disaster in Tanzania. *Oryx* 16(1): 59–66, p. 59.

121. Ibid.

122. Ibid., p. 60.

123. Ibid., p. 61.

124. Victor A. Runyoro et al. (1995) Long-term trends in the herbivore population of the Ngorongoro Crater, Tanzania. In A.R.E Sinclair and Peter Arcese (eds) *Serengeti II Dynamics: Management and conservation of an ecosystem*, pp. 146–68. Chicago: University of Chicago Press, p. 165.

125. Borner (1981), p.62.

126. Martin (1982), p. 50.

127. Roderick P. Neumann (1998) *Imposing Wilderness. Struggle over Livelihood and Nature Preservation in Africa.* Berkeley: University of California Press, p. 7.

128. Personal communication from guides who wished to remain anonymous.

129. Neumann (1998), p. 202.

130. Ibid., p. 63.

131. Ibid., pp. 64–5.

132. Martin (1982), p. 50.

133. Ibid.; personal communication with Esmond Bradley Martin.

134. Cumming, Du Toit and Stuart (1990), p. 56.

135. Hillman (1980), p. 23.

136. Runyoro et al. (1995), p. 165.

137. Patricia D. Moehlman, George Amato and Victor Runyoro (1996) Genetic and demographic threats to the black rhinoceros population in the Ngorongoro Crater. *Conservation Biology* 10(4): 1107–11, p. 1107.

138. Runyoro et al. (1995), p. 165.

139. Ibid.

140. Robert D. Fyumagwa and Julius W. Nyahongo (2010) Black rhino conservation in Tanzania: translocation efforts and further challenges. *Pachyderm* 47: 59–65, p. 59.

141. Emslie and Brooks (1999), p. 57.

142. Anthony Mills et al. (2006) Managing small populations in practice: black rhino *Diceros bicornis michaeli* in the Ngorongoro Crater, Tanzania. *Oryx* 40(3): 319–23, p. 319.

143. Ibid., p. 20.

144. Richard Emslie, IUCN AfRSG (2006) Ngorongoro black rhino population dynamics: what does the data tell us? In Mills et al., p. 8.

145. Titus Mlengaya, Chief Veterinary Officer, Tanzania Wildlife Authority (2003) The status of the black rhino in the Serengeti National Park. In Mills et al. 2006, p. 10.

146. Robert D. Fyumagwa and Julius W. Nyahongo (2010) Black rhino conservation in Tanzania: translocation efforts and further challenges. *Pachyderm* 47: 59–65, p. 59.

147. Vigne, Martin and Okita-Ouma (2007), p. 74.

148. Sidney (1961), pp. 56–7.

149. A.J.E. Cave (1963) The white rhinoceros in Uganda. *Oryx* 1: 28.

150. Ibid., p. 29.

151. *Oryx* I, 1963, 2, p. 80.

152. Sidney (1961), p. 58.

153. John Savidge (1961) The introduction of white rhinoceros into the Murchison Falls National Park, Uganda. *Oryx* 3: 184–6, 189.

154. Roger J. Wheater and Ian S.C. Parker (2019) The fate of Uganda's Northern white rhino translocated to Murchison Falls National Park in 1961 and 1964. *Pachyderm* 60: 112–17, p. 112.

155. Eric L. Edroma (1982) White rhino extinct in Uganda. *Oryx* 16(4): 352–55, p. 353.

156. Ibid., p. 354.

157. Ian Parker (2001) The darting trip. In Parker and Bleazard, *An Impossible Dream*, 181–9, p. 181.

158. Martin (1982), p. 43.

159. Edroma (1982), p. 352.

160. Hillman (1980), p. 23.

161. Emslie and Brooks (1999), p. 57.

162. Vigne, Martin and Okita-Ouma (2007), pp. 74–5.

163. Western and Vigne (1985), p. 218.

164. Emslie and Brooks (1999), p. 9.

165. John R. Platt (2013) How the western black rhino went extinct. *Scientific American*, 13 November. https://blogs.scientificamerican.com/extinction-countdown/how-the-western-black-rhino-went-extinct/. Accessed 7 June 2023.

166. Lagrot, Lagrot and Bour (2007), p. 25.

167. Ibid.
168. Hillman (1980), p. 23.
169. World Wildlife Fund (1984) African Rhino Conservation Project, WWF Yearbook 1983/84, http://www.rhinoresourcecenter.com/pdf_files/127/1276415467.pdf, p. 314. Accessed 12 April 2023.
170. Western and Vigne (1985), p. 216.
171. Emslie and Brooks (1999), p. 44.
172. Hillman (1980), p. 23.
173. Martin (1982), p. 52; Emslie and Brooks (1999), p. 44.
174. Emil K. Urban and Leslie H. Brown (1968) Wildlife in an Ethiopian valley. *Oryx* 9(6): 342–53, p. 346.
175. Emslie and Brooks (1999), p. 46.
176. Ibid., p. 51.
177. D.C. Happold (1966) The future for wildlife in the Sudan. *Oryx* 8(6): 360–73, p. 360.
178. Kes Hillman-Smith, Mankoto ma Oyisenzoo and Fraser Smith (1986) A last chance to save the northern white rhino? *Oryx* 20(1): 20–6, p. 20.
179. Ibid.
180. World Wildlife Fund (1984), 314.
181. Ibid., p. 315.
182. Emslie and Brooks (1992), p. 55.
183. J. Haezaert (1959) The black rhinoceros is brought back to Ruanda. *Oryx* 3: 96.
184. Emslie and Brooks (1992), p. 50
185. World Wildlife Fund (1984), p. 314.
186. Hillman, Oyisenzoo and Smith (1986), p. 20.
187. Rookmaaker (2013), pp. 453–4.
188. Kes Hillman-Smith (1987) Developing strategies for the northern white rhino. *Pachyderm* 9: 19–22. http://www.rhinoresourcecenter.com/pdf_files/117/1175860990.pdf, pp. 19–20. Accessed 8 April 2023.
189. Emslie and Brooks (1999), p. 44.
190. Ibid.
191. Hillman-Smith (1987), pp. 19–20.
192. Charles Mackie (Garamba Rehabilitation Project) (1987) *Pachyderm* 9: 22.
193. Ibid.
194. Kes Hillman-Smith (1988) Northern white rhinos born at Garamba. *Pachyderm* 10(1).
195. Kes Smith et al. (1993) Garamba National Park project. General aerial count, https://portals.iucn.org/library/node/17852, p. 8. Accessed 21 September 2013.
196. Fraser Smith and Kes Hillman Smith (1998) *Garamba National Park Project, Conservation of the Northern White Rhinoceros, Democratic Republic of the Congo.* World Wide Fund for Nature (WWF) Project No: ZR 0009. http://www.rhinoresourcecenter.com/pdf_files/133/1334459045.pdf, p. 1. Accessed 11 April, 2023.
197. Ibid., p. 11.
198. Ibid., pp. 17–18.
199. Kes Hillman Smith (2001) Status of northern white rhinos and elephants in Garamba National Park, Democratic Republic of Congo, during the wars. *Pachyderm* 31: 79–81, p. 79.
200. K. Hillman-Smith et al. (2003) Poaching upsurge in Garamba National Park, Congo. *Pachyderm* 35: 147–50, p. 149.
201. Kes Hillman Smith and Jerome Amube Ndey (2005) Post-war effects on the rhinos and elephants of Garamba National Park. *Pachyderm* 39: 106–10, p. 107.
202. Kes Hillman Smith (2006) Past population dynamics and individual information on possible surviving northern white rhinos in Garamba National Park and surrounding reserves. *Pachyderm* 40: 107–15, p. 115.
203. IUCN Red List (n.d.) Northern white rhino. https://www.iucnredlist.org/species/4183/45813838. Accessed 18 July 2023.
204. Martin Brooks (2009) AfRSG Report. *Pachyderm* 45: 8–15, p. 9.

Chapter 7: Southern Africa, 1960–2005:
armed conflict, crime and conservation

1. Mr Justice M.B. Kumleben (1996) *Commission of Inquiry into the Alleged Smuggling and Illegal Trade in Ivory and Rhinoceros Horn in South Africa*, http://www.rhinoresourcecenter.com/pdf_files/131/1311074532.pdf accessed 5 January 2023.

2. Michael 't Sas Rolfes et al. (2022) Legal hunting for conservation of highly threatened species: the case of African rhinos. *Conservation Letters* e12877: 1–9, p. 2. https://doi.org/10.1111/conl.12877. Accessed 31 July 2023.

3. D. Western and L. Vigne (1985) The deteriorating status of African rhinos. *Oryx* 19: 215–20, p. 218.

4. Rademeyer (2012), loc. 1838.

5. Western and Vigne (1985), p. 218.

6. Barry Dalal-Clayton and Brian Child (2003) *Lessons from Luangwa: The story of the Luangwa Integrated Resource Development Project, Zambia*. London: International Institute for Environment and Development, Wildlife and Development Series No. 13, p. 8. https://www.iied.org/sites/default/files/pdfs/migrate/9079IIED.pdf. Accessed 17 January 2023.

7. Milliken, Nowell and Thomsen (1993), p. 1.

8. Emslie and Brooks (1999), p. 51.

9. Cumming, Du Toit and Stuart (1990), p. 9.

10. M. Karsten et al. (2011) The history and management of black rhino in KwaZulu-Natal: a population genetic approach to assess the past and guide the future. *Animal Conservation* 14: 363–70, p. 364.

11. P.A. Lindsey, S.S. Romanach and H.T. Davies-Mostert (2009) The importance of conservancies for enhancing the value of game ranch land for large mammal conservation in southern Africa. *Journal of Zoology* 277: 99–105, p. 99–100.

12. Ibid.

13. J.L. Anderson (1993) Management of translocated white rhino in South Africa. In Ryder (ed.) *Rhinoceros Biology and Conservation*, pp. 287–93, p. 287.

14. Emslie and Brooks (1999), p. 10.

15. Milliken and Shaw (2012), p. 93.

16. Emslie and Brooks (1999), p. 39.

17. Stephen Ellis (1994) Of elephants and men: politics and nature conservation in South Africa. *Journal of Southern African Studies* 20(1): 53.

18. Somerville (2017), pp. 322–3.

19. Keith Somerville (1990) *Foreign Military Intervention in Africa*. London: Pinter, pp. 99–100.

20. Jan Breytenbach (1997) *Eden's Exiles. One soldier's fight for paradise*. Cape Town: Quellerie, 1997, p. 103.

21. Ibid., p. 242.

22. Ibid., p. 252.

23. Kumleben (1996).

24. Ian Parker (1979) *The Ivory Trade*. Washington: US Fish and Wildlife Service, pp. 117–18 (Ian Parker Collection of East African Wildlife Conservation: the Ivory Trade http://ufdc.ufl.edu/AA00020117/00001/3j University of Florida, accessed 5 March 2015, p. 138.

25. Rademeyer (2012), loc. 792–801.

26. Ellis (1994), p. 56.

27. Kumleben (1996), pp. 20–1.

28. Ibid., pp. 100–4.

29. Personal communication, November 1990.

30. Breytenbach (1990), p. 69.

31. Personal communication from Breytenbach and Stephen Ellis.

32. Environmental Investigation Agency (1989) A system of extinction: the African elephant disaster. London: EIA. https://eia-international.org/wp-content/uploads/A-System-of-Extinction.pdf, pp. 25–6. Accessed 1 March 2023.

33. Kumleben (1996), pp. 48–9.
34. Huntley (2023), p. 357.
35. Huntley (2017), p. 26.
36. Bradley Martin and Bradley Martin (1982), p. 94.
37. Ibid., p. 52.
38. Emslie and Brooks (1999), p. 42.
39. Hillman (1980), p. 23.
40. Keith Somerville (2020) Botswana's rhino poaching crisis: COVID-19 increases the pressure. *Global Geneva*, 11 April. https://global-geneva.com/botswanas-rhino-poaching-crisis-covid-19-increases-the-pressure/. Accessed 8 August 2023.
41. Martin (1982), p. 52.
42. Emslie and Brooks (1999), p. 42.
43. Ibid., pp. 42–3.
44. Personal communication with Botswana's Foreign Minister Gaositwe Chiepe, October 1993.
45. Dan Henk (2007) *The Botswana Defense Force in the Struggle for an African Environment*. New York: Palgrave Macmillan, p. 2.
46. Personal communication with community members and leaders in the Okavango region and the Kalahari Conservation Society in Gaborone, October 1993, and Chief Timex Moalosi of Sankuyo in Maun, May 2018.
47. Henk (2007), p. 24.
48. Ibid.
49. Ibid., p. 54.
50. Ibid., p. 61; personal communication with politicians in Zambia and Namibia in 1990 and 1991.
51. Henk (2007), p. 67.
52. Personal communication with Derek de la Harpe, a senior executive with Wilderness Safaris, 7 August 2023, and with wildlife veterinarian Erik Verreynne, April 2020 and September 2023. See also Erik Verreynne (2020) Botswana's rhinos are under siege: It's time to learn from historical mistakes. *Daily Maverick*, 10 January. https://www.dailymaverick.co.za/article/2020-01-10-botswanas-rhinos-are-under-siege-its-time-to-learn-from-historical-mistakes/. Accessed 8 November 2023.
53. Ole-Gunnar Støen (2009) Same-site multiple releases of translocated white rhinoceroses *Ceratotherium simum* may increase the risk of unwanted dispersal. *Oryx* 43(4): 580–5, p. 580.
54. Verreynne (2020).
55. Vera Pfannerstil et al. (2022) Effects of age and sex on site fidelity, movement ranges and home ranges of white and black rhinoceros translocated to the Okavango Delta, Botswana. *African Journal of Ecology* 60: 344–56, p. 346.
56. Annette A. LaRocco and Emmanuel Mogende (2022) Fall from grace or back down to earth? Conservation and political conflict in Africa's 'miracle' state. *Environment and Planning E: Nature and Space* Online First, 1 June, no page numbers.
57. Brian T.B. Jones (2009) Community benefits from safari hunting and related activities in Southern Africa. In B. Dickson, J. Hutton and W.M. Adams (eds) *Recreational Hunting, Conservation and Rural Livelihoods: Science and practice*, pp. 157–77. London: Wiley-Blackwell, p. 158.
58. Personal communication with Chief Timex Moalosi of Sankuyo, May 2018; Jones (2009), p. 161.
59. The Kingdom of Swaziland's Big Game Parks (1996) Rhino gift to Swaziland. *Oryx* 30(1): 12–13.
60. Emslie and Brooks (1999), p. 55.
61. Ibid., p. 48.
62. Hillman (1980), p. 23.
63. Personal communication with Hugo Jachman, May 2015 and November 1981.

64. Emslie and Brooks (1999), p. 48.
65. Ibid.
66. Wildlife Society of Malawi (2000) Mammal re-introduction into Liwonde National Park. *Oryx* 34(2): 94.
67. Paul Taylor (2004) The first privatization of a wildlife area in Malawi. *Oryx* 38(2): 134.
68. Krisztián Gyöngyi and Morten Elmeros (2017) Forage choice of the reintroduced black rhino and the availability of selected browse species at Majete Wildlife Reserve, Malawi. *Pachyderm* 58: 40–50, p. 41.
69. Ellis (1994), p. 55.
70. Kumleben (1996), pp. 79–80.
71. Hillman (1980), p. 23.
72. Emslie and Brooks (1992), p. 49.
73. Martin (1982), p. 52.
74. Emslie and Brooks (1992), p. 49.
75. Ibid.
76. Environmental Investigation Agency (1989), p. 29.
77. Joubert (1971), p. 38.
78. Ibid., pp. 38–40.
79. Ibid.
80. Ibid.
81. Ibid., p. 42.
82. Anthony Hall-Martin (1979) Black rhinoceros in southern Africa. *Oryx* 15(1): 27–32, p. 30.
83. Ibid.
84. Cited by Rademeyer (2012), loc. 1002.
85. Hillman (1980), p. 23.
86. !Uri‡khob (2004), p. 18.
87. Milliken, Nowell and Thomsen (1993), p. 45.
88. Emslie and Brooks (1999), p. 49.
89. Owen-Smith (2010), p. 3.
90. Ibid., p. 99.
91. Ibid.
92. Ibid., p. 174.
93. Ibid.
94. Ibid., p. 224.
95. Kumleben (1996), pp. 80–4.
96. Jan Breytenbach (1997) *Eden's Exiles: One Soldier's Fight for Paradise*. Cape Town: Quellerie, p. 28.
97. Ibid.
98. Ibid., pp. 70–1.
99. Owen-Smith (2010), pp. 320–1.
100. Ibid., p. 376.
101. Ibid., p. 386.
102. Ibid., p. 393.
103. Ibid., p. 507.
104. Sullivan and Muntifering (2020), pp. 11–12; Owen-Smith (2010), p. 474.
105. Milliken, Nowell and Thomsen (1993), p. 38.
106. Sullivan and Muntifering (2021), p. 13.
107. Ibid., p. 14.
108. Sullivan et al. (2021), pp. 15–16.
109. Christo Botha (2005) People and the environment in colonial Namibia. *South African Historical Journal* 52: 170–190, p. 182.
110. Ibid.
111. Botha (2005), p. 188.
112. !Uri‡khob (2004), pp. 16–17.

113. Personal communication with Derek de la Harpe of Wilderness Safaris, 7 August 2023.
114. Jones (2009), p. 161.
115. The High-Level Panel of Experts for the Review of Policies, Legislation and Practices on Matters of Elephant, Lion, Leopard and Rhinoceros Management, Breeding, Hunting, Trade and Handling (2020) *High-Level Panel Report for Submission to the Minister of Environment, Forestry and Fisheries*, 15 December, p. 48. https://www.dffe.gov.za/sites/default/files/reports/2020-12-22_high-levelpanel_report.pdf. Accessed 22 August 2023.
116. Emslie and Brooks (1999), pp. 49–50.
117. Ibid.
118. Ibid.
119. Save the Rhino Trust (n.d.) Our mandate. https://www.savetherhinotrust.org/about-us.html. Accessed 20 November 2023.
120. Personal communications with SRT guides and Derek de la Harpe of Wilderness.
121. !Uri‡khob (2004), p. 10.
122. Ibid., p. 11.
123. Sullivan and Muntifering (2020), p. 14.
124. !Uri‡khob (2004), p. 10.
125. Ibid., p. 19.
126. Ibid., p. 35.
127. Emslie and Brooks (1999), p. 51.
128. Ibid.
129. Milliken and Shaw (2012), p. 24.
130. Emslie and Brooks (1999), p. 51.
131. Sidney (1961).
132. Caroline Reid et al. (2007) Habitat changes reduce the carrying capacity of Hluhluwe-Umfolozi Park, South Africa, for Critically Endangered black rhinoceros *Diceros bicornis*. *Oryx* 41(2): 247–54, p. 247.
133. Player (1972), p. 15.
134. Ian Player (1967) Translocation of White Rhinoceros in South Africa. *Oryx* 9(2): 137–50, p. 140.
135. Ibid., pp. 145–7.
136. Ibid.
137. Player (1972), p. 228.
138. Pienaar, Bothma and Theron (1993), p. 641.
139. Player (1972), p. 228.
140. S.M. Ferreira, J.M. Botha and M.C. Emmett (2012) Anthropogenic influences on conservation values of white rhinoceros. *PLoS ONE* 7(9): 5.
141. Reid et al. (2007), p. 247.
142. Rademeyer (2012), loc. 1826; Orenstein (2013), p. 71.
143. Milliken and Shaw (2012), pp. 45–6.
144. Ibid.
145. Personal communication with John Hanks on a press trip with him to South Africa and eSwatini in late August 2016.
146. John Hanks (2015) *Operation Lock and the War on Rhino Poaching*. London: Penguin, p. 13.
147. D. Buys (1987) A summary of the introduction of white rhino on to private land in the Republic of South Africa: Report for Rhino and Elephant Foundation. http://www.rhinoresourcecenter.com/pdf_files/129/1291956323.pdf. Accessed 21 November 2023.
148. Michael 't Sas Rolfes M. (2012) *Saving African Rhinos: A Market Success Story*. Bozeman: PERC Case Studies, pp. 3–4. https://www.perc.org/wp-content/uploads/2011/08/Saving-African-Rhinos-final.pdf. Accessed 21 August 2023.
149. Ibid., p. 4.
150. Milliken and Shaw (2012), pp. 45–6.
151. High-Level Panel (2020), p. 48.

152. Martin (1982), p. 96.

153. Milliken and Shaw (2012), p. 11.

154. Kumleben (1996), p. 162.

155. Emslie and Brooks (1999), pp. 51–2.

156. Hrabar and du Toit (2005), p. 260.

157. Sam M. Ferreira and Cathy Greaver (2016) Re-introduction success of black rhinoceros in Marakele National Park. *African Journal of Wildlife Research* 46(2): 135–8, pp. 135–6. Ferreira is Scientific Officer for the AfRSG and a leading rhino specialist and mammal scientist with SANParks.

158. Emslie and Brooks (1999), p. 52.

159. Ibid.

160. High-Level Panel (2020), pp. 48–9.

161. Emslie and Brooks (1999), p. 53.

162. Ibid. (1999), p. 52.

163. Chansa Chomba and Wigganson Matandikohansa (2011) Population status of black and white rhinos in Zambia. *Pachyderm* 50: 50–5, p. 50.

164. Emslie and Brooks (1999), p. 57.

165. Hanks (2015), p. 12.

166. Clark C. Gibson (1999) *Politicians and Poachers. The political economy of wildlife policy in Africa*. Cambridge: Cambridge University Press, p. 27.

167. Ibid., p. 32.

168. Dalal-Clayton and Child (2003), p. 7.

169. Ibid.

170. S. L. Atkins (1984) The socio-economic aspects of the Lupande Game Management Area. In D.B. Dalal-Clayton (ed.) *Proceedings of the Lupande Development Workshop*. Lusaka: Government Printer, p. 52.

171. Simon Kumwenda (2021) Whose heritage? The state, local communities and game in South Luangwa National Park (SLNP) of Eastern Zambia, 1890–2001, p. 1. University of Zambia. http://dspace.unza.zm/bitstream/handle/123456789/7292/Main%20Document.PDF?sequence=1. Accessed 23 January 2023.

172. Republic of Zambia (1970) *Debates of the Second Session of the Second National Assembly, 7th January-25th March 1970*. Lusaka: Government printer, pp. 550–1.

173. Emslie and Brooks (1999), p. 57.

174. Hillman (1980), p. 23.

175. Save the Rhino (2023) Zambia: North Luangwa Conservation Programme, https://www.savetherhino.org/programmes/north-luangwa-conservation-programme/. Accessed 19 January 2023.

176. Chomba and Matandikohansa (2011), p. 50.

177. Dalal-Clayton and Child (2003), p. 8.

178. Gibson (1999), p. 36.

179. Personal communication with Richard Bell.

180. Gibson (1999), p. 35.

181. Ibid., and personal communication with Richard Bell in Maun, Botswana, November 1993, and also with Guy Scott of the MMD, who was Minister of Agriculture, Food and Fisheries in the first MMD government and later acting President in 2014-5, in Lusaka, February 1991.

182. Zambian Department of Game and Fisheries (1970) *Annual Report, 1968*. Lusaka: Government Printer, p. 43.

183. Personal communication with Richard Bell.

184. Personal communication with Zambian President Kenneth Kaunda at State House, Lusaka, 22 February 1991.

185. Gibson (1990), p. 53.

186. Ibid.

187. Milliken, Nowell and Thomsen (1993), p. 25.

188. Ibid.
189. Personal communication with Richard Bell.
190. Personal communication with Richard Bell.
191. Gibson (1999), p. 57.
192. Ibid.
193. Zambian National Assembly (1982) Parliamentary debates, August, p. 4112.
194. *Times of Zambia*, 28 November 1986. Journalists working on the *Times of Zambia* and others working for the BBC African Service told the author in Lusaka in February 1991 that the levels of corruption in NPWS were well known, but because of ministerial, senior civil service, police and army involvement, nothing was done to combat it.
195. Gibson (1999), p. 56.
196. Personal communication with Richard Bell.
197. *Times of Zambia*, 8 January 1991.
198. Hanks (2015), p. 28.
199. Ibid.
200. Parker (2004), pp. 344–6.
201. Kumleben (1996), p. 139.
202. Rademeyer (2012), loc. 75.
203. This was made clear in the unpublished research notes and writings of the late Stephen Ellis, which I had access to through his publisher.
204. Hanks (2015), p. 29.
205. Ibid., p. 44.
206. N. Leader-Williams (1985) Black rhino in South Luangwa National Park: their distribution and future protection. *Oryx* 19(1): 27–33, pp. 28–9.
207. National Parks and Wildlife Service (1979) *1979 Annual Report*, p. 2.
208. Dalal-Clayton and Child (2003), p. 8.
209. Gibson (1999), p. 56.
210. Ibid., p. 6.
211. Personal communication with Richard Bell.
212. Ibid.
213. Dalal-Clayton and Child (2003), p. 11.
214. Ibid., p. 11.
215. Interviews with Fred Chiluba, MMD leader and then president, and senior MMD politician Vernon Mwaanga, Lusaka, February 1991.
216. Emmanuel Koro (2021) Zambian community's wildlife conservation pays off 45 years later. *Conservation Force*, 2 January. https://www.conservationforce.org/zambian-community-s-wildlife-conser. Accessed 24 January 2021.
217. Save the Rhino (2023).
218. Ibid.
219. World Wide Fund for Nature (2019) Walking with Rhinos in Zambia. https://www.worldwildlife.org/magazine/issues/summer-2018/articles/walking-with-white-rhinos-in-zambia. Accessed 25 January 2023.
220. Chansa Chomba and Wigganson Matandikohansa (2011) Population status of black and white rhinos in Zambia. *Pachyderm* 50: 50–5, p. 50; Save the Rhino (2023) Zambia: North Luangwa Conservation Programme, https://www.savetherhino.org/programmes/north-luangwa-conservation-programme/. accessed 25 January 2023.
221. Milliken, Nowell and Thomsen (1993), p. 17.
222. H.H. Roth (1967) White and black rhinoceros in Rhodesia. *Oryx* 9: 217–231, pp. 233, 226.
223. Player (1967), pp. 147–8.
224. Janet L. Rachlow and Joel Berger (1998) Reproduction and population density: trade-offs for the conservation of rhinos in situ. *Animal Conservation* 1: 101–6, p. 102.
225. Roth (1967), p. 230.
226. Ibid.

227. Michael A. Kerr and Rupert Fothergill (1971) Black rhinoceros in Rhodesia. *Oryx* 11 (2–3): 129–34, p. 129.
228. Ibid., p. 133.
229. Martin (1982), p. 93.
230. Ibid., p. 94.
231. Gibson (1990), p. 45.
232. Ibid.
233. Somerville (2019), p. 176. Some of the information presented there is from personal communication with ivory and rhino horn trade investigators, given under conditions of anonymity, and communication with former ZANU guerrilla commanders in Harare in April 1982 and February 1991.
234. Cumming, Du Toit and Stuart (1990), p. 56.
235. Personal communication with wildlife officials and ZANU ministers in Harare during research trip to Zimbabwe, April 1982.
236. Emslie and Brook (1999), p. 57.
237. Personal communication with Bubye Valley Conservancy head Blondie Leatham, July 2018, and Brian Child, London, June 2019.
238. Milliken, Nowell and Thomsen (1993), p. 45.
239. Savé Valley Conservancy. https://savevalleyconservancy.org/. Accessed 2 September 2023.
240. http://chilogorge.com/about/. Accessed 2 June 2015.
241. Milliken, Nowell and Thomsen (1993), p. 45.
242. Esmond Bradley Martin (1984) Zimbabwe's ivory carving industry. *Traffic Bulletin* 6 (2).
243. Environmental Investigation Agency (1992) *Under Fire: Elephants in the Front Line.* London: EIA, p. 21.
244. Duffy (2000), p. 63.
245. Rosaleen Duffy (1999) The role and limitations of state coercion: anti-poaching policies in Zimbabwe. *Journal of Contemporary African Studies* 17(1): 97–121, p. 109.
246. Rademeyer (2012), loc. 386.
247. Hanks (2015), p. 30.
248. Rademeyer (2012), loc. 404–416.
249. Ibid.
250. Milliken, Nowell and Thomsen (1993), pp. 19–20.
251. Ibid., pp. 20–3.
252. Ibid., pp. 25–6.
253. Cited in ibid., pp. 25–6.
254. Ibid., p. 27.
255. Rademeyer (2012), loc. 386.
256. Ibid., loc. 429–41.
257. Personal communication with Raoul du Toit, 5 September 2023.
258. Milliken, Nowell and Thomsen (1993), p. 28.
259. Ibid. p. 27.
260. Hanks (2015), pp. 45–7, 93–4 and 139.
261. Rosaleen Duffy (2000) *Killing for Conservation.* Oxford: James Currey/The International African Institute, p. 54.
262. Ibid., p. 55.
263. TRAFFIC (1992) *Bulletin* 13 (2), p. 47.
264. For further details of land reform, see S.T. Williams et al. (2016) The impact of land reform on the status of large carnivores in Zimbabwe. *PeerJ* 4: 1–21.
265. Peter Makumbe et al. (2022) Human–wildlife conflict in Save Valley Conservancy: residents' attitude toward wildlife conservation. *Scientifica* 2022: no page numbers.
266. AfRSG (2007) Current rhino numbers and trends, recommended conservation strategies and the EAZA Rhino Campaign, pp. 58–9. http://www.rhinoresourcecenter.com/pdf_files/117/1175858257.pdf. Accessed 28 November 2023.

Chapter 8: East and Central Africa, 2006–24

1. Richard H Emslie, Tom Milliken and Bibhab Talukdar (2010) *African and Asian Rhinoceroses – Status, Conservation and Trade: A report from the IUCN Species Survival Commission (IUCN/SSC) African and Asian Rhino Specialist Groups and TRAFFIC to the CITES Secretariat pursuant to Resolution Conf. 9.14 (Rev. CoP15).* https://rhinos.org/wp-content/uploads/2020/10/final-cop16-rhino-rpt.pdf. Accessed 22 December 2023.

2. Milliken and Shaw (2012), pp. 18, 24.

3. IUCN Red List (n.d.).

4. Rhino poaching dataset supplied by Emme Pereira, Save the Rhino, 17 October 2023.

5. Knight, Mosweub and Ferreira (2023) AfRSG chair report. *Pachyderm* 64: 13–30, pp. 13–16.

6. CITES (2022) Nineteenth meeting of the Conference of the Parties, Panama City (Panama), 14–25 November 2022. Species specific matters RHINOCEROSES (RHINOCEROTIDAE SPP.), pp. 2–3.

7. Ibid.

8. CITES (2022), pp. 2–3.

9. Ibid., p. 2.

10. IUCN (2023) African rhino numbers are increasing despite poaching. Press release, 21 September. https://www.iucn.org/press-release/202309/african-rhino-numbers-are-increasing-despite-poaching#:~:text=South%20Africa%20is%20still%20home,to%2047%20the%20previous%20year. Accessed 4 December 2023.

11. Save the Rhino (2023) World Rhino Day 2023: new numbers and challenges. https://www.savetherhino.org/africa/world-rhino-day-2023-new-numbers-and-challenges/?sfmc_id=203184. Accessed 23 September 2023.

12. Personal communication, 11 December 2023.

13. CITES (2022), p. 3.

14. Mike Knight, Keitumetse Mosweub and Sam M. Ferreira (2022) AfRSG chair report. *Pachyderm* 63: 17–32, pp. 18–19.

15. US Attorney's Office (2022) Wildlife trafficker sentenced to 57 months for large-scale trafficking of rhinoceros horns and elephant ivory, 14 December. https://www.justice.gov/usao-sdny/pr/wildlife-trafficker-sentenced-57-months-large-scale-trafficking-rhinoceros-horns-and. Accessed 21 September 2023.

16. Knight, Mosweub and Ferreira (2023), pp. 18–19.

17. Kristopher Carlson (2015) *In the Line of Fire: Elephant and rhino poaching in Africa.* https://www.researchgate.net/publication/277598416_In_the_Line_of_Fire_Elephant_and_Rhino_Poaching_in_Africa, pp. 21–22. Accessed 6 September 2023.

18. Ibid.

19. Ibid.

20. Ibid.

21. Keith Somerville (2023) South African white rhinos relocated to DRC's Garamba National Park. *Commonwealth Opinion*, 16 June. https://commonwealth-opinion.blogs.sas.ac.uk/2023/south-african-white-rhinos-relocated-to-drcs-garamba-national-park/. Accessed 4 December 2023.

22. Keith Somerville (2023) African Parks buys John Hume's 2000 rhinos to rewild them. *Commonwealth Opinion*, 7 September. https://commonwealth-opinion.blogs.sas.ac.uk/2023/african-parks-buys-john-humes-2000-rhinos-to-rewild-them/. Accessed 18 September 2023.

23. Graeme Green (2023) Sold: 2,000 captive southern white rhino destined for freedom across Africa. *Guardian*, 1 December, https://www.theguardian.com/environment/2023/dec/01/sold-2000-captive-southern-white-rhino-destined-for-freedom-across-africa-aoe. Accessed 4 December 2023.

24. Susan Anyango Oginah, Paul O. Ang'ienda and Patrick O. Onyango (2020) Evaluation of habitat use and ecological carrying capacity for the reintroduced Eastern black rhinoceros (*Diceros bicornis michaeli*) in Ruma National Park. *African Journal of Ecology* 58: 34–45, p. 35.

25. Save the Rhino (2018) All rhinos translocated to Tsavo East National Park have died. https://www.savetherhino.org/africa/kenya/all-rhinos-translocated-to-tsavo-east-national-park-have-died/#:~:text=On%2026%20July%202018%2C%20the,in%20Tsavo%20East%20National%20Park. Accessed 8 December 2023.

26. Knight, Mosweub and Ferreira (2023), pp. 18–19.

27. Hayley S. Clements, Dave Balfour and Enrico Di Minin (2023) Importance of private and communal lands to sustainable conservation of Africa's rhinoceroses. *Frontiers in Ecology and the Environment*, 21(3)1–8: 2.

28. CITES-based figures courtesy of Emme Pereira of Save the Rhino.

29. Personal communication with Pelham Jones, chair of PROA, 6 December 2023.

30. Clements, Balfour and Di Minin (2023), p. 5.

31. Ibid., p. 6.

32. Jamie Gaymer (2023) Laikipia Rhino-Range Expansion. *The Horn*, December, pp. 10–11. https://media.savetherhino.org/prod/uploads/2023/11/SR3498_TheHorn23-v6.1-ISSU-Eds.pdf?sfmc_id=203184. Accessed 6 December 2023.

33. Ibid.

34. Mike Knight (2016) AfRSG report. *Pachyderm* 57: 17.

35. Don Pinnock (2023) Canada blocks import of ivory, rhino horn and trophies. *Daily Maverick*, 21 November. https://www.dailymaverick.co.za/article/2023-11-21-canada-blocks-import-of-ivory-rhino-horn-and-trophies/. Accessed 5 December 2023.

36. Clements, Balfour and Di Minin (2023), p. 6.

37. Save the Rhino (2023) *The Horn*, December, p. 5. https://media.savetherhino.org/prod/uploads/2023/11/SR3498_TheHorn23-v6.1-ISSU-Eds.pdf?sfmc_id=203184. Accessed 6 December 2023.

38. Sam M. Ferreira and Benson Okita-Ouma (2012) A proposed framework for short-, medium- and long-term responses by range and consumer states to curb poaching for African rhino horn. *Pachyderm* 51: 52–9, p. 53.

39. Milliken, Nowell and Thomsen (1993), p. 41.

40. Chris Barichievy (2021) A demographic model to support an impact financing mechanism for black rhino metapopulations. *Biological Conservation* 257: 1–11, p. 5.

41. Cedric Khayale et al. (2020) Management progress on the Kenya Black Rhino Action Plan 2017–21. *Pachyderm* 61: 109–19, pp. 110–11.

42. Ibid., p. 118.

43. Carlson (2015), p. 7.

44. John Mbaria and Mordecai Ogada (2016) *The Big Conservation Lie: The untold story of wildlife conservation in Kenya*. Nairobi: John Mbaria and Mordecai Ogada, p. 74.

45. Carlson (2015), p. 7.

46. Save the Rhino (n.d.) Kenya: Ol Pejeta Conservancy. https://www.savetherhino.org/programmes/kenya-ol-pejeta-conservancy/#:~:text=Since%20then%2C%20enhanced%20protection%20efforts,complacency%20is%20not%20an%20option. Accessed 28 August 2023.

47. Gaymer (2023).

48. Ibid.

49. Margaret F. Kinnaird and Timothy O'Brien (2012) Effects of private-land use, livestock management, and human tolerance on diversity, distribution, and abundance of large African mammals. *Conservation Biology* 26(6): 1026–39, p. 102.

50. Personal communication with Marc Dupuis-Désormeaux, August 2023.

51. Marc Dupuis-Désormeaux et al. (2016) Usage of specialized fence-gaps in a black rhinoceros conservancy in Kenya. *African Journal of Wildlife Research* 46(1): 22–32, pp. 23–4.

52. Personal communication with Kenyan conservationist Mordecai Ogada, 26 August 2023; Mbaria and Ogada (2016), p. 69; and Anuradha Mittal, Zahra Moloo and Frederic Mousseau (2021) *Stealth Game: 'Community' conservancies devastate land & lives in northern Kenya*. Oakland, CA: Oakland Institute, p. 4.

53. Northern Rangelands Trust (n.d.) Transforming lives, conserving nature. https://www.nrt-kenya.org/. Accessed 24 August 2023.

54. See, for example, Mittal, Moloo and Mousseau (2021).

55. Mordecai Ogada (2019) *Standards of Conservation Practice and the Status of Community Natural Resource Rights in Kenya's Rangelands*. UN Department of Economic and Social Affairs, Division for Inclusive Social Development, Indigenous Peoples and Development Branch – Secretariat of the Permanent Forum on Indigenous Issues. https://www.un.org/development/desa/indigenouspeoples/wp-content/uploads/sites/19/2019/01/UNEP-Paper.pdf, p. 1. Accessed 24 August 2023.

56. Mittal, Moloo and Mousseau (2021), p. 5.

57. Ibid., pp. 5–6.

58. Personal communication with Mordecai Ogada, August 2023.

59. Cedric Khayale et al. (2021) Kenya's first white rhino conservation and management action plan. *Pachyderm* 62: 112–18, p. 112; CITES (2022), pp. 2–3.

60. Khayale et al. (2021), p. 114.

61. Save the Rhino (2018) All rhinos translocated to Tsavo East National Park have died. https://www.savetherhino.org/africa/kenya/all-rhinos-translocated-to-tsavo-east-national-park-have-died/#:~:text=We%20understand%20that%20two%20of,on%20arrival%20in%20Tsavo%20East. Accessed 13 December 2023.

62. Richard Leakey on Twitter/X, 27 July 2018. https://twitter.com/safariwithselle/status/1022801557820203010.

63. SEEJ-Africa (2021) Your rangers are in the crosshairs – an open letter to Brigadier (Rtd.) John Waweru, Director General KWS, Seej-Africa, 10 August. https://www.seej-africa.org/2021/10/08/your-rangers-are-in-the-crosshairs/. Accessed 28 April 2022.

64. Houghton Irungu (2021) Wildlife defender Kofa did not deserve a brutal death. *Standard*. https://www.standardmedia.co.ke/houghton-irungu/article/2001423856/wildlife-defender-kofa-did-not-deserve-a-brutal-death. Accessed 11 December 2023.

65. Ibid.

66. Ibid.

67. Chris Morris (2018) The appeal acquittal of Feisal Mohamed Ali: a victory for rule of law, a process corrupted, or both? (commentary). *Mongabay*, 17 August. https://news.mongabay.com/2018/08/the-appeal-acquittal-of-feisal-mohamed-ali-a-victory-for-rule-of-law-a-process-corrupted-or-both-commentary/. Accessed 11 Dedember 2023.

68. Xinhua (2022) Tanzania plans to introduce 30 white rhinos to boost conservation. *Xinhua*, 3 June. https://english.news.cn/africa/20220604/404d07966ae6420ba468c1fa1cc56e10/c.html#:~:text=According%20to%20statistics%20by%20the,2018%20and%20190%20in%202020. Accessed 11 December 2023.

69. Shetler (2007), p. 225.

70. CITES (2022), pp. 2–3.

71. Personal communication with Sam Ferreira, 11 December 2023.

72. Grumeti Fund (2019) Impact report 2019. http://grumetifund.org/wp-content/uploads/2020/04/Grumeti-Fund_Impact-Report-2019-1.pdf. Accessed 18 September 2019.

73. Ibid.

74. T. Michael Anderson et al. (2020) The burning question: does fire affect habitat selection and forage preference of the black rhinoceros *Diceros bicornis* in East African savannahs? *Oryx* 54(2): 234–43, p. 235.

75. Siringit (n.d.) Black rhino of the Serengeti. https://www.siringit.co.tz/rhinos-of-serengeti/. Accessed 12 December 2023.

76. Air Tanzania (n.d.) Black rhinos are back in Mkomazi National park. https://issuu.com/landmarine/docs/air-tanzania-twiga_11/s/13630934. Accessed 12 December 2023.

77. Gideon A. Mseja et al. (2020) Dry season wildlife census in Mkomazi National Park, 2015. In Jeffrey O. Durrant et al. (eds) *Protected Areas in Northern Tanzania: Local Communities, Land Use Change, and Management Challenges*, pp. 133–45. Cham: Springer Nature, p. 143.

78. Ibid.

79. Tanzania Times (2021) Knock! Knock? Black Rhinos return at the foot of Mount Meru. *Tanzania Times*, 6 August. https://tanzaniatimes.net/knock-knock-knock-black-rhinos-may-soon-return-to-the-foot-of-mount-meru/. Accessed 20 September 2023.

80. D. Western and L. Vigne (1985) The deteriorating status of African rhinos. *Oryx* 19: 215–20, p. 216; Milliken, Nowell and Thomsen (1993), p. 5; Emslie and Brooks (1999), pp. 8–9.

81. Felix J. Patton et al. (2012) Dispersal and social behaviour of the three adult female white rhinos at Ziwa Rhino Sanctuary in the immediate period before, during and after calving. *Pachyderm* 52: 66–71, p. 67.

82. Felix Patton and Angie Genade (2013) Ziwa Rhino Sanctuary – the first 10 years. *Pachyderm* 54: 75–9, p. 78.

83. Felix J. Patton, Petra E. Campbell and Angie Genade (2018) Observations on the 24-hour clock, reproduction and gestation periods of the white rhinoceros at Ziwa Rhino Sanctuary, Uganda. *Pachyderm* 59: 103–8, p. 103.

84. Reuters (2021) Conservancy rekindles hope for Uganda's rhino population. Reuters, 9 December. https://www.reuters.com/business/environment/conservancy-rekindles-hope-ugandas-rhino-population-2021-12-09/. Accessed 1 August 2023.

85. Pamela Amia (2022) Uganda to reintroduce rhinos into the wild, *Independent*, 20 August. https://www.independent.co.uk/voices/campaigns/giantsclub/uganda-reintroduce-rhinos-wild-b2149292.html. Accessed 11 September 2023.

86. Derrick Wandera (2022) Police intercept Shs1.4b of rhino horn at airport. *Monitor*, 30 May.

87. African Parks (n.d.) Zakouma National Park. https://www.africanparks.org/the-parks/zakouma. Accessed 12 December 2023.

88. African Parks (2018) The governments of the Republic of South Africa and Chad together with African Parks and SANparks confirm the discovery of two additional black rhino carcasses in Zakouma National Park in Chad. Press release, 2 November. https://www.africanparks.org/press-release/rhinos-zakouma-chad. Accessed 12 December 2023.

89. Personal communication, June 2023.

90. Personal communication, June 2023.

91. Ed Stoddard (2023) Conserving and rewilding John Hume's rhinos may cost R1bn or more. *Daily Maverick*, 12 September. https://www.dailymaverick.co.za/article/2023-09-12-conserving-and-rewilding-john-humes-rhinos-may-cost-r1bn-or-more/. Accessed 18 December 2023.

92. Graeme Green (2023) VIP passengers: the five black rhinos flown 2,700 miles on a mission to repopulate Chad. *Guardian*, 6 December.

93. IUCN Red List (n.d.).

94. Lawrence Anthony and Graham Spence (2012) *The Last Rhinos*. London: Pan Books, p. 125.

95. Ibid., pp. 201–31.

96. African Parks (n.d.) Garamba National Park. https://www.africanparks.org/the-parks/garamba. Accessed 12 November 2023.

97. Ibid.

98. Personal communication, 11 December 2023.

99. Jim Tan (2023) African Parks to rewild 2,000 rhinos from controversial breeding program. *Mongabay*, 11 September. https://news.mongabay.com/2023/09/african-parks-to-rewild-2000-rhinos-from-controversial-breeding-program/#:~:text=On%20Sept.,simum)%20that%20came%20with%20it. Accessed 12 December 2023.

100. Phys.org (2023) White rhinos reintroduced to DR Congo national park, Phys.org, 11 June. https://phys.org/news/2023-06-white-rhinos-reintroduced-dr-congo.html. Accessed 12 December 2023.

101. Mike Knight (2017) AfRSG report. *Pachyderm* 58: 32.

102. African Parks (n.d.) Rhinos return to Rwanda 2017. https://www.africanparks.org/campaign/rhinos-return-rwanda/about. Accessed 6 September 2023.

103. Personal communication with Ladis Ndahiriwe of African Parks, 29 November 2021.

104. Ed Stoddard (2022) Rwanda's rhinos are safer than its dissidents. *Daily Maverick*, 8 June. https://www.dailymaverick.co.za/article/2022-06-08-rwandas-rhinos-are-safer-than-its-dissidents/?fbclid=IwAR2DHkFR00JMAlCpN7B7FEYFUVvYT38KpnJMCv4KUUB-HRT-wc8Znz2a_8LE. Accessed 24 August 2023.

105. Personal communication with Michaela Wrong, 24 August 2023.

Chapter 9: Southern Africa's poaching storm

1. CITES (2022) Nineteenth meeting of the Conference of the Parties, Panama City (Panama), 14–25 November 2022. https://cites.org/sites/default/files/documents/E-CoP19-Inf-81.pdf, pp. 2–3. Accessed 27 September 2023.

2. Ibid. and Keith Somerville (2022) Kruger National Park's rhino losses mirror continental African trend. University of Kent, 13 December. https://www.kent.ac.uk/anthropology-conservation/news/8752/kruger-national-parks-rhino-losses-mirror-continental-african-trend. Accessed 27 December 2023.

3. Africa Geographic (2023) Kruger rhino population update – the losses continue. *Africa Geographic*, 24 October. https://africageographic.com/stories/kruger-rhino-population-update-losses-continue/#:~:text=Despite%20investing%20in%20access%20control,195%20rhinos%20lost%20in%202021. Accessed 27 December 2023.

4. Forestry, Fisheries and Environment on Rhino Poaching (2024).

5. MEFT (2023) *National Report: Wildlife protection and law enforcement in Namibia for the year 2022*. Windhoek: Ministry of Environment, Forestry and Tourism, p. 2.

6. IUCN/TRAFFIC (2022) Analyses of the proposals to amend the CITES Appendices at the 19th Meeting of the Conference of the Parties, IUCN Species Survival Commission/ TRAFFIC 14–25 November 2022, p. 21.

7. Personal communication with Pelham Jones, 6 December 2023.

8. R. Emslie (2020) *Diceros bicornis* ssp. *bicornis*. *The IUCN Red List of Threatened Species* 2020: e.T39318A45814419. https://dx.doi.org/10.2. Accessed 27 September 2023.

9. HALO (n.d.) Landmine clearance for conservation. https://www.halotrust.org/latest/special-projects/protecting-the-okavango/. Accessed 12 December 2023; personal communication with Chris Pym of HALO in 2015.

10. Personal communication with Peter Fearnhead, African Parks CEO, 20 June 2023.

11. Priya Tekriwal (2023) Iona National Park takes a step forward in its transformation. African Parks, 1 November. https://www.africanparks.org/iona-national-park-takes-step-forward-its-transformation. Accessed 13 December 2023.

12. Somerville (2019), pp. 265–6.

13. CITES (2022), pp. 9, 21.

14. Vera Pfannerstil et al. (2022) Effects of age and sex on site fidelity, movement ranges and home ranges of white and black rhinoceros translocated to the Okavango Delta, Botswana. *African Journal of Ecology* 60: 344–56, p. 346.

15. Personal communication with conservationists and wildlife officials in Botswana, June 2015 and September 2018. Because of the bitter Khama–Masisi conflict, the interviewees wished to remain anonymous.

16. Somerville (2020).

17. Emslie et al. (2019).

18. Keith Somerville (2016) Botswana's elephants and conservation – are things starting to fall apart? *Talking Humanities*, 6 October. https://talkinghumanities.blogs.sas.ac.uk/2016/10/06/botswanas-elephants-and-conservation-are-things-starting-to-fall-apart/. Accessed 18 December 2023.

19. Personal communication with Amos Ramokati and Michael Flyman in Maun in June 2015.

20. Ibid. and personal communication with Babolaki Autlwetse, Gaborone, June 2015.

21. Jones (2009), p. 158.

22. Mbongeni Mguni (2017) Botswana banks on fierce reputation as anti-poaching funds shrink. *Mmegi* 17. https://www.mmegi.bw/features/botswana-banks-on-fierce-reputation-as-anti-poaching-funds-shrink/news. Accessed 18 December 2023.

23. Oscar Nkala (2017) Botswana budget cuts imperil anti-poaching. Conservation Action Trust, 19 April. https://www.conservationaction.co.za/botswana-budget-cuts-imperil-anti-poaching/. Accessed 18 December 2023.

24. Somerville (2020). Much of the information is based on interviews with a number of conservationists, wildlife officials and community leaders in Botswana in 2018, who wished to remain anonymous.

25. Ibid.

26. Bhejane Trust (2020).

27. CITES (2022), p. 6.

28. Spencer Mogapi (2020) Botswana fast losing the war on rhino poaching. *Sunday Standard* (Botswana). https://www.sundaystandard.info/botswana-fast-losing-the-war-against-rhino-poachers/. Accessed 18 December 2023.

29. Bhejane Trust (2020) The on-going rhino massacre in Botswana. Facebook, 14 March. https://www.facebook.com/bhejanetrust/photos/a.854716111277976/29328663867962 61/?type=3&source=57&paipv=0&eav=AfaVru-8_Ht-gFEB62mpNryQpbU9EJ9nAwg GcJqeUMtfiO52X05qwfBFAqtQia3bpxg&_rdr. Accessed 18 December 2023.

30. Ibid. This is borne out by information supplied to the author on condition of anonymity by wildlife conservation specialists in Botswana.

31. Simon Espley (2021) Rhino poaching in Botswana – why the smoke and mirrors? *Africa Geographic*, 10 June. https://africageographic.com/stories/rhino-poaching-in-botswana-why-the-smoke-and-mirrors/. Accessed 8 August 2021.

32. Ibid.

33. Mmegi Online (2020) Soldier, poacher killed in rhino hotspot shootout, 11 March. https://www.mmegi.bw/news/soldier-poacher-killed-in-rhino-hotspot-shootout/news. Accessed 18 December 2023.

34. Somerville (2020).

35. Oscar Nkala, Facebook posts for Oxpeckers, 12 April 2022 and 12 April 2023.

36. Ibid.; personal communication with wildlife conservationists in Botswana.

37. Verreyne (2020).

38. Ibid.

39. Environmental Investigation Agency (2022) The rise of rhinoceros poaching in Botswana, briefing document for delegates to CITES SC74 (Lyon, March 2022). https://us.eia.org/wp-content/uploads/2022/03/EIA-SC74-Botswana-Rhino-Briefing.pdf, p. 1. Accessed 7 September 2023.

40. Thobo Motlhoka (2022) Rhino killed at Khama Sanctuary. *Sunday Standard* (Botswana), 19 August. https://www.sundaystandard.info/rhino-killed-at-khama-sanctuary/. Accessed 14 September 2023.

41. Mompati Tlhankane (2023) Khama Sanctuary no longer safe haven. *Mmegi*, 24 February. https://www.mmegi.bw/news/khama-sanctuary-no-longer-safe-haven/news. Accessed 26 September 2023.

42. Erik Verreynne, personal communication, 19 September 2023.

43. Mqondisi Dube (2022) Rhino poaching way down in Botswana. Voice of America, 16 November. https://www.voanews.com/a/rhino-poaching-way-down-in-botswana-/6837159.html. Accessed 19 September 2023.

44. Thobo Motlhoka (2023) Botswana presents intensified rhino anti-poaching efforts at global conference for trade in wild species. *Independent*, 7 February. https://www.independent.co.uk/voices/campaigns/giantsclub/botswana/botswana-rhino-anti-poaching-b2277469.html. Accessed 21 September 2023.

45. CITES (2022), p. 6.

46. IUCN/TRAFFIC (2022), p. 28.

47. Personal communication with Ted Reilly, head of eSwatini's national parks and reserves, August 2016, at Mlilwane Sanctuary.
48. CITES (2022), pp. 2–3.
49. Mwangi S. Kimenyi (2012) The human development cost of the King of Swaziland's lifestyle and his 'bevy' of wives. Brookings Institution, 21 September. https://www.brookings.edu/articles/the-human-development-cost-of-the-king-of-swazilands-lifestyle-and-his-bevy-of-wives/. Accessed 27 December 2023.
50. Mike Knight (2017) AfRSG report. *Pachyderm* 58: 20–1.
51. Ibid.
52. Keith Somerville (2016) Swaziland's rhino horn trade bid defeated at CITES – but what are the alternatives? Talking Humanities, 25 October. https://talkinghumanities.blogs.sas.ac.uk/2016/10/25/swazilands-rhino-horn-trade-bid-defeated-at-cites-but-what-are-the-alternatives/. Accessed 10 August 2023.
53. Personal communication with Ted Reilly.
54. Save the Rhino (2016) Will Travers and John Hume debate the legalisation of the rhino horn trade. Save the Rhino, 24 August. https://www.savetherhino.org/poaching-crisis/will-travers-and-john-hume-debate-the-legalisation-of-the-rhino-horn-trade/#:~:text=Hume%20argued%20breeding%20rhinos%20and,this%20debate%20reached%20far%20wider. Accessed 19 December 2023.
55. Paul Taylor (2004) The first privatization of a wildlife area in Malawi. *Oryx* 38(2): 134.
56. Krisztián Gyöngyi and Morten Elmeros (2017) Forage choice of the reintroduced black rhino and the availability of selected browse species at Majete Wildlife Reserve, Malawi. *Pachyderm* 58: 40–50, p. 41.
57. CITES (2022), pp. 2–3.
58. African Parks (n.d.) The Parks. https://www.africanparks.org/campaign/rhinos-return-malawi/about. Accessed 6 September 2023.
59. Ibid.
60. Milliken and Shaw (2012), p. 85.
61. CITES (2022), pp. 2–3.
62. Milliken and Shaw (2012), p. 85.
63. Rademeyer (2012), loc. 3493–505.
64. D.V. Roque et al. (2022) Historical and current distribution and movement patterns of large herbivores in the Limpopo National Park, Mozambique. *Frontiers in Ecology and Evolution*, https://doi.org/10.3389, pp. 8–9, p. 16.
65. Personal communication with Ken Maggs, Skukuza, August 2016.
66. Rademeyer (2012), loc. 3717.
67. Hübschle (2017).
68. Ibid., pp. 432–4.
69. 702 News (2022) White rhino reintroduced to Mozambique. http://www.702.co.za/articles/449382/white-rhino-reintroduced-to-mozambique. Accessed 24 August 2023.
70. IUCN/TRAFFIC (2022), p. 23.
71. Etosha National Park (n.d.) Collaboration is key for Kunene rhino conservation. https://www.etoshanationalpark.org/news/collaboration-is-key-for-kunene-rhino-conservation. Accessed 6 September 2023; Keith Somerville (2023) Namibia: rhino poaching almost doubles in 2022 with Etosha hit hard. *Commonwealth Opinion*, 6 February. https://commonwealth-opinion.blogs.sas.ac.uk/2023/namibia-rhino-poaching-almost-doubles-in-2022-with-etosha-hit-hard/. Accessed 30 August 2023.
72. The Staff Reporter (2022) Rhino poachers shift focus from national parks to custodianship and private farms. *Namibia Economist*, 5 August. https://economist.com.na/72471/environment/rhino-poachers-shift-focus-from-national-parks-to-custodianship-and-private-farms/. Accessed 7 September 2023.
73. Joe-Chinta Garises (2022) Poachers kill 11 black rhinos in Etosha in two weeks. *Namibia Daily News*, 15 June. https://namibiadailynews.info/poachers-kill-11-black-rhinos-in-etosha-in-two-weeks. Accessed 24 August 2022.

74. MEFT (2023), p. 2.
75. Somerville, 2023n.
76. Personal communication, Maxi Pia Louis, Director of NACSO (National Association of CBNRM Organisations), February 2023.
77. Xinhua (2023) 16 rhinoceros, 4 elephants poached in Namibia since January: official. https://english.news.cn/africa/20230523/d332c67b50b0419f963e555affab8166/c.html. Accessed 27 September 2023.
78. The Namibian (2022) Government probes workers for poaching. *The Namibian*, 21 June. https://www.namibian.com.na/government-probes-workers-for-poaching/. Accessed 5 February 2023.
79. Personal communication, Kenneth Heinrich Uiseb, 31 January 2023.
80. MEFT (2023), p. 60.
81. Etosha National Park (n.d.).
82. Personal communication with Derek de la Harpe of Wilderness, 7 August 2023, and SRT guides at desert Rhino Camp in September 2019.
83. Knight, Mosweub and Ferreira (2023), p. 25.
84. NACSO (n.d.) Registered communal conservancies. https://www.nacso.org.na/conservancies. Accessed 14 September 2023.
85. Sian Sullivan (2002) How sustainable is the communalizing discourse of 'new' conservation? The masking of difference, inequality and aspiration in the fledgling 'conservancies' of Namibia. In D.E. Chatty and M. Colchester (eds) *Conservation and Mobile Indigenous Peoples: Displacement, forced settlement and sustainable development*. Oxford: Berghahn, p. 159; Sian Sullivan (2023) 'Hunting Africa': how international trophy hunting may constitute neocolonial green extractivism. *Journal of Political Ecology* 30(1): 1–31.
86. Jeff J.R. Muntifering et al. (2017) Harnessing values to save the rhinoceros: insights from Namibia. *Oryx* 51(1): 98–105, p. 101.
87. Ibid., p. 101.
88. NACSO (2022) Natural Resource Report. https://www.nacso.org.na/sites/default/files/Torra%20Audit%20Report%202021.pdf. Accessed 7 August 2023.
89. Personal communication, Derek de la Harpe.
90. Knight, Mosweub and Ferreira (2023), pp. 25, 26.
91. Sullivan et al. (2021), p. 18.
92. Chris Brown and Gail C. Potgeiter (2020) How hunting black rhino contributes to conservation in Namibia. *Conservation Frontlines*. https://www.conservationfrontlines.org/2020/01/how-hunting-black-rhino-contributes-to-conservation-in-namibia/#:~:text=The%20direct%20benefits%20from%20wildlife,without%20negatively%20affecting%20each%20other. Accessed 28 December 2023.
93. Ibid.
94. Ibid.
95. Conservation Namibia (n.d.) The Game Products Trust Fund factsheet. https://conservationnamibia.com/factsheets/game-products-trust-fund.php. Accessed 18 December 2023.
96. Ibid.
97. AFP/Guardian (2015) Hunter pays $350,000 to shoot black rhino: 'I believe in survival of species'. *Guardian* 21 May. https://www.theguardian.com/environment/2015/may/21/hunter-paid-225000-shoots-black-rhino-i-believe-survival-of-species. Accessed 27 September 2023.
98. M. Hannis (2016) *Killing nature to save it? Ethics, economics and rhino hunting in namibia*. Future Pasts Working Paper 4, pp. 1–17, p. 3. https://www.futurepasts.net/fpwp4-hannis-2016. Accessed 6 September 2023.
99. Personal communication with John Jackson II, 8 September 2023.
100. CITES (2022), p. 26.
101. Ibid., p. 27.
102. Hannis (2016), p. 5.

103. B. Kötting (2020) Namibia's black rhino custodianship program. *Conservation Frontlines*, 2 April. https://www.conservationfrontlines.org/2020/04/namib ias-black-rhino-custodianship-program/. Accessed 1 February 2023.

104. Jeff J.R. Muntifering et al. (2021) Black rhinoceros avoidance of tourist infrastructure and activity: planning and managing for coexistence. *Oryx* 55(1): 150–9, p. 150.

105. Ibid., p. 155.

106. Jeff J.R. Muntifering et al. (2019) Sustainable close encounters: integrating tourist and animal behaviour to improve rhinoceros viewing protocols. *Animal Conservation* 22: 189–97, p. 192.

107. Personal communication with rhino guides and conservancy members, Torra Conservancy, 19 September 2019.

108. Save the Rhino Trust (2023) Strategic Plan 2023–2028, p. 6. https://www.scribd.com/document/639609552/srt-strategic-plan-final-print-file?secret_password=5xjM21vLmg OxoQSPaGSU. Accessed 7 January 2023.

109. Ibid., p. 16.

110. Ibid., p. 29.

111. Victoria Schneider (2023) Namibian community protects its rhinos from poaching but could lose them to mining. *Mongabay*, 13 March. https://news.mongabay.com/2023/03/namibian-community-protects-its-rhinos-from-poaching-but-could-lose-them-to-mining/. Accessed 26 August 2023.

112. Ibid.

113. Personal communication, Tristan Cowley, 28 December 2023.

114. Personal communication, Kenneth Uiseb, 28 December 2023.

115. Knight (2016), p. 17.

116. CITES (2022), pp. 2–3.

117. International Rhino Foundation (2022) 2022 State of the Rhino Report. https://rhinos.org/wp-content/uploads/2022/09/IRF-State-of-the-Rhino-2022.pdf. Accessed 3 January 2024.

118. Department of the Environment, Forestry and Fisheries, South Africa (n.d.) Statistics. https://www.facebook.com/photo?fbid=10157389946172827&set=a.10150931650187827. Accessed 27 December 2023; Team AG (2022) Another year of loss – an update on Kruger's rhino populations. *Africa Geographic*, 6 December. https://africageographic.com/stories/another-year-of-loss-an-update-on-krugers-rhino-populations/#:~:text=Posted%20on%20December%206%2C%202022,the%20News%20Desk%20post%20series.&text=There%20are%20now%20an%20estimated,a%2050%25%20reduction%20since%202013. Accessed 27 December 2023.

119. Ibid., and Keith Somerville (2022).

120. Personal communication with Pelham Jones, 6 December 2023.

121. Forestry, Fisheries and Environment on Rhino Poaching (2024).

122. International Rhino Foundation (n.d.) Updated poaching numbers from Kruger National Park. https://rhinos.org/blog/updated-poaching-numbers-from-kruger-national-park/. Accessed 21 December 2023; Africa Geographic (2022) Another year of loss – an update on Kruger's rhino populations, Africa Geographic, 6 December. https://africageographic.com/stories/another-year-of-loss-an-update-on-krugers-rhino-populations/#:~:text=Posted%20on%20December%206%2C%202022,the%20News%20Desk%20post%20series.&text=There%20are%20now%20an%20estimated,a%2050%25%20reduction%20since%202013. Accessed 3 January 2023.

123. Personal communication with Cedric Coetzee, Hlulhuwe, September 2016.

124. Personal communication with wildlife officials and private rhino owners in February and March 2018 – for obvious reasons they spoke on condition of anonymity. See Keith Somerville (2018) Corruption hindering South Africa's anti-poaching operations. *Talking Humanities*, 20 March. https://talkinghumanities.blogs.sas.ac.uk/2018/03/20/corruption-hindering-south-africas-anti-poaching-operations/. Accessed 3 January 2024.

125. John Ledger (2022) Crisis in the cradle of rhino conservation. *African Hunting Gazette*, 9 June. https://africanhuntinggazette.com/crisis-in-the-cradle-of-rhino-conservation/. Accessed 24 August 2023.
126. TRAFFIC (2018) South Africa: rhino poaching in 2017 almost matches 2016 figure, with KwaZulu Natal now bearing the brunt. https://www.traffic.org/news/south-africa-rhino-poaching-in-2017-almost-matches-2016-figure-with-kwazulu-natal-now-bearing-the-brunt/. Accessed 3 January 2024.
127. Personal communication with Cathy Deam, then CEO of Save the Rhino, March 2018.
128. South African Government News Agency (2022) SA records decline in rhino poaching, 8 February. https://www.sanews.gov.za/south-africa/sa-records-decline-rhino-poaching. Accessed 1 January 2024.
129. Knight, Mosweub and Ferreira (2023), pp. 13–16.
130. Personal communication, 11 December 2023; Ferreira, who also works as a large mammal scientist with SANParks, was speaking in his AfRSG capacity.
131. See Rademeyer (2012) for a very detailed examination of the pseudo-hunts and those involved.
132. Milliken and Shaw (2012), pp. 79–80.
133. Rademeyer (2012), loc. 2408.
134. Ibid., loc. 2239–42.
135. Rademeyer (2012).
136. Ibid., loc. 1678–84.
137. Milliken and Shaw (2012), p. 34.
138. Ibid., p. 56.
139. Ibid.
140. Rademeyer (2012), loc. 2091.
141. Emslie et al. (2019), pp. 14–15.
142. Ibid., pp. 79–80.
143. Keith Somerville (2012) Rhino poaching in South Africa: organised crime and economic opportunity driving trade. *African Arguments*, 29 October.
144. Ibid.
145. Orenstein (2013), p. 87.
146. Fiona MacLeod (2012) Vets charged for illegal use of tranquillisers. *Mail and Guardian*, 2 March. https://mg.co.za/article/2012-03-02-vets-charged-for-illegal-use-of-tranquillisers/. Accessed 19 June 2023.
147. Cited by ibid.
148. Rademeyer (2012), loc. 2097–2117.
149. Ibid., loc. 2194.
150. Ibid., loc. 2431.
151. Keith Somerville (2019) Will South Africa's national strategy stop the rhino poaching crisis? Talking Humanities, 21 February. https://talkinghumanities.blogs.sas.ac.uk/2019/02/21/will-south-africas-national-strategy-stop-the-rhinos-poaching-crisis/. Accessed 4 April 2023.
152. Marvin Charles (2021) Two Pretoria men jailed for selling boat that didn't belong to them. *News24*, 22 May. https://www.news24.com/news24/southafrica/news/pretoria-magistrates-court-sentences-2-men-to-41-years-for-fraud-and-theft-20210522. Accessed 5 January 2024.
153. Don Pinnock (2021) Rhino 'kingpin' arrested again for dealing in horn. *Daily Maverick*, 22 July. https://www.dailymaverick.co.za/article/2021-07-22-rhino-kingpin-arrested-again-for-dealing-in-horn/. Accessed 5 September 2021.
154. Zelda Venter (2022) Rhino horn trade case delayed again 12 years after arrest of suspects. *Pretoria News*, 12 April. https://rhinoreview.org/rhino-horn-trade-case-delayed-again-12-years-after-arrest-of-suspects/. Accessed 4 January 2024.
155. Tzaneen Voice (2023) American national arrested after found with 26 unreported rhino carcasses and horn in Gravellote. *Tzaneen Voice*, 27 December. https://tzaneenvoice.co.za/

american-national-arrested-after-found-with-26-unreported-rhino-carcasses-and-horn-in-gravellote/. Accessed 5 January 2024.
156.   Emma Ogao (2023) US national arrested in South Africa on wildlife trafficking, weapons charges. ABC News, 29 December. https://abcnews.go.com/International/us-national-arrested-south-africa-wildlife-trafficking-weapons/story?id=105995418. Accessed 1 January 2024.
157.   Black Rock Rhino. https://www.blackrockrhinoconservation.com/general-8-2. Accessed 5 January 2024.
158.   Milliken and Shaw (2012), pp. 79–80.
159.   Ibid.
160.   Rademeyer (2012), loc. 65.
161.   Kim Willsher (2017) Rhino shot dead by poachers at French zoo. Guardian, 7 March. https://www.theguardian.com/world/2017/mar/07/rhino-shot-dead-by-poachers-at-french-zoo. Accessed 5 January 2017.
162.   Simon Bloch (2023) 'Kick in the gut' – thieves escape with 51 rhino horns from North West Parks Board HQ. Daily Maverick, 27 June. https://www.dailymaverick.co.za/article/2023-06-27-thieves-escape-with-51-rhino-horns-from-north-west-parks-board/. Accessed 21 September 2023.
163.   Tshepiso Moche (2023) Hawks mum on stolen rhino horn stockpile. SABC News, 3 August. https://www.sabcnews.com/sabcnews/hawks-mum-on-stolen-rhino-horn-stockpile/. Accessed 4 September 2023.
164.   Nicole McCain (2023) Hawks make arrest after theft of 50 rhino horns from North West stockpile. News 24, 3 July. https://www.news24.com/news24/southafrica/news/hawks-make-arrest-after-theft-of-50-rhino-horns-from-north-west-stockpile-20230703. Accessed 5 January 2023.
165.   Milliken and Shaw (2012), p. 70.
166.   Emslie et al. (2019), p. 3.
167.   Johan Jooste with Tony Park (2022) Rhino War. South Africa: Ingwe Publishing, p. 44.
168.   Ibid., pp. 100–3.
169.   Ibid., p. 122.
170.   Ibid., pp. 19, 21.
171.   Ibid.
172.   Jasper Humphreys and M.L.R. Smith (2014) The 'rhinofication' of South African security. International Affairs 90(4): 795–818, p. 805.
173.   Jooste (2022), pp. 131–2.
174.   Ibid., p. 149.
175.   Personal communication with Xolani Nicholus Funda, KNP, August 2016.
176.   Africa Geographic (2023) Kruger rhino population update – the losses continue. Africa Geographic, 24 October. https://africageographic.com/stories/kruger-rhino-population-update-losses-continue/#:~:text=Despite%20investing%20in%20access%20control,195%20rhinos%20lost%20in%202021. Accessed 27 December 2023.
177.   Jooste (2022), p. 145.
178.   Vigne and Bradley Martin (2018), p. 9.
179.   Ibid., p. 41.
180.   Ibid.
181.   Don Pinnock (2022) Kruger National Park's rhinos are headed for extinction, we must declare emergency. Daily Maverick, 30 November. https://www.dailymaverick.co.za/article/2022-11-30-kruger-national-parks-rhinos-are-headed-for-extinction-we-must-declare-emergency/. Accessed 28 September 2023.
182.   Club of Mozambique (2022) South Africa turns to lie detectors in anti-poaching war. Club of Mozambique, 13 December. http://clubofmozambique.com/news/south-africa-turns-to-lie-detectors-in-anti-poaching-war-231008/?fbclid=IwAR3m5kd4dKQZKWiSdhntwzC_JioHB_8BaMP1Ag9ZmHFDZKjwC8LCXClqcxs. Accessed 21 September 2023.

183. Zoliswa N. Nhleko et al. (2022) Poaching is directly and indirectly driving the decline of South Africa's large population of white rhinos. *Animal Conservation* 25: 151–63, p. 151.
184. Ibid., pp. 159–60.
185. A.J. Selier and E. Di Minin (2022) How to reverse the rhino poaching crisis: a commentary on Nhleko et al. *Animal Conservation* 25(2): 164–5, p. 165; Z.N. Nhleko et al. (2022) Poaching is directly and indirectly driving the decline of South Africa's large population of white rhinos. *Animal Conservation* 25(2): 151–63.
186. Ibid.
187. Rademeyer (2012), loc. 3680.
188. Julian Rademeyer (2023) Landscape of fear: crime, corruption and murder in greater Kruger. *Enact* 36: 15.
189. Buks Viljoen (2023) Alleged rhino poaching kingpin nabbed again. *Lowvelder*, 29 May. https://www.citizen.co.za/lowvelder/news-headlines/2023/05/29/alleged-rhino-poaching-kingpin-nabbed-again/. Accessed 27 September 2023.
190. Lindile Sifile (2023) Slain chief Mnisi was to be prosecuted as middleman in rhino poaching syndicate. *Sowetan*, 5 April. https://www.sowetanlive.co.za/news/south-africa/2023-04-05-slain-chief-mnisi-was-to-be-prosecuted-as-middleman-in-rhino-poaching-syndicate/. Accessed 21 September 2023.
191. Rademeyer (2023), p. 1.
192. Ibid., p. 3.
193. BBC (2020) Leroy Brewer: South Africa hunts rhino-poaching investigator's killers. BBC News, 18 March. https://www.bbc.co.uk/news/world-africa-51942957. Accessed 25 September 2023.
194. Tembile Sgqolana (2022) The violent death of Timbavati ranger Anton Mzimba mourned by conservationists worldwide. *Daily Maverick*, 29 July. https://www.dailymaverick.co.za/article/2022-07-29-the-violent-death-of-timbavati-ranger-anton-mzimba-mourned-by-conservationists-worldwide/. Accessed 5 September 2023.
195. Cited by Rademeyer (2023), pp. 9–10.
196. IOL (2023) Rogue Kruger National Park rangers: 14th suspect arrested for rhino poaching case, 7 February. https://www.iol.co.za/news/crime-and-courts/rogue-kruger-national-park-rangers-14th-suspect-arrested-for-rhino-poaching-case-1c33c02f-e67d-48fd-ba45-577d2a612bbc.
197. Sheree Bega (2023) Creecy: just five of 87 ranger posts at Kruger Park have been filled. *Mail and Guardian*, 22 April. https://mg.co.za/environment/2023-04-22-creecy-just-five-of-87-ranger-posts-at-kruger-park-have-been-filled/. Accessed 26 September.
198. Jamie Joseph (2018) War on rhino: gambling on corruption (victory! Nzimande suspended). Saving the Wild, 3 October. https://www.savingthewild.com/news/war-on-rhino-gambling-on-corruption/. Accessed 7 January 2024.
199. Gaddafi Zulu (2023) Mtubatuba court finds rhino poaching 'kingpin' accused not guilty. *Zululand Observer*, 28 July. https://www.citizen.co.za/zululand-observer/news-headlines/local-news/2023/07/28/mtubatuba-court-finds-rhino-poaching-kingpin-accused-not-guilty/. Accessed 8 January 2024.
200. Personal communication with Pelham Jones, chair of PROA, 6 December 2023.
201. Ed Stoddard (2023) Saving private rhino — non-government owners of the animals succeed in stemming poaching carnage. *Daily Maverick*, 14 February. https://www.dailymaverick.co.za/article/2023-02-14-saving-private-rhino-non-government-owners-of-the-animals-succeed-in-stemming-poaching-carnage. Accessed 25 September 2023.
202. Personal communication with Pelham Jones, 6 December 2023.
203. Knight (2016), p. 17.
204. Hayley S. Clements et al. (2020) Private rhino conservation: diverse strategies adopted in response to the poaching crisis. *Conservation Letters*, 15 June. https://conbio.onlinelibrary.wiley.com/doi/full/10.1111/conl.12741. Accessed 9 January 2024.
205. Keith Somerville (2017) Has the tide turned for South Africa's rhino poaching crisis? *Talking Humanities*, 21 March. https://talkinghumanities.blogs.sas.ac.uk/2017/03/21/has-the-tide-turned-for-south-africas-rhino-poaching-crisis/. Accessed 10 August 2023.

206. CITES (2022), p. 28.
207. Clements, Balfour and Di Minin (2023), p. 4.
208. The High-Level Panel of Experts for the Review of Policies, Legislation and Practices on Matters of Elephant, Lion, Leopard and Rhinoceros Management (2020) *High-Level Panel Report – For Submission to the Minister of Environment, Forestry and Fisheries*. https://www.dffe.gov.za/sites/default/files/reports/2020-12-22_high-levelpanel_report.pdf, p. 306. Accessed 9 January 2024.
209. Tan (2023).
210. Elena C. Rubino and Elizabeth F. Pienaar (2018) Understanding South African private landowner decisions to manage rhinoceroses. *Human Dimensions of Wildlife* 23(2): 160–95, p. 160.
211. Clements, Balfour and Di Minin (2023), p. 4.
212. Ibid., pp. 161–2.
213. Ibid., p. 162.
214. Oliver Thomas Wright, Georgina Cundill and Duan Biggs (2016) Stakeholder perceptions of legal trade in rhinoceros horn and implications for private reserve management in the Eastern Cape, South Africa. *Oryx* 52(1): 1–11, p. 1.
215. Biological Management Working Group, African Rhino Specialist Group, Species Survival Commission, International Union for the Conservation of Nature (2023) The conservation contribution of the Platinum Rhino southern white rhino Captive Breeding Organization. https://media.savetherhino.org/prod/uploads/2023/04/AfRSG-2023-The-conservation-contribution-of-the-Platinum-Rhino-white-rhino-Captive-Breeding-Organization.pdf, p. 2. Accessed 5 July 2023.
216. Personal communication with John Hume, Buffalo Dreams ranch, 29 August 2016.
217. News 24 (2022) SA breeder to release 100 rhinos back into the wild every year. *News 24 (South Africa)*, 7 June. https://www.news24.com/news24/southafrica/news/sa-breeder-to-release-100-rhinos-back-into-the-wild-every-year-20220607#:~:text=A%20private%20rhino%20farmer%20in,rhinos%20were%20killed%20in%202021. Accessed 24 August 2023.
218. Knight (2023), p. 1.
219. Ed Stoddard (2023) The opening bid for world's largest white rhino population is $10m. *Daily Maverick*, 8 March. https://www.dailymaverick.co.za/article/2023-03-08-the-opening-bid-for-worlds-largest-white-rhino-population-is-10m/. Accessed 9 March 2023.
220. Tan (2023).
221. African Parks (2023) 2,000 southern white rhino to be released into the wild over next 10 years. https://www.africanparks.org/2000-southern-white-rhino-be-released-wild-over-next-10-years. Accessed 9 January 2023.
222. Keith Somerville (2023) African Parks buys John Hume's 2000 rhinos to rewild them. *Commonwealth Opinion*, 7 September. https://commonwealth-opinion.blogs.sas.ac.uk/2023/african-parks-buys-john-humes-2000-rhinos-to-rewild-them/. Accessed 18 September 2023.
223. Ed Stoddard (2023) Conserving and rewilding John Hume's rhinos may cost R1bn or more. *Daily Maverick*, 12 September. https://www.dailymaverick.co.za/article/2023-09-12-conserving-and-rewilding-john-humes-rhinos-may-cost-r1bn-or-more/. Accessed 18 September 2023.
224. Personal communication with Pelham Jones, 6 December 2023.
225. Chansa Chomba and Wigganson Matandikohansa (2011) Population status of black and white rhinos in Zambia. *Pachyderm* 50: 50–5, p. 50; Save the Rhino (2023) Zambia: North Luangwa Conservation Programme. https://www.savetherhino.org/programmes/north-luangwa-conservation-programme/. Accessed 25 January 2023.
226. Chansa Chomba, David Squarre and Harvey Kamboyi (2012) Notes on black rhino mortalities in North Luangwa National Park, Zambia. *Pachyderm* 52: 88–90, p. 88.
227. Save the Rhino (2023) Zambia: North Luangwa Conservation Programme. https://www.savetherhino.org/programmes/north-luangwa-conservation-programme/. Accessed 19 January 2023.

228. Chansa Chomba and Wigganson Matandikohansa (2011) Population status of black and white rhinos in Zambia. *Pachyderm* 50: 50–5, p. 55.

229. CITES (2022), pp. 2–3.

230. AfRSG (2007), pp. 58–9.

231. CITES (2022), pp. 2–3.

232. Rademeyer (2012), loc. 158.

233. Ibid.

234. Ibid., loc. 285.

235. Cited by Rademeyer (2012), loc. 202.

236. Ed Warner (2016) *Running with Rhinos*. Austin, TX: Greenleaf, p. 61.

237. Peter Lindsey and Andrew Taylor (2011) *A Study on the Dehorning of African Rhinoceroses as a Tool to Reduce the Risk of Poaching*. Pretoria: South African Department of Environmental Affairs, p. 26.

238. Ibid., p. 20.

239. Ibid., p. 12.

240. Keith Somerville (2017) Wave of rhino killings points to shifting poaching patterns in South Africa. *The Conversation*. https://theconversation.com/wave-of-rhino-killings-points-to-shifting-poaching-patterns-in-south-africa-77806. Accessed 10 August 2023.

241. Knight, Mosweub and Ferreira (2022).

242. Mike B. Ball et al. (2019) What does it take to curtail rhino poaching? Lessons learned from twenty years of experience at Malilangwe Wildlife Reserve, Zimbabwe. *Pachyderm* 60: 96–104, p. 97.

243. Ibid., p. 96.

244. Ibid., p. 97.

245. Ibid., pp. 99–100.

246. Ibid., p. 100.

247. Reuters (2021) Zimbabwe re-introduces rhinos in Gonarezhou park after three decades. Reuters, 13 May. https://www.reuters.com/business/environment/zimbabwe-re-introduces-rhinos-gonarezhou-park-after-three-decades-2021-05-13/. Accessed 27 September 2023.

248. Natasha Anderson (2022) The long road to Gonarazhou. International Rhino Foundation. https://rhinos.org/blog/the-long-road-to-gonarazhou/. Accessed 2 January 2024.

249. Hayley S. Clements, Dave Balfour and Enrico Di Minin (2023) Importance of private and communal lands to sustainable conservation of Africa's rhinoceroses. *Frontiers in Ecology and the Environment*, doi:10.1002/fee.2593: 1–8, p. 2.

250. Personal communication with Raoul du Toit of the International Rhino Foundation and Lowveld Trust, 5 September 2023.

251. International Rhino Foundation (2020) Black rhino births in Zimbabwe happy news after poaching crisis. https://rhinos.org/blog/black-rhino-births-in-zimbabwe-happy-news-after-poaching-crisis/. Accessed 4 September 2023.

252. Personal communication with Blondie Leathem of BVC, 11 September 2023.

253. *Bhejane Trust Newsletter* (April 2023).

254. Personal communication with Blondie Leathem and Raoul du Toit.

255. Personal communication with Blondie Leathem and Raoul du Toit.

Chapter 10: What next?

1. Kes Hillman-Smith, Mankoto ma Oyisenzoo and Fraser Smith (1986) A last chance to save the northern white rhino? *Oryx* 20(1): 20–6, pp. 23–4.

2. Ibid.

3. M. Svitalsky, J.Vahala and P. Spala (1993) Breeding experience with northern white rhinos (*Cerotherium simum cottoni*) at Zoo Dvur Kralove. In Oliver A. Ryder (ed.) *Rhinoceros Biology and Conservation*, pp. 282–5, p. 283.

4. Emslie and Brooks (1999), p. 22.

5. Ruth Appeltant and Suzannah Williams (2020) Only two northern white rhinos remain, and they're both female – here's how we could make more. *The Conversation*, 15 October. https://theconversation.com/only-two-northern-white-rhinos-remain-and-theyre-both-female-heres-how-we-could-make-more-147608. Accessed 9 March 2023.

6. Kes Hillman-Smith (2022) Northern White rhino – extinction or survival. *Biorescue*, October-December 2022, 29–34, p. 30.

7. Petr David Josek (2019) Scientists fine-tune method to save rhinos. *Phys.org*, 15 February. https://phys.org/news/2019-02-scientists-fine-tune-method-rhinos.html. Accessed 11 January 2024.

8. International Rhino Foundation (2019) Advancements in assisted reproductive techniques – northern white rhino embryos. International Rhino Foundation, 11 September 2019. https://rhinos.org/tough-issues/advancements-in-assisted-reproductive-techniques-northern-white-rhino-embryos/#:~:text=The%20IRF%20applauds%20the%20efforts,subspecies%20that%20is%20functionally%20extinct. Accessed 6 September 2023.

9. Leibniz-Institut für Zoo-und Wildtierforschung (2023) Five new embryos and new surrogate mothers added to northern white rhino rescue project. Phy.org, 17 July. https://phys.org/news/2023-07-embryos-surrogate-mothers-added-northern.html. Accessed 13 September 2023.

10. Emslie and Brooks (1999), pp. 64–5.

11. Ibid. and personal communication with Sam Ferreira, scientific officer of the AfRSG, 11 December 2023.

12. Emslie et al. (2019), p. 19.

13. Personal communication with Sam Ferreira, 11 December 2023.

14. Personal communication with Peter Kernhead, 8 June 202.

15. Lindsey and Taylor (2011), pp. 2–3.

16. TRAFFIC (1992) *Bulletin* 13(2): 47.

17. Milliken and Shaw (2012), pp. 76–7.

18. Cited by Lindsey and Taylor (2011), p. 20.

19. Ibid., pp. 12, 15–16.

20. Ibid., p. 29.

21. Ibid., p. ix.

22. Samuel G. Penny et al. (2020) Does dehorning lead to a change in inter-calf intervals in free-ranging white rhinoceros? *Pachyderm* 61: 191–3, p. 192.

23. Lucy C. Chimes et al. (2022) Effects of dehorning on population productivity in four Namibia sub-populations of black rhinoceros (*Diceros bicornis bicornis*). *European Journal of Wildlife Research* 68 (58): 1–17, p. 6.

24. J. Berger and C. Cunningham (1994) Phenotypic alterations, evolutionarily significant structures, and rhino conservation. *Conservation Biology* 8(3): 833–840, p. 835.

25. Chimes et al. (2022), pp. 13–14.

26. Vanessa Duthé et al. (2023) Reductions in home-range size and social interactions among dehorned black rhinoceroses (*Diceros bicornis*). PNAS, 12 June. https://doi.org/10.1073/pnas.2301727120. Accessed 28 September 2023, no page numbers; Erik Stokstad (2023) Cutting off rhino horns to prevent poaching makes them homebodies. *Science.org*, 12 June. https://www.science.org/content/article/cutting-rhino-horns-prevent-poaching-makes-them-homebodies. Accessed 28 September 2023.

27. Duthé et al. (2023).

28. Penny et al. (2022), p. 7.

29. Keith Somerville (2016) Taking the bull by the horns – dehorning rhinos to protect them. *Commonwealth Opinion*, 6 September. https://commonwealth-opinion.blogs.sas.ac.uk/2016/taking-the-bull-by-the-horns-dehorning-rhinos-to-protect-them/. Accessed 10 August 2023.

30. Mike Knight (2019) AfRSG report. *Pachyderm* 60: p. 21.

31. Sam Ferreira et al. (2014) Chemical horn infusions: a poaching deterrent or an unnecessary deception? *Pachyderm* 55: 54–61, p. 55.

32. Ibid., p. 56.

33. Rhino rescue project (n.d.) About the project. https://rhinorescueproject.org/about-the-project/. Accessed 13 January 2024; Ferreira (2014), p. 60.

34. Save the Rhino (n.d.) Reducing illegal horn trade. https://www.savetherhino.org/what-we-do/reducing-illegal-trade/. Accessed 13 January 2024.

35. Save the Rhino/TRAFFIC (2013) Tackling the demand for rhino horn. https://www.savetherhino.org/thorny-issues/tackling-the-demand-for-rhino-horn/. Accessed 13 January 2024.

36. Giovanni Ortolani (2017) Reducing Asia's hunger for rhino horn. *Mongabay*, 9 March. https://news.mongabay.com/2017/03/reducing-asias-hunger-for-rhino-horn/#:~:text=The%20vast%20majority%20of%20rhino%20horn%20is%20bound,Vietnam%2C%20the%20primary%20destinations%20for%20poached%20rhino%20horn. Accessed 13 January 2024.

37. Wild Aid (n.d.) Sir Richard Branson-Nail Biters Ad-English. https://wildaid.org/richard-branson-speaks-out-against-rhino-horn-trade/sir-richard-branson-nail-biters-ad-english/ and https://wildaid.org/maggie-q-nail-biters/. Accessed 13 January 2024.

38. Hoai Nam Dang Vua and Martin Reinhardt Nielsen (2021) Evidence or delusion: a critique of contemporary rhino horn demand-reduction strategies. *Human Dimensions of Wildlife* 26(4): 390–400, p. 390.

39. Kristin Nowell (2012) Assessment of rhino horn as a traditional medicine: a report prepared for the CITES Secretariat, TRAFFIC. http://www.rhinoresourcecenter.com/pdf_files/138/1389957235.pdf, p. 1. Accessed 16 August 2023.

40. Ibid., pp. 25–6.

41. Hoai Nam Dang Vua and Martin Reinhardt Nielsen (2018) Understanding utilitarian and hedonic values determining the demand for rhino horn in Vietnam. *Human Dimensions of Wildlife* 23(5): 417–32, p. 419.

42. Anh Ngoc Vu (2023) Demand reduction campaigns for the illegal wildlife trade in authoritarian Vietnam: ungrounded environmentalism. *World Development* 164: 1–14, p. 1.

43. Ibid., pp. 2, 4.

44. Ibid., p. 9.

45. Michael Scott Smith (2018) Framing rhino horn demand reduction in Vietnam. Dismissing medical use as voodoo. *Pacific Journalism Review* 24(2): 241–56, p. 251.

46. Cited by Milliken, Nowell and Thomsen (1993), p. 2.

47. Emslie and Brooks (1999), p. 71.

48. Duan Biggs et al. (2013) Legal trade of Africa's rhino horns. *Science* 339: 1038–9, p. 1038.

49. 't Sas Rolfes (2012); Michael Eustace (2015) Smart Trade, Committee of Enquiry, 25/26 March. https://www.dffe.gov.za/sites/default/files/docs/paper_readbymichaeleustace.pdf, p. 1; Jane Wiltshire (2021) It's time for a meaningful discussion on the future of rhinos and the trade in rhino horn. *Daily Maverick*, 7 June. https://www.dailymaverick.co.za/opinionista/2021-06-07-its-time-for-a-meaningful-discussion-on-the-future-of-rhinos-and-the-trade-in-rhino-horn/, p. 8. Accessed 20 June 2023.

50. 't Sas Rolfes (2012), p. 15.

51. Ibid.

52. Save the Rhino (2016) Will Travers and John Hume debate the legalisation of the rhino horn trade. Save the Rhino, 24 August. https://www.savetherhino.org/poaching-crisis/will-travers-and-john-hume-debate-the-legalisation-of-the-rhino-horn-trade/#:~:text=Will%20Travers%20and%20John%20Hume%20debate%20the%20legalisation%20of%20the%20rhino%20horn%20trade,-24%20Aug%202016&text=On%20Wednesday%203%20August%202016,horn%20trade%20should%20be%20legalised. Accessed 18 December 2023. The author attended the event on 3 August 2016.

53. 't Sas Rolfes (2012), p. 16.

54. Ferreira and Okita-Ouma (2012), p. 56.

55. Milliken and Shaw (2012), pp. 102–3.

56. Orenstein (2013), pp. 171–2.

57. Personal communication with John Hume, Michael Eustace and Jane Wiltshire, August 2016.
58. Biggs et al. (2013), p. 1038.
59. Eustace (2015), p. 1.
60. Personal communication with Pelham Jones, 6 and 12 December 2023.
61. Nikolaj Bichel and Adam Hart (2023) *Trophy Hunting*. Springer Nature Singapore. Kindle edition.
62. IUCN (2016) IUCN Briefing Paper: Informing Decisions on Trophy Hunting. https://www.iucn.org/sites/dev/files/iucn_sept_briefing_paper_-_informingdecisionstrophyhunting.pdf. Accessed 6 January 2020; Bichel and Hart (2023), p. 16.
63. Milliken and Shaw (2012), p. 8.
64. Ibid.
65. Nigel Leader-Williams (2009) Conservation and hunting: friends or foes. In Barney Dickson, Jon Hutton and William M. Adams (eds) *Recreational Hunting, Conservation and Rural Livelihoods: Science and practice*, pp. 9–24. London: ZSL/Wiley-Blackwell, p. 12.
66. Milliken and Shaw (2012), pp. 45–6.
67. 't Sas Rolfes et al. (2022) Legal hunting for conservation of highly threatened species: the case of African rhinos. *Conservation Letters* e12877. https://doi.org/10.1111/conl.12877, pp. 1–9. Accessed 31 July 2023.
68. Ibid., p. 1.
69. Ibid., p. 5.
70. Economists at Large. The $200 million question. How much does trophy hunting really contribute to African communities? https://www.ecolarge.com/work/the-200-million-question-how-much-does-trophy-hunting-really-contribute-to-african-communities/. Accessed 30 July 2024.
71. Bichel and Hart (2023), loc. 10822; 't Sas Rolfes et al. (2022), p. 4.
72. Vernon Booth (2010) *Contribution of Wildlife to National Economies*. CIC Technical Series Publication No. 8, joint publication of FAO and CIC. http://www.cic-wildlife.org/wp-content/uploads/2012/12/Technical_series_8.pdf. Accessed 6 February 2020.
73. Robin Naidoo et al. (2016) Complementary benefits of tourism and hunting to communal conservancies in Namibia. *Conservation Biology* 30(3): 628–38, p. 630.
74. Ibid., p. 632.
75. Ibid., p. 634.
76. Save the Rhino (2019) Trophy hunting and sustainable use: rhinos. https://www.savetherhino.org/thorny-issues/trophy-hunting-and-sustainable-use-rhinos/. Accessed 4 September 2023.
77. Keith Somerville (2022) Conservationists look for new ways of combating rhino poaching as South African poachers go back to work. *Commonwealth Opinion*, 13 April. https://commonwealth-opinion.blogs.sas.ac.uk/2022/conservationists-look-for-new-ways-of-combating-rhino-poaching-as-south-african-poachers-go-back-to-work/. Accessed 10 August 2023.
78. Chris Barichievy (2021) A demographic model to support an impact financing mechanism for black rhino metapopulations. *Biological Conservation* 257: 1–11, p 2.
79. Glen Jeffries et al. (2019) The Rhino Impact Investment Project – a new, outcomes-based finance mechanism for selected AfRSG-rated 'key' black rhino populations. *Pachyderm* 60: 88–95, p. 89.
80. ZSL (n.d.) Rhino Impact Investment Project, Zoological Society of London. https://www.zsl.org/what-we-do/projects/rhino-impact-investment-project/. Accessed 28 July 2023.
81. WFA News (2023) A dedicated tax incentive for rhino and OECMS in South Africa. Wilderness Foundation Africa, 13 November. https://wildernessfoundation.co.za/a-dedicated-tax-incentive-for-rhino-and-oecms-in-south-africa/. Accessed 15 November 2023.
82. Rowan Martin (1992) *The Influence of Governance on Conservation and Wildlife Utilization: Alternative Approaches to Sustainable Use*. Harare: DNPWLM, 1992, pp. 3–8.
83. Save the Rhino Trust (2023).

84. Cited by Raymond Bonner (1993) *At the Hand of Man: Peril and Hope for Africa's Wildlife*. London: Simon & Schuster, pp. 145–6.

85. Vigne and Bradley Martin (2018), p. 9.

86. Ibid., pp. 46–7.

87. Siyabona Africa (2012) Kruger National Park field rangers strike called off. https://www.krugerpark.co.za/kruger-park-news-kruger-national-park-field-rangers-strike-25499.html. Accessed 31 August 2023.

88. Vanda Felbab-Brown (2017) *The Extinction Market: Wildlife Trafficking and How to Counter It*. London: Hurst and Co., p. 118.

89. Duffy (2022), p. 3.

90. Muntifering et al. (2017), pp. 98–9.

91. Ibid., p. 99.

92. S. Sullivan (2006) The elephant in the room? Problematising 'new' (neoliberal) biodiversity conservation. *Forum for Development Studies* 33: 105–135, p. 112.

Abbreviations

$ – dollars. US dollars unless otherwise specified

ADMADE – Zambian Administrative Management Design for Game Management programme

AfRSG – African Rhino Specialist Group of the IUCN

ANC – African National Congress of South Africa

APLRS – Association of Private and Community Land Rhino Sanctuaries (Kenya)

APU – anti-poaching unit

AWF – African Wildlife Foundation (based in the USA)

BCE – Before Common Era (formerly BC)

BDF – Botswana Defence Force

BP – years before present: the date for 'present' is 1950

BSAC –British South Africa Company

BVC – Bubye Valley Conservancy (Zimbabwe)

CAMPFIRE – Communal Areas Management Programme for Indigenous Resources

CAR – Central African Republic

CBNRM – community-based natural resource management

CE – Common Era (formerly AD)

CEO – chief executive officer

CID – Criminal Investigation Department (of the Kenyan Police)

CITES – Convention on International Trade in Endangered Species of Wild Fauna and Flora

CoP – Conference of the Parties to CITES

DRC – Democratic Republic of Congo (formerly Zaire)

DWNP – Department of Wildlife and National Parks (Botswana)

EIA – Environmental Investigation Agency

ENV – Education for Nature Vietnam

EWT – Endangered Wildlife Trust (South Africa)

FAO – Food and Agriculture Organization (of the UN)

FNLA – National Front for the Liberation of Angola

GEF – Global Environment Facility (a multilateral environmental fund that provides grants and finance for projects related to biodiversity, climate change, international waters, land degradation, persistent organic pollutants, mercury, sustainable forest management, food security, and sustainable cities in developing countries; it is the largest source of multilateral funding for biodiversity, and distributes more than $1 billion a year to address interrelated environmental challenges)

GMA – game management area

GR – game reserve

GSU- General Service Unit (Kenyan paramilitary unit)

IBEAC – Imperial British East Africa Company

ICCN – Institut Congolais pour la Conservation de la Nature (DRC)

IFAW – International Fund for Animal Welfare

IPZ – intensive protection zone (used in both Zimbabwe and South Africa to provide maximum protection for rhinos)

IUCN – International Union for the Conservation of Nature

KANU – Kenyan African National Union

KAS Enterprises – private military security company established by SAS founder David Stirling

KNP – Kruger National Park (South Africa)

KSh – Kenyan shillings

KWS – Kenya Wildlife Service

kya – thousand years ago

KZN – KwaZulu-Natal province (South Africa)

LGMA – Lupande Game Management Area (Zambia)

LIRDP – Luangwa Integrated Resource Development Project (Zambia)

LNP – Limpopo National Park (Mozambique, Greater Limpopo Transfrontier Park)

LRA – Lord's Resistance Army (Ugandan rebel movement)

MEFT – Ministry of Environment, Forestry and Tourism (Namibia)

MPLA – Popular Movement for the Liberation of Angola

mya – million years ago

NCA – Ngorongoro Conservation Area (Tanzania)

NFD – Northern Frontier District (Kenya)

NGO – non-governmental organisation

NNP – Nairobi National Park

NORAD – Norwegian Agency for Development Cooperation

NP – national park

NPWS – National Parks and Wildlife Service (Zambia)

NRT – Northern Rangelands Trust (Kenya)

p.a. – per annum (per year)

PHASA – Professional Hunters' Association of South Africa
PROA – Private Rhino Owners Association (South Africa)
R – rand (South African currency)
Renamo – Mozambique National Resistance Movement (rebel movement)
RMG – Rhino Management Group (of the SADC)
SADC – Southern African Development Community
SADF – South African Defence Force (pre-1994)
SANDF – South African National Defence Force (post-1994)
SANParks – South African National Parks
SAP – South African Police
SEEJ – Saving Endangered Species through Education and Justice
    (Kenyan-based organisation)
SLNP – South Luangwa National Park
SNP – Serengeti National Park (Tanzania)
SPLA – Sudan People's Liberation Army
SRT – Save the Rhino Trust (there were trusts in Zambia and Namibia, in
    addition to the London-based Save the Rhino International)
SWAPO – South West African People's Organization (Namibian liberation
    movement)
TCM – traditional Chinese medicine
UN – United Nations
UNEP – United Nations Environment Programme
UNIP – United National Independence Party (Zambia)
UNITA – National Union for the Total Independence of Angola
VOC – Dutch East India Company
WCMD – Wildlife Conservation and Management Department (Kenya)
WWF – World Wide Fund for Nature
ZNA – Zimbabwe National Army
ZSL – Zoological Society of London

Index

www.ingramcontent.com/pod-product-compliance
Lightning Source LLC
Chambersburg PA
CBHW020813300125
21007CB00002B/1